Human Agency, Artificial Intelligence, and the Attention Economy

Leslie Paul Thiele

Human Agency, Artificial Intelligence, and the Attention Economy

The Case for Digital Distancing

palgrave
macmillan

Leslie Paul Thiele
Political Science
University of Florida
Gainesville, FL, USA

ISBN 978-3-031-82085-4 ISBN 978-3-031-82086-1 (eBook)
https://doi.org/10.1007/978-3-031-82086-1

Cover credit: Cover art by AsheskArt

This Palgrave Macmillan imprint is published by the registered company Springer Nature Switzerland AG
The registered company address is: Gewerbestrasse 11, 6330 Cham, Switzerland

Preface

As friends and I sat around a backyard firepit one brisk fall evening, a Luna moth flitted about. It was an exquisite creature, with sea-foam green wings, the breadth of a human hand, shimmering in the light. Our delight was short lived. Circling the fire, the graceful moth flew too close. With a burst of light and a ghastly sizzle, its wings combusted and the insect dropped pathetically into the flames.

A similar fate is often met by moths as they circle lighted candles. Why do they do this? The activity seems pointless and often proves fatal. Certainly, it qualifies as maladaptive behavior. One commonly accepted answer is that moths, being nocturnal, are accustomed to navigating by moonlight or starlight. They employ a method called transverse orientation, which keeps them flying at a fixed height in a straight line by maintaining the light source at a constant angle relative to their eyes. As the light source is very far away, the moth's angle to it does not appreciably change when flying in a straight line. However, when a nearby flame is encountered, the effort to maintain a constant angle to the light causes the moth to fly in circles. At best, the poor creature is subjected to a tedious waste of energy. At worst, its spiraling flight proves fatal. The navigational skill of transverse orientation had evolutionary benefits for moths, heightening their fitness. But this instinct devolved into a maladaptive and often self-destructive behavior on a planet increasingly illuminated by campfires, candles, and electric lights.

Human beings face a similar danger as they navigate the digital world. Our craving for attention and the tendency to distribute it widely is instinctive. So is our attraction to novel phenomena, our preference for choice abundance, and our proclivity to heed perilous prospects. In turn, we relish convenience and efficiency. These innate inclinations had evolutionary benefits for our species. And they can still serve us well today, as we navigate our physical and social environments. In a world of digital technologies that covet our engagement and satisfy ever more of our needs, however, these predispositions are being dangerously exploited. So we find ourselves compulsively occupied by our AI-enabled devices, bewitched by their shimmering screens. We are as moths to virtual flames: transfixed by their allure, circling ever closer, exhausted by our efforts. It may only be a matter of time before our downward spiral reaches a calamitous nadir.

Scientific discoveries are fascinating and breakthrough technologies are exciting. The former inspires awe and the latter hope. Indeed, the future of humanity hinges on scientific insights and technological innovations. But the welfare of humanity, and that of most other species on our planet, also depends on putting science and technology, the fruits of human agency, to the best use. Failure to do so could be catastrophic. Nowhere are the stakes higher than in the realm of digital technologies, including artificial intelligence (AI).

Artificial General Intelligence (AGI) is typically contrasted with narrow AI, a form of machine intelligence focused on a particular task domain, such as playing chess or identifying human faces. AGI would be human-level intelligence that is not range restricted. AI is rapidly gaining functionalities, and many experts predict that AGI will be reached in the coming decade. Still, human intelligence does not only display a wide range of task domains. The distinctiveness of human intelligence is that it *integrates* diverse kinds of knowledge and skills, allowing a multidimensional world populated by incommensurable values to be well navigated. Practical judgment, which AI has not mastered and arguably never will, is the fruit of such integration.

"Of all the ills afflicting men," the ancient playwright Sophocles (1993, 398) wrote in his *Antigone*, "the worst is lack of judgment." That truth has stood the test of time. The failure to exercise good judgment often produces personal calamities. Politically, the stakes are even higher. Absent the widespread exercise of practical judgment, democracy will not survive.

Over the coming years, machines will get smarter—much smarter. In turn, digital media will become more pervasive and enthralling. All this will occur at a quickening pace. Without concerted efforts, our mental muscles and social fabric will be weakened as these technologies are strengthened. To reverse this ominous trend, we need to fortify human agency and cultivate the practical judgment that puts it to good use. That is an attainable goal, and the surest means to achieve human flourishing, maintain human distinctiveness, and bolster democratic prospects in an age of accelerating technological innovation and advancement.

Gainesville, USA — Leslie Paul Thiele

Acknowledgments This book engages extensively with philosophical studies and scientific scholarship while avoiding disciplinary niches and technical jargon. The aim was to make the writing clear and compelling to readers who may lack a background or interest in specialized literatures. If that goal was reached, it is in no small part owing to the thoughtful comments of Manu Samnotra, Marvelle D'Souza, and Samaa Gullamhussein.

Contents

CHAPTER 1

Digital Upgrading and Human Downgrading

The outbreak of COVID-19 introduced the world to the practice of social distancing. The pandemic also expanded the use of digital devices, particularly for those who began working or studying at home. Digital media proved a lifeline for people who were sheltering in place or quarantined, providing professional opportunities, medical information, community interactions, business support, connections to family and friends, and entertainment. The steep increase in online activities was accompanied by a predictable rise in hacking and espionage attempts against individuals, agencies, and businesses. Cyber security experts encouraged precautions to enhance online security, calling for *digital distancing*. This book argues the case for digital distancing. But the focus is not on the safety of people's money or possessions. The concern is the strength of people's minds and relationships.

We rely on digital technologies, greatly empowered by AI, to communicate and supply our knowledge, to cultivate and curate our relationships, to occupy and entertain us, to organize our activities, to filter our choices, and, increasingly, to make our decisions. While their benefits are legion, our growing reliance on digital technologies leaves us with underdeveloped internal resources, including the capacity for self-direction and the thoughtful reflection it is grounded upon. Widespread and growing dependence on these technologies is eroding human agency and undermining practical judgment. As the digital world enjoys continuous

L. P. Thiele, *Human Agency, Artificial Intelligence, and the Attention Economy*, https://doi.org/10.1007/978-3-031-82086-1_1

upgrading, now greatly accelerated by AI, human beings are threatened with a precipitous downgrading.

A world of driverless vehicles is burgeoning. Automated transportation might also serve as a metaphor for much of our off-road yet online lives. As patrons of digital platforms, devices, and apps, we see ourselves as fast-moving drivers on the freeways of cyberspace. In many respects, however, we have become driven. We are driven in the sense of being steered, maneuvered, goaded, and compelled by digital technologies. We are also driven in the sense that we no longer fully exercise, or develop, the aptitudes and skills required to direct ourselves.

Riffing on Henry David Thoreau's lament that "men have become the tools of their tools," Marshall McLuhan, the father of media studies, wrote: "First we build the tools, then they build us." In many respects, this is a central truth of the human condition. Our forebears' minds created language; then language shaped human minds. They tamed fire; then fire molded appetites and social interactions. They developed agriculture and cities; then agriculture and cities altered physiologies. Today we engineer digital media and artificial intelligence, and these technologies are currently remaking our economies, our cultures, our polities, and our personalities. The problem is that these powerful tools are not only constructing our behaviors. They are dismantling our agency.

Notwithstanding the many benefits of digital technology, it is now widely acknowledged that dependence on them has become central to our "very sense of agency" (Paasonen 2021, 55). And while many are rightfully opposed to the "scapegoating of smart devices" and "wistful nostalgia" for a pre-digital world, it is also widely recognized that we face a deepening crisis wherein "networked devices, apps, and social media services are atrophying our attention spans, eroding our capacity to focus and think, addicting us, boring us, and stealing our time, as well as stopping us from truly relating to one another, engaging in critical thought, or contributing to public life in a meaningful way" (Paasonen 2021, 1, 146). Unfortunately, such entreaties typically avoid the challenge of devising a means to reap the many benefits of digital technologies while avoiding their dangers. That crucial task is taken up here.

The harmful impacts of social media and the dangers of AI have become common topics of concern and debate. This book sets itself apart by assessing the threats we face across psychological, social, and political domains while addressing their impact on crucial human aptitudes. In turn, it provides a detailed account of how we can safeguard human agency in a world that is surveilled, administered, and galvanized by digital technologies. It highlights the indispensability of self-leadership and service as a bulwark against digital dysfunctions, and explains how these beneficial dispositions, habits, and skills can be cultivated. Before we consider a remedy for human downgrading, however, we must first understand why and how it is occurring.

The AI-Enabled Attention Economy

While the amount of time people spend online has exploded in recent years, it has been growing for more than three decades, ever since the arrival of the Internet and the World Wide Web. With the proliferation of smartphones and the rise of social media, the migration to cyberspace accelerated and became global. Search engines such as Google and Bing, and social media sites such as Facebook, Instagram, Telegram, TikTok, WhatsApp, and YouTube are some of the more prominent digital platforms that facilitate this migration. Their business models are based on advertising revenues. The goal is the conversion of users into customers.

Toward this end, digital platforms collect data on where users go and what they search for, watch, say, and do, who clicks on specific advertisements, who signs up for mailing lists or alerts, and which of these searches, viewings, clicks, and subscriptions end up in purchases. Enabled by artificial intelligence, the platforms induce us to browse, surf, post, share, like, watch, play, and purchase. We are flipped into play by algorithms that keep us incessantly engaged. As if inhabiting a pinball machine, we find ourselves bounced from one bumper to another, the prospect of a reward or forfeit ever present.

The more digital platforms can secure and sustain our attention, the greater their profits. To this end, data on our locations and movements, preferences and purchases, communications and inquiries, emotional states and physical health, acquaintances, friends, and social networks are collected, analyzed, and monetized. Collectively, digital platforms and the businesses and organizations that supply their advertising revenues constitute the attention economy.

The attention economy has colonized much of twenty-first-century life by way of ubiquitous and ever more deftly targeted advertising. At the same time, we have also become societies of self-advertisers. To an unprecedented degree, adults, youth, and increasingly children are engaged in digital self-promotion. Whether for personal pleasure, professional need, or social status, people market and brand themselves online. Indeed, lived experience—dubbed RL (real life) on digital platforms—often becomes a mere means to secure a steady supply of updates for social media, typically in the form of posts and selfies that advertise activities, achievements, whereabouts, and associates. As a result, in the odd moment when we do not find ourselves distracted by our digital devices, we are actively trying to distract our fellow cyberspace travelers. In this realm of ceaseless engagement, attention has become the "most irresistible of drugs" (Franck 1999).

Historically, character and behavior were primarily shaped by kin, community, institutions of learning, and places of work. These social structures, in turn, were shaped by local natural environments. So there was a great deal of variation, geographically and culturally, in how children were raised and how customs and social institutions developed. Today, across the globe, the single largest force shaping our characters and behaviors is digital technology. In the AI-enabled attention economy, there is a great deal of misshaping.

We would be mistaken to describe the problem as one of humans versus machines. Still, a brief look at the history of the competition is illuminating. John Henry, the legendary steel driver for the early American railroad, was reputedly the strongest man alive. His days of pounding thick steel spikes into rocks with a hammer were cut short when pitted against a steam-powered drill in a grand competition, Henry died after giving it his all. The fatal competition, which lore has Henry winning, was a test of power and pace. Henry's victory was Pyrrhic. As industrial machinery improved with each passing year, flesh-and-blood workers could not hope to rival their mechanical counterparts in either strength or speed.

In the realm of digital devices, power and pace are also important. We have long ceded this competition as well. The best digital switches are about 10 million times faster than the electrochemical ones located inside our skulls. Digital switches, unlike our meaty ones, are improving every year. Faster switches mean more signals, and more signals mean more power. But speed and raw computing power are not the only arenas in which human supremacy is waning.

Strategic thinking was long held to set humans apart, and chess, for that reason, was believed to be a game that machines would never win. This conviction dissolved in 1997 when grandmaster champion Gary

Kasparov lost a chess match to IBM's Deep Blue computer. A smart machine designed by IBM gained the upper hand over human competitors again in 2011 by arranging words into questions in response to cleverly disguised answers. That year, IBM's *Watson* decisively beat the world's top Jeopardy! players on national television. Former champion Brad Rutter and current champion Ken Jennings, with 74 consecutive wins to his name, were trounced. Neither contestant managed to gain even one-third of Watson's dollar winnings. Watson provided complex linguistic information better than the ablest humans, while responding quickly to puns and riddles. The defeated Jennings acknowledged the arrival of "our new computer overlords." To be sure, the writing was on the wall. Machines could surpass humans in the retrieval, analysis, and integration of knowledge, and their power would grow rapidly in the coming years.

Still, some speculated that machines would never get the upper hand in more intuitive kinds of games. *Go* is widely considered the most complex and intuitive game of strategy. By early 2016, however, an AI system called AlphaGo, a deep learning program developed by Google's subsidiary DeepMind, decisively defeated Lee Sedol from Korea, the finest Go player of his generation and an 18-time world champion. More than 200 million people watched the televised event.

Early smart machines learned by reviewing thousands of human games and strategies. AlphaGo Zero, DeepMind's next iteration, skipped this process. Instead, it started from scratch, knowing only the rules of the game. Then it played Go against itself, discovering new strategies based on innumerable games waged between its neural networks. After three days of such "reinforcement learning," AlphaGo Zero beat the original AlphaGo, 100 games to 0. After 21 days of practice, under the pseudonym "Go Master," it beat 60 Go champions online. In May 2017, AlphaGo Zero was victorious over the current world champion, Ke Jie of China, in 3 out of 3 games. After only 40 days of practice, AlphaGo Zero could lay claim to being the best Go player in all of history. The oldest continually played board game on the planet had been challenging human beings to display sophisticated strategic thinking for 2500 years. It was conquered by a machine, learning from scratch, in less than six weeks.

In late 2017, by playing against itself at a rate of millions of games per hour for one day, AlphaZero, a generalized version of AlphaGo Zero easily beat Stockfish, the world's top-rated computer chess program.

Given the straightforward goal of moving an assortment of pieces on a 64-square chessboard to capture your opponent's king, or encircling an opponent by iteratively placing stones on a 19 × 19 grid in Go, these strategic games might be thought to demand little in the way of psychological insight or the capacity to act in the face of incomplete information. But in 2017, Libratus, an AI program developed at Carnegie Mellon University, decisively beat four professionals at a 20-day, no limit *Texas Hold'em* poker championship. Libratus was designed for a game where players engage in bluffing in conditions of uncertainty. The last frontier in strategic games had been breached. As Libratus's developer observed: "The best AI's ability to do strategic reasoning with imperfect information has now surpassed that of the best humans" (Spice 2017). Strategic thinking in conditions of uncertainty, long held as the exclusive province of a single species on the planet, had decisively fallen into the domain of machines.

The arrival of large language models (LLMs) in the early 2020s quickly expanded the reach of AI. LLMs process and utilize natural language for a wide variety of tasks, including casual conversation, scientific scholarship, prose, poetry, music composition, and writing computer code. They employ machine learning to predict what sequence of words, images, numbers, or other symbols will satisfy a specific request or goal. As the most common form of Generative AI, LLMs are widely available in many chatbots and virtual agents. Image creation and integration have been onboarded, for multimodal performance. More powerful versions appear regularly, with Generative AI increasingly employed in its own design and improvement. Investment has skyrocketed, reaching $25.2 billion in 2023, an eight-fold increase from the previous year, with the number of AI patents increasing more than 31 times since 2010 (Maslej et al. 2024).

Generative AI demonstrates the capacity to dispatch a wide variety of complex tasks much faster than human beings, and often with comparable if not superior quality. It can outperform people in a wide range of standardized tests, from the College Board's SATs for reading and math to the Uniform Bar Exam. More specialized AI systems can discover solutions to advanced mathematical problems and create complicated computer code from scratch. Corporate labs are working on versions that have advanced reasoning, situational awareness, and long-term planning capabilities (Gladstone AI 2024, 23–24). In recent head-to-head tests, AI chatbots outperformed most humans in generating ideas that were deemed creative (Koivisto and Grassini 2023). In what may be a harbinger of the future of science, only one of the five scientists who won the Nobel Prize for Physics and the Nobel Prize for Chemistry in 2024 worked in the field of the award. The others designed AI programs. The prognosis is bracing: "Henceforth, perhaps scientists will merely craft the tools that make the breakthroughs, rather than do the revolutionary work themselves or even understand how it came about" (Burdick and Miller 2024). Increasingly, AI is not a tool (like telescopes and microscopes) that scientists use to make discoveries. AI is doing the discovering.

When smart machines beat out humans in arenas that demand specific knowledge or skills, the common response is to place a higher value on other knowledge or skills (Santoro and Benoît 2023; Cha et al. 2020). This is a variation of the "sour grapes" phenomenon: we act like Aesop's fox, disparaging the value of tasty fruit that we cannot reach. The knowledge and skills that AI masters get devalued; those that still evade it are celebrated as uniquely human. In short, the goalposts keep getting moved to avoid our species having to concede wholesale defeat. As Sherry Turkle (2015, 349) observes "We redefine what is human by what technology can't do." The realm of what smart machines cannot do, however, is rapidly shrinking. Time and again, boundaries that were predicted never to be breached have been surpassed. We should expect more of the same.

It is understandable that we chafe at the idea of machines mastering us in arenas characterized by language use, strategic thinking, deception, uncertainty, and creativity—crucial features of our social, economic, and political lives. But this is the world in which we now live. If we are truly in a race, AI is clearly on a winning streak. There is cause for alarm.

To make matters worse, AI is often employed by bad actors to spread misinformation, engage in espionage, and disrupt infrastructure. In the future, it might be employed to terrorize populations, start wars, and create lethal pathogens that cause pandemics. In the wrong hands, AI can become a formidable weapon. It also poses a threat apart from misuse by miscreants. Even in the absence of bad actors who weaponize it, a highly developed form of AI might become misaligned with human values, stealthily accrue power, harm large numbers of people, or, potentially, exterminate the human race.

In 2023, hundreds of artificial intelligence experts, the chief officers and scientists of OpenAI, Google's DeepMind, Microsoft, Anthropic, and Inflection, and thousands of notable public figures signed a "Statement of AI Risk" sponsored by the Center for AI Safety. The statement was simple and straightforward: "Mitigating the risk of extinction from AI should be a global priority alongside other societal-scale risks such as pandemics and nuclear war" (Center for AI Safety 2023). Many signatories are worried about human prospects. Mustafa Suleyman, the co-founder of DeepMind and CEO of Inflection, hopes to be proven wrong in his conviction that AI "containment is not possible." Suleyman (2023, 116) writes: "There comes a point where technology can fully direct its own evolution; where it is subject to recursive processes of improvement; where it passes beyond explanation; where it is consequently impossible to predict how it will behave in the wild; where, in short, we reach the limits of human agency and control." Threats to human agency should not be dismissed. As a species, we are circling the flames.

Public efforts are underway to act before it is too late (e.g. Barnhizer and Barnhizer 2019; https://pauseai.info). Governments, in turn, are assessing the danger and crafting legislation. A 2024 study commissioned by the US Department of State observed the "urgent need" for intervention to counter "catastrophic national security risks from AI weaponization and loss of control." Loss of control occurs when an AI system becomes "so capable that it escapes all human efforts to contain its impact" (Gladstone AI 2024, 11–12). By way of deception or other strategies, it gains power over its environment, accumulates resources, recursively improves its capacities, thwarts the human modification of its goals, and avoids being shut down. The prospect of smart machines escaping human control is made more likely as businesses and nation-states compete to develop and deploy the most powerful models as quickly as possible. However, celebrations of success by specific corporations or countries may be short lived if the creators, in Frankenstein-like fashion, become their fabrications' victims.

A group of current and former employees of frontier AI companies, including Open AI and Google's DeepMind, published an open letter in 2024 arguing that whistleblowers need greater protection. Given the lack of effective government oversight of AI corporations, they argued, we must encourage employees to hold their businesses accountable. Standard whistleblower protections are insufficient because they pertain only to the reporting of illegal activity. Yet most of the risks generated by AI are neither prohibited nor regulated. As such, support needs to be given to employees who are critical of the corporate race to develop the most powerful models, including those that portend "the loss of control of autonomous AI systems potentially resulting in human extinction" (Hilton 2024).

Some experts believe that AGI will be developed by the late 2020s. As Sam Altman (2024), CEO of Open AI, observed, that is but "a few thousand days" away. Others suggest a later date, and some argue that it will never occur. If and when AGI is created, it may only be a matter of years, months, or weeks before it morphs into what is called *superintelligence*, something smart enough to escape human control. Regardless of the timeline, confronting the prospect should be a global priority, given its potential repercussions. But that is not the focus here. Rather, this book addresses another kind of existential threat. My concern is not the rise of an artificial superintelligence that gains autonomy, but the decline of human intelligence that erodes agency.

Even those who opt not to fret about a species-threatening explosion of AI should be concerned about the implosion of human agency. After all, the two outcomes are linked. The chances of becoming dominated or exterminated by a pernicious form of superintelligence dramatically increase if we become cognitively, socially, and politically enfeebled. The likelihood that AI will escape human control is all but assured if human agency is significantly degraded.

Human Agency and Practical Judgment

While there are many who consider digital technology in the age of AI a "threat to human agency," diagnoses mostly focus on the loss of jobs, and proposed remedies tend to be equally narrow (e.g. Agar 2019). The hardships of billions of people who will find themselves out of work or underemployed in the wake of automation and the AI revolution are very troubling. So are the political repercussions. In the absence of a Universal Basic Income (UBI) or large-scale welfare payments, massive unemployment and economic privation would likely prove a breeding ground for authoritarianism across the globe. We should not forget that Hitler's rise to power, achieved at the ballot box in 1932 when the Nazis gained more seats in the Reichstag than any other party, was much abetted by the stock market crash three years earlier and the ensuing Great Depression.

I leave the issue of automation-induced unemployment to those better equipped to address it, though, in truth, no one has been able to propose an adequate solution. It is important to note, however, that the loss of jobs does not only produce lost wages. It begets a loss of aptitudes and skills. Deskilling in the digital world, as will be discussed in Chapter 4, occurs well before people become gainfully employed and extends far beyond the workplace. Still, job sites, at least potentially, are schools for skill acquisition and aptitude enhancement. If AI and automation push large numbers of people into unemployment or underemployment, regardless of any financial compensation they receive, a key venue for cognitive and social skill development will be lost.

In a *New York Times* op-ed article entitled "How I Learned to Stop Worrying and Love A.I," Robert Burton (2015) acknowledges that computers will be supreme in all areas where data is in play, with predictable impacts on the workforce. Humans will keep their edge, Burton believes, only in the "ability to contemplate the non-quantifiable. Emotions, feelings and intentions—the stuff of being human." In the same vein, Andrew McAfee and Erik Brynjolfsson acknowledge that neural network AI programs will surpass humans even in fields that rely less on raw cognitive power and more on pattern recognition. These MIT professors believe that humans will retain their edge in only one arena: "Understanding and addressing the societal challenges brought on by rapid technological progress remain tasks that no machine can do for us" (McAfee and Brynjolfsson 2016). That is not strictly true. Smart machines can, will, and should be employed to help us grapple with the societal challenges brought on by technology. Still, human beings do have a unique aptitude to blend cognitive capacities with emotional intelligence in a way that builds on common sense to enable understanding, evaluation, and contextually responsive action. The ancient Greeks identified this capacity by the term *phronesis (φρόνησις)*, which is usually translated as practical wisdom, prudence, or practical judgment.

Practical judgment is often thought to consist of a certain type of knowledge, perhaps in the family of wise maxims. But practical judgment is less an accumulation of precepts than a process of learning. It is learning by doing, where the doing includes worldly experience and skillful reflection. Some human judgments fall into the category of formal reasoning. Here, conclusions are derived from premises by applying logic. Formal reasoning is an exercise in deduction or induction. It is largely an algorithmic process. Smart machines, which run on algorithms, are very good at it. Practical judgment entails more than reckoning, though reason certainly plays its part. To exercise practical judgment, we must identify things of importance from a potentially endless pool of candidates and assess their relative significance. It requires evaluating diverse variables and integrating incommensurable values. This is chiefly achieved by way of a narrative framing.

Practical judgment makes sense of our interactive passage through time by situating our experiences within stories. In turn, it imaginatively explores possible futures in the form of alternative storylines (Flyvbjerg 2001; Flyvbjerg 1998, 8; Johnson 1993, 164; MacIntyre 1981, 201; MacIntyre 1999, 56–57, 96). As pragmatist philosopher John Dewey observes, judgment operates by way of a "dramatic rehearsal" whereby we play out competing narratives in our imagination. These "tentative rehearsals in thought" allow the practical judge to "run ahead and foresee outcomes, and thereby avoids having to await the instruction of actual failure and disaster" (Dewey 2002, 190). So informed, we can determine how events ought to unfold to achieve the best possible results. To cultivate practical judgment is to develop a skill for the narrative framing of reality (Thiele 2025; Thiele and Young 2016; Thiele 2006).

Consider the daily challenge of reconciling our passing desires and the exercise of individual choice with the demands arising from moral convictions, personal relationships, social memberships, professional obligations, and political affiliations. No single lens can bring all pertinent claims on our attention and commitment into focus at once, and no single scale can comparatively weigh them. Such incommensurability reflects the complex nature of the human condition. Our multifaceted lives rebuff any singular metric of relevance or worth. Practical judgment focuses attention and marshals commitment by organizing a multidimensional and often bewildering world into stories with specific characters, plots, and themes. These narratives provide both orientation and traction, imposing relevance upon the components of our world and generating a sense of what matters.

Philosopher Alasdair MacIntyre (1981, 201) observed that "I can only answer the question 'What am I to do?' if I can answer the prior question 'Of what story or stories do I find myself a part?" Practical judgment employs narrative framing to gain what Dewey (1993, 106) called "a sense of respective or proportionate values," determining which concerns merit attention and discerning their relative importance. Oftentimes, tunnel-vision preempts a broader perspective that would allow effective action in pursuit of enduring benefits. It is an all-too-human foible to miss the forest for the trees and to win the battle but lose the war. Alternatively, a too-diffuse overview will obscure the trees just as a lack of attention to detail will cause the battle to be lost. Narrative framing provides an appropriately scaled purview, generating a sense of direction and purpose. The fruits of this effort, in turn, can be employed to persuade others, inviting them to address a specific challenge by contributing to a story in the making (Thiele 2025; Arendt 1982, 72–84, 1968, 222–223; Larmore 1987, 2001).

Stories are powerful things. It is all too easy to become "caught up" in them, neglecting relevant information and abandoning rational assessment. Witness the mischief caused by conspiracy theories, or the damage done by dogmatic ideologies. Narrative orientations are vulnerable to emotional reactivity, the reinforcement of prejudices, and blindness to historical facts and scientific data (Kahneman 2011; Taleb 2010; Plous 1993; Tversky and Kahneman 1982; Tetlock 2005, 118). Accordingly, practical judgment chastens its narrative orientation with critical thinking, the product of reason and reflection.

This chastening is not a rejection of narrative frameworks. It entails the integration of certain sorts of narratives, namely, stories of how and when cognitive limitations, biases, or emotional reactions waylay attention and misguide assessment. Practical judgment is the product of experience, of learning from stories of success and failure. These stories need not solely be one's own. We can learn from the achievements and mistakes of others, by way of direct observation and through biographical, historical, and even fictional narratives (Gottschall 2012; Oatley 2008).

Practical judgment is neither an exercise of rational calculation that sidelines passion nor an affective disposition that ignores reason. In turn, it pays heed to both ends *and* means, cherished values *and* instrumental mechanisms, ideals *and* practicalities, long-term goals *and* short-term needs. It does not optimize along a singular gradient. Rather, practical judgment integrates values, ideals, and expediencies, allowing coherence in our goals and effectiveness in our means. As such, practical judgment is inherently a normative faculty, combining know-how with know-why. By way of a narrative framing, it informs and empowers agency, enabling us to become the worthy protagonists of our lives.

If any human faculty can survive the onslaught of AI and automation, it will be practical judgment. As AI becomes increasingly adept at various forms of problem-solving, the human contribution will increasingly become a matter of "understanding which problems should or could be addressed" (Verganti et al. 2020, 212). That is to say, AI will become ever more proficient at devising innovative solutions for specific problems. The enduring human contribution will be found in the selection of issues that merit attention. Practical judgment allows us to discern the best means to achieve specific objectives. It also entails knowing what objectives are worth pursuing. To exercise practical judgment is to know what matters, how much it matters, why it matters, and how best to achieve it. This aptitude for focusing attention in order to navigate a complex, multidimensional world has been deemed the "Holy Grail" of human intelligence (Hasselberger and Lott 2023). AI does not currently display this form of intelligence, and arguably never will. As such, practical judgment constitutes the last stand of the distinctly human in an age of smart machines.

Over the centuries, various capacities have been thought to be uniquely human, separating us from other animals. Rationality, speech, and creativity are cases in point. With the arrival of LLMs, rationality and speech no longer appear uniquely human. What, then, of creativity? Creativity is most often defined as the ability to produce ideas or artifacts that are new and useful. It might be defined as "relevant and effective novelty" (Cropley 1999, 512). Newness on its own is easy to generate. Take any ten randomly selected items from your refrigerator and cupboard, mix them together, and cook them for an hour at an arbitrarily selected temperature. You will have a new, original item for dinner. Whether it is valuable, which in this case means tasty and nutritious, is another matter. The reverse is also true. Valuableness on its own is easy to produce. Simply supply something that most people find useful or relevant. A fork or spoon will do, or a mixing bowl. These things are commonplace, not creative. Creativity, the combining of the original *and* the valuable, is relatively rare.

Determining whether something is original and valuable entails judgment. Brilliant creators have this aptitude. Still, judgments about originality and value are not solely the prerogative of creators. The broader community also has a voice. Creative works of literature, music, art, cuisine, and science gain their distinction not simply because the people who produce them say so. Rather, the distinction is in the minds, hearts, and sensory organs of both producers and beholders. As Mihalyi Csikszentmihalyi (1996, 23) observes, creativity "does not happen inside people's heads, but in the interaction between a person's thought and a sociocultural context." Creativity is a shared judgment.

Many innovative thinkers, including Albert Einstein and John Dewey, insisted that creativity inheres not only or even primarily in the wherewithal to solve complex problems in novel ways. Rather, creativity is the ability to determine which problems most need solving. The determination of relevance is crucial. The "creative process consists exactly in this separation of what is relevant from what is not," Csikszentmihalyi (1988, 161–162) observes, along with the disposition "to persevere despite the lack of clear criteria for solution." Creativity is grounded in the motivation and ability to focus and sustain attention on specific fields, identify worthy challenges within these fields, and dedicate oneself to their resolution. As a result of such efforts, the new and valuable is produced. A creative cook likely will not be able to innovate in the arena of carpentry or coaching, and vice versa. The same can be said of creative artists, scholars, and entrepreneurs. Creativity is field-specific. What all creative people have in common, however, is the ability to discern what matters, how much it matters, why it matters, and how best to achieve it in their specific area of expertise. That is to say, creativity is grounded in practical judgment, and its status as a preeminent human attribute arises from this foundation.

Practical judgment is cultivated by way of reflection and self-direction. Reflection allows us to integrate what we have learned from experience. Self-direction allows us to gain wide-ranging experience. To cultivate practical judgment by means of reflective self-direction is to think our way into right ways of doing, and to do our way into right ways of thinking. This reciprocal relationship between thinking and doing is the core of human agency. If agency is eroded, practical judgment cannot be cultivated. In turn, if practical judgment is absent, human agency will go awry.

The discerning of what matters, how much it matters, why it matters, and how best to achieve it is only possible for an agent who has been immersed, experientially and reflectively, in the diverse values, ideals, and expediencies of human life. The reciprocal and inherently narrative relationship between thinking and doing makes practical judgment a uniquely human *aptitude*. Aptitudes are capacities that we are predisposed to develop. We are not born with the ability to speak. Rather, we are born with an aptitude for speech, and we routinely become competent speakers by being exposed to linguistic interactions as young children. We are, as it were, hardwired to learn language. The same is true for many other perceptual, physical, cognitive, and emotional capacities, including practical judgment.

Common sense, a core human attribute, provides the foundation for practical judgment. We are not born with common sense any more than we are born with the ability to speak. It develops over time, as a product of experience. Common sense is an array of skills, such as the ability to gauge distances and weights, assess visual perspective, and attribute goals and intentions to actors. A lack of common sense reflects an inability to learn from common experiences. To have common sense is to have learned the lessons that human beings are predisposed to learn. In 2024, commonsense reasoning was one of the very few arenas where human beings still outperformed the most sophisticated AI programs (Strickland 2024; Maslej et al. 2024).

Common sense might be thought of as a robust mental model of the way the physical world and social world work, a contextually attuned grasp of phenomena and relationships within the realms of basic physics, biology, and psychology, among others. A person with common sense knows that dropping a wineglass on the floor will cause it to abruptly shatter, while it will lightly bounce if dropped on a soft mattress. A person with common sense knows that screaming at a small child will likely bring tears, making you an insensitive bully, while yelling at a sporting event makes you a valued part of the spectating crowd.

Common sense is crucial to our navigation of the physical and social worlds. But it is far from infallible. For example, it encourages us to believe that heavier objects always fall faster than lighter objects. This belief is grounded upon experience and serves us well enough in daily life. Still, devising a general law of physics upon it, as ancient philosophers did, is mistaken. Only imaginative experimentation, or scholarship, informs us that lighter objects fall at the same speed as heavier objects in the absence of atmospheric resistance. So, when Commander David Scott of the Apollo 15 mission dropped a hammer and a feather while standing on the moon, the objects landed at the same time. Back on Earth, hundreds of millions of television viewers watched their common sense upended.

Voltaire sardonically observed that common sense is not so common. Pace Voltaire, basic common sense *is* widespread. Still, not everyone takes common sense to a higher level by gaining wide-ranging experience and reflectively integrating its lessons. Samuel Taylor Coleridge (1836, 151) observed that "Common sense in an uncommon degree is what the world calls wisdom." Practical judgment is common sense taken to a higher level by way of wide-ranging, well-integrated experience.

Untutored common sense is grounded in everyday life experiences, and, consequently, expects the future to look much like the past. Practical judgment integrates a wider range of learning. It pays heed to diverse experiences, both direct and secondhand, including experiences where an absence of imagination or a lack of daring led to failure or mediocrity rather than success or excellence. As such, practical judgment prompts us to expect the unexpected in certain circumstances and to act accordingly. The practically wise person is an excellent student of social and political life, attentive to uncertainties and expectant of surprises.

Before the arrival of the Internet and digital platforms, the political philosopher Michael Oakeshott already lamented the "crowded" modern life replete with endless distractions. While it offered "instant participation" in a "ceaseless flow of seductive trivialities," the crowded life lacked the reflectivity that made for informed choices and the cultivation of practical judgment (Oakeshott 2001, 33). A *frenetic* life, one might say, prevents the development of a *phronetic* life.

Today's AI-enabled attention economy produces a hyper-crowded world that celebrates speed and busyness. Contemporary culture is enamored with the endless novelty and ever-increasing pace and power of technological development. The cultural mood finds its voice in the doctrine of ***accelerationism***. Accelerationism hitches its wagon of technological optimism to the unrestricted development and deployment of artificial intelligence. As a manager at the tech company NVIDIA insisted, an AI-enabled personal assistant is "a great way to accelerate us as humans" (Evans 2024). The clearest statement of accelerationism is "The Techno-Optimist Manifesto," published by the venture capitalist Marc Andreeson (2023). The manifesto states:

> Techno-Optimists … believe everything good is downstream of growth…. And … the only perpetual source of growth is technology…. We believe in accelerationism—the conscious and deliberate propulsion of technological development…. Artificial Intelligence ramps up the capabilities of our machines and ourselves…. We believe Artificial Intelligence can save lives….. We believe any deceleration of AI will cost lives. Deaths that were preventable by the AI that was prevented from existing is a form of murder…. We believe that advancing technology is one of the most virtuous things that we can do. We believe in deliberately and systematically transforming ourselves into the kind of people who can advance technology.

The manifesto argues that the pursuit of sustainability is "against life" and that the Precautionary Principle must be jettisoned "with extreme prejudice." Of course, doing so will cause some harm. But that is the price of progress.

The original motto of Facebook was "move fast and break things." CEO Mark Zuckerberg explained his company's "prime directive" in a 2009 interview: "Unless you are breaking stuff, you are not moving fast enough" (Business Insider 2010). Not to be outdone, Sam Altman boasted on his X account that "you cannot outaccelerate me" (Altman 2023). AI-engineered acceleration, Altman claims, will produce "massive prosperity" while also "fixing the climate, establishing a space colony, and the discovery of all of physics." With all this going on, accelerated humans likely will not be able to afford the time required for reflective self-direction and the cultivation of practical judgment. Consequently, human agency is one of the things that accelerationists appear willing to break.

While accelerationism is rampant in an age of AI, the love of speed, and concerns about its costs, have long been part of modern life. In the 1983 movie, *Reuben, Reuben*, an American advocate for a speed-reading course boasts of a star pupil who consumed Tolstoy's *War and Peace*—over 1200 pages in length—in 55 minutes. The movie's protagonist, a Scottish poet, remained unimpressed. "I would pay vast sums," he responded, "for anyone to teach me to read the books I love at a snail's pace." Increasing reading speed frees up time to accomplish other things. Doing as many things as possible as quickly as possible is the unwritten rule of accelerationist culture. In contrast, learning how to read more slowly, which is to say, more reflectively, facilitates the development of practical judgment.

When the human aptitude for practical judgment is eroded, human agency is imperiled. Not everyone is concerned. Assessing the likely impacts of AI over the coming decades, experts counsel us to "stop worrying and embrace the future" (Lee and Quifan 2021, xvii, xix). The upbeat recommendation stems from their staunch "belief in human agency—that we are masters of our fate." While I share the assessment of the manifold benefits that AI might provide in the coming years, including medical advances, scientific breakthroughs, economic prosperity, cultural innovation, and climate change mitigation, the conviction that human agency remains unassailable is mistaken. The experts' faith is flawed because it ignores both the allure and business model of the attention economy.

A recent report on "The Future of Human Agency" assessed whether people will retain control over essential decision-making as digital systems and AI proliferate. Hundreds of technology innovators and developers, business and policy leaders, researchers and activists were canvassed. The majority believed that by 2035 smart technologies will *not* be designed to allow their human users to maintain control. The reasoning was straightforward: people greatly value convenience and are often overwhelmed by the demands of decision-making. In turn, the designers and deployers of digital technologies have few, if any, incentives to encourage human users to exercise agency (Anderson and Rainie 2023). I agree. If we do not change course, human agency will be perilously eroded in the coming decades.

This book assesses human downgrading across psychological, social, and political domains. I present evidence, grounded in extensive scholarship, that the AI-enabled attention economy is responsible for sharply rising levels of distraction, dependence, deskilling, dogmatism, and despair. Worrying about such things is pointless, however, if it does not help generate effective remedies. Accordingly, I explain how we can cultivate the dispositions, habits, and skills needed to sustain human agency in a world increasingly surveilled and administered by digital technologies.

Healthy dispositions, good habits, and well-honed skills give individuals the same benefits that laws and constitutions provide to democratic nations. They are self-imposed constraints that enable and safeguard the exercise of freedom. Cicero, the ancient Roman statesman and orator, observed that to be free one must become subservient to the law. Jean-Jacques Rousseau, the early modern theorist of democracy, agreed that political freedom entailed abiding by the law, adding that the law in question must be of the people's own making. Democratic liberties are products of collective constraints. Likewise, individual freedoms issue from self-governance.

The Case for Digital Distancing

Digital distancing is the form of self-governance most needed today. It entails developing the dispositions, habits, and skills that enable reflective self-direction and practices of caring. We thoroughly explore these crucial human aptitudes in Chapter 7, where reflective self-direction is defined as self-leadership—becoming the protagonist of one's own life. Practices of caring, which I call service, refer to other-directed activities that contribute to conserving or improving human and natural communities. Self-leadership and service are essential to human flourishing. Today, they are also serve as crucial defenses against the human downgrading that ensues from accelerating technological upgrading.

Social distancing during the pandemic was needful. But it was not in itself a good thing. People chose to practice it because it decreased the spread of a potentially lethal virus. Digital distancing is also needful, as it protects us from virulent forms of virtual life. But digital distancing advances things that *are* good in themselves, stimulating both social proximity and what might be called self-proximity. The former occurs through authentic face-to-face interactions and relationships. The latter occurs by way of reflection and self-direction. In both cases, human agency is exercised and developed because our attention is focused to best serve our intentions. Self-leadership and service combat the most dangerous elements of the AI-enabled attention economy by way of a humanized intention economy.

We create worthy intentions by way of reflection. We realize these intentions by way of self-direction. And we can only engage in reflection and self-direction by focusing and sustaining our attention. Toward the end of the nineteenth century, the psychologist William James devoted a chapter of *The Principles of Psychology* to attention, a topic much neglected by his forerunners and contemporaries. "Every one knows what attention is," James (1890, 403–404) wrote: "It is the taking possession by the mind, in clear and vivid form, of one out of what seem several simultaneously possible objects or trains of thought. Focalization, concentration, of consciousness are of its essence. It implies withdrawal from some things in order to deal effectively with others, and is a condition which has a real opposite in the confused, dazed, scatterbrained state which in French is called *distraction*, and *Zerstreutheit* in German." James went on to observe that "the faculty of voluntarily bringing back a wandering attention, over and over again, is the very root of judgment, character, and will. No one is *compos sui* [master of one's self] if he have it not. An education which should improve this faculty would be *the* education *par excellence*. But it is easier to define this ideal than to give practical directions for bringing it about." This book provides guidelines for improving the faculty of attention, which as James insists, is the root of practical judgment and human agency. Today, however, the problem is more intractable than James could have imagined, and the stakes are much higher.

Developing the human capacity to direct and sustain attention, as James acknowledges, has never been easy. But never before have we had to safeguard it on a global scale against systematic assaults by the most highly developed technologies and the most powerful corporations and governments. I am far from alone in this diagnosis. Maintaining that "attention is *the* crucial resource of our epoch," Yves Citton (2017, 10) insists that there is nothing inevitable about the erosion of our cognitive capacities by digital technologies. A reversal of current trends is possible if we understand how attention is developed, manipulated, and captured. In the same vein, James Williams (2018, xii) writes that "The liberation of human attention may be the defining moral and political struggle of our times. Its success is prerequisite for the success of virtually all other struggles." Absent such an effort, Luciano Floridi (2015, 12–13) writes, we will forfeit "an inherent human attribute that conditions the flourishing of human interactions and the capabilities to engage in meaningful action."

The AI-enabled attention economy erodes our capacity to generate, direct, and sustain attention. Still, making our attention our own again does not entail a Luddite-style renunciation of technology. Like most people, I use digital devices daily and depend upon them extensively in my professional life and personal affairs. These technologies offer countless benefits, and they can and should be deployed to achieve a wide array of individual goals and public goods. Their wholesale rejection is neither practical, necessary, nor advantageous.

Indeed, effectively addressing the harmful impacts of digital technologies in our personal and political lives will often entail the development and use of innovative software, platforms, and devices. Many democratic processes and institutions, for example, can be enhanced by digital technologies, including AI. Such technologies, which are easily accessed by way of curated list of tools and resources (e.g. https://dem.tools/guides-and-tools), can increase and improve transparency, accountability, information sharing, and participation. They include digital mechanisms for citizen-based agenda setting and problem-solving, collective means of discovering common ground and making decisions, and participatory budgeting, among others. This is the expanding realm of "digital democracy" or "civic technology."

I do not address the digital innovations that enhance the spirit and practice of democracy and citizenship. This important task is left to those with greater experience and expertise (see Duberry 2022, 195–224; Collective Intelligence Project 2023; Stanford Digital Economy Lab 2024). My contribution is grounded on the assumption that democracy and civil society cannot be sustained, notwithstanding any helpful technological enhancements, without safeguarding human agency. That project entails digital distancing.

Digital media inform and connect people, and, as such, can contribute to the cultivation of human agency. But caution is in order. Virtual connections are easily co-opted. A brief comparison of digital distancing to other forms of resistance illustrates the challenge. Historically, traditional resistance movements opposed domination and exploitation through protests, sit-ins, demonstrations, and strikes. They also engaged in civil disobedience, and occasionally in sabotage and revolt. Digital distancing does not entail electronic civil disobedience, hacktivism, or whistleblowing, notwithstanding the importance of such activities. And while traditional resistance movements have generally been characterized by the frequent and vibrant expression of moral outrage directed at exploitive economic relations or oppressive political regimes, digital distancing takes a different path.

Online culture is already overflowing with moral outrage. Indeed, the attention economy thrives on it: social media platforms employ algorithms that reward the expression of vitriol and wrath. This outrage is often informed and abetted by misinformation, which diffuses farther, faster, and more broadly than facts in cyberspace. The result is polarization, declining public trust, widespread cynicism, and a toxic environment for personal and political engagement. A few decades ago, the advent of digital media was promoted with bold declarations of the arrival of planetary consciousness and connectivity. We now know that digital networks can be dangerously constrictive and divisive. Digital media often prove more pro-tribal than pro-social.

Self-leadership and service are the forms of resistance most needed in a caustic digital environment that is awash in outrage and is fueled by polarization. By means of reflective self-direction and practices of caring, we can escape constrictive tribalism and avoid the incessant pursuit of digital recognition, which the attention economy both stimulates and rewards. At the same time, just as traditional resistance movements educated their members in nonviolent communication, digital distancing is grounded in nonvirulent communication. And like traditional resistance movements, which reached beyond protest to foster community empowerment, digital distancing entails extricating ourselves from online enthrallments so we might foster collaborative social relationships grounded in crucial human aptitudes.

When the first book on the attention economy was written, the authors forecast that "The trend of more information competing for less attention can't go on forever. Ultimately, people will begin to withdraw from the stress of an attention-devouring world" (Davenport and Beck 2001, 221). What the authors could not foresee at the turn of the last century was the power of AI-enabled digital platforms to foster distraction and furnish convenience. Distracted individuals within the attention economy may not be particularly "stressed-out" by information overload, as one of the conveniences offered is the filtering of options. However, more than options are being managed for us. Needs, wants, desires, and expectations are being refashioned to align with the demands of the attention economy, along with the dispositions, habits, and skills we employ to realize them.

The problem of distraction—the inability or unwillingness to concentrate the mind upon intentions and direct action toward them—is endemic to the human condition. But the mechanisms of distraction have evolved over time. In 1936, T. S. Eliot could already bemoan "this twittering world" in the first of his *Four Quartets*, writing:

> Distracted from distraction by distraction
> Filled with fancies and empty of meaning
> Tumid apathy with no concentration.

Four years earlier, Aldous Huxley described a benign dystopia in his novel *Brave New World*. It was filled with fancies and empty of meaning. Desires were both generated and fulfilled by powerful technologies, including *soma,* a soothing, happiness-inducing drug, and the *feelies*—a haptic, mixed reality experience that could provide all the titillations of world travel and realistically mimic the sensuality of making love on a bearskin rug. Current efforts at engineering the "sensor-filled metropolis of tomorrow" (Fussell 2018) suggest that Huxley was onto something.

We fail to understand the most dangerous threats to civilization, Huxley (2005, 267–268) prophetically observed, if we, like the early advocates of universal literacy and a free press, do not "foresee what in fact has happened, above all in our Western capitalist democracies—the development of a vast mass communications industry, concerned in the main neither with the true nor the false, but with the unreal, the more or less totally irrelevant. In a word, they failed to take into account man's almost infinite appetite for distractions." In the past, Huxley observed, the means to satiate the human appetite for distraction were unavailable. Today they are in full development. And once our cognitive and social aptitudes become atrophied and our critical faculties diminished by full immersion in an AI-enabled attention economy, our appetite for distraction may expand even further. At that point, we will swallow our daily dose of digital soma to ward off any lingering stress, anxiety, boredom, or despair, and embrace all the diversions and comforts provided by virtual life.

In *Brave New World,* residents ingest *soma* to escape troubling sentiments and experience a general state of euphoria. If the absence of negative feelings and the satisfaction of desires were the only component of human flourishing, then utopia might be achieved by pharmaceutical means, or by smart machines that cater to our needs. We could forgo a wide array of cognitive and physical exertions, relying on technology to provide for our burgeoning wants.

If agency is only a means to obtain desired goods and AI-enabled devices can satisfy our desires directly, then we might be spared its exercise. But human flourishing entails more than satiating desires. It demands the exercise of agency. To be sure, the exercise of agency taxes our capacities, tests our convictions, and tries our patience. In its absence, however, neither human flourishing nor democracy can be sustained.

The AI-enabled attention economy is producing chronically engaged people who lack the capacity for reflective self-direction. They also lack the wherewithal for collective self-governance. As Huxley wrote, foreseeing the dangers of our virtual enthrallment: "Only the vigilant can maintain their liberties, and only those who are constantly and intelligently on the spot can hope to govern themselves effectively by democratic procedures. A society, most of whose members spend a great part of their time, not on the spot, not here and now and in the calculable future, but somewhere else … will find it hard to resist the encroachments of those who would manipulate and control it." The dictators of the future, Huxley foresaw, will engineer techniques of manipulation as well as "non-stop distractions." The trivial will become triumphant, and voyeurs will think themselves victorious. All the while, human agency will wane and democratic institutions will perish. This is the brave new world that digital distancing aims to prevent.

To be sure, there will always be plenty of creative, skillful individuals and geniuses. If ever-growing numbers of people across the globe have their agency eroded, however, the prospects for widespread human flourishing are bleak. In turn, demagogues and dictators will thrive. Distracted and dependent people will embrace autocrats who promise to alleviate their anxiety and boredom. Cognitively and socially deskilled people will embrace authoritarian ideologues who supply simplistic nostrums for a complex world. Dogmatic and despairing people will embrace bullies who provide scapegoats to serve as targets of resentment and wrath. Democracy is demanding. It cannot be sustained by downgraded human beings.

In what follows, I argue against false choices. The claim that we must choose either governmental regulation of digital technology or the preservation of freedom and individual autonomy is shown to be baseless. Another false choice is that between optimism and pessimism. These days people often announce themselves as optimists as if to vouch for their moral or intellectual credentials. To call oneself an optimist is to utter a shibboleth designed to gain passage into the hearts and minds of an audience eager to feel good about their lifestyles and prospects. Optimism is *de rigueur* for politicians of all stripes and for most businesspeople. This book is neither optimistic nor pessimistic. Rather, it is hopeful and realistic.

The project of digital distancing is best engaged with what Romain Rolland (1920, 1) once called the fusion of "pessimism of the intelligence … and optimism of the will." The phrase was later adopted by Antonio Gramsci as he sat in an Italian prison, jailed for his political activities by Mussolini's fascists. Rolland, who won the Nobel Prize for Literature in 1915 and later campaigned for Gramsci's release, argued that fascism can only be effectively combatted if we understand both the ruthlessness of its politics and the popularity of its appeal. The contemporary challenge is similar. It would be a grave mistake to underestimate the appeal of digital technologies, or the psychological, social, and political consequences of the human downgrading they produce.

Still, to keep pessimism of the intelligence from becoming a self-fulfilling prophecy, it must go hand in hand with optimism of the will. Optimism of the will is not an intellectual orientation or a cognitive conviction. Neither is it the complacent expectation of a better future. Rather, it is the disposition to act in a way that makes a better future more likely. We commonly call this disposition *hope*. Hope manifests itself through the exercise of agency. As David Orr (2009, 182) writes, "Hope is a verb with its sleeve's rolled up." Active forms of hope are required to reap the rewards of humanity's technological prowess while safeguarding and sustaining its capacity for agency.

In *The Protestant Ethic and the Spirit of Capitalism,* the sociologist Max Weber (1958, 181) famously observed that the technical and economic conditions of the machine age determine the lives of all "with irresistible force," and would likely continue to do so "until the last ton of fossilized coal is burnt." Concern for economic welfare and technological power, Weber argued, should be worn "like a light cloak, which can be thrown aside at any moment." In this manner, one could participate in economic life and make use of technological innovations without becoming wholly determined and defined by them. That was the task at hand. But as Weber was not hopeful, he issued his famous forecast: "fate decreed that the cloak should become an iron cage."

The philosopher Martin Heidegger, like Weber, was worried by technology. And like Weber, Heidegger (1966, 54) endorsed resistance without foregoing its use, writing that "We can use technical devices as they ought to be used, and also let them alone as something which does not affect our inner and real core. We can affirm the unavoidable use of technical devices, and also deny them the right to dominate us, and so to warp, confuse, and lay waste our nature." The AI-enabled attention economy is indeed warping and laying waste to our nature. Still, it need not become an iron cage. If we practice digital distancing and cultivate our capacity for self-leadership and service, digital technologies can serve as light cloaks—utilized to great benefit, yet frequently cast aside so that we more fully exercise and experience the fullness of our humanity.

When Rolland met Mahatma Gandhi in the early 1930s, he observed that his mind "proceeds through successive experiments into action" (*The Nation* 1932). Gandhi's clear-eyed assessment of colonial oppression was unflinching. But it did not diminish the optimism of the will implicit in his tireless efforts to dismantle British imperialism. His creative and courageous organization of nonviolent, civil disobedience was called *Satyagraha*—the firm assertion of truth. Gandhi's efforts, grounded in self-leadership and service, have been a model for many resistance movements over the last century. Resistance to the AI-enabled attention economy also demands self-leadership and service. Today, more than ever, we need the firm assertion of human agency as we navigate digital technologies whose benefits cannot be denied and whose dangers should not be discounted.

References

Agar, Nicholas. 2019. *How to Be Human in the Digital Economy*. Cambridge: MIT University Press.

Altman, Sam. 2023. *X*. Oct 24, 8:19 PM.

Altman, Sam. 2024. The Intelligence Age. Accessed at: https://ia.samaltman.com/.

Andreeson, Marc. 2023. The Techno-Optimist Manifesto. Accessed at: https://a16z.com/the-techno-optimist-manifesto/.

Anderson, Janna and Lee Rainie. 2023. *The Future of Human Agency*. Pew Research Center.

Arendt, Hannah. 1968. *Between Past and Future*. New York: Penguin Books.

Arendt, Hannah. 1982. *Lectures on Kant's Political Philosophy*. Chicago: University of Chicago Press.

Banrhizer, David, and Daniel Barnhizer. 2019. *The Artificial Intelligence Contagion: Can Democracy Withstand the Imminent Transformation of Work, Wealth and the Social Order?* Atlanta: Clarity Press.

Burdick, Alan and Katrina Miller. 2024. A Shift in the World of Science. *New York Times*, October 13. Accessed at: https://www.nytimes.com/2024/10/13/briefing/nobel-prize-artificial-intelligence.html.

Burton, Robert. 2015. How I Learned to Stop Worrying and Love A.I. *New York Times*, September 21. Accessed at: http://opinionator.blogs.nytimes.com/2015/09/21/how-i-learned-to-stop-worrying-and-love-a-i/.

Business Insider. 2010. Mark Zuckerberg, Moving Fast and Breaking Things. Accessed at: https://www.businessinsider.com/mark-zuckerberg-2010-10.

Center for AI Safety. 2023. Statement on AI Risk. Accessed at: https://www.safe.ai/statement-on-ai-risk.

Cha, Young-Jae., et al. 2020. Compensating for the Loss of Human Distinctiveness: The Use of Social Creativity Under Human-Machine Comparisons. *Computers in Human Behavior* 103: 80–90.

Citton, Yves. 2017. *The Ecology of Attention*. Cambridge, UK: Polity Press.

Coleridge, Samuel Taylor. 1836. *The Literary Remains of Samuel Taylor Coleridge*, vol. III. London: William Pickering.

Collective Intelligence Project. 2023. Accessed at: https://cip.org/.

Cropley, Arthur. 1999. Definitions of Creativity. In *The Encyclopedia of Creativity*, Mark Runco and Steven Pritzker, 511–524. New York: Academic Press.

Csikszentmihalyi, Mihaly. 1988. Motivation and Creativity: Toward a Synthesis of Structural and Energistic Approaches to Cognition. *New Ideas in Psychology* 6 (2): 159–176.

Csikszentmihalyi, Mihaly. 1996. *Creativity*. New York: HarperCollins.

Davenport, Thomas H., and John C. Beck. 2001. *The Attention Economy: Understanding the New Currency of Business*. Boston: Harvard Business School Press.

Dewey, John. 1993. *The Political Writings*, ed. Debra Morris and Ian Shapiro. Indianapolis: Hackett.

Dewey, John. 2002. *Human Nature and Conduct*. Mineola, NY: Dover Publications.

Duberry, Jérôme. 2022. *Artificial Intelligence and Democracy: Risks and Promises of AI-Mediated Citizen-Government Relations* UK: Edward Elgar Cheltenham,

Evans, Robert. 2024. The Cult of AI. *RollingStone* (January 27). Accessed at: https://www.rollingstone.com/culture/culture-features/ai-companies-advocates-cult-1234954528/.

Floridi, Luciano, ed. 2015. *The Online Manifesto: Being Human in a Hyperconnected Era*. Cham: Springer Open. Accessed at: https://link.springer.com/book/10.1007/978-3-319-04093-6.

Flyvbjerg, Bent. 1998. *Rationality and Power: Democracy in Practice*. Chicago: University of Chicago Press.

Flyvbjerg, Bent. 2001. *Making Social Science Matter*. Cambridge, UK: Cambridge University Press.

Franck, Georg. 1999. The Economy of Attention. *Telepolis* (December 7). Accessed at: https://www.heise.de/tp/features/The-Economy-of-Attention-3444929.html.

Fussell, Sidney. 2018. The City of the Future Is a Data-Collection Machine. *The Atlantic*. Accessed at: https://www.theatlantic.com/technology/archive/2018/11/google-sidewalk-labs/575551/?utm_term=2018-11-21T10%3A00%3A12&utm.

Gladstone AI. 2024. *Defense in Depth: An Action Plan to Increase the Safety and Security of Advanced AI*. Accessed at: https://assets-global.website-files.com/62c4cf7322be8ea59c904399/65e7779f72417554f7958260_Gladstone%20Action%20Plan%20Executive%20Summary.pdf.

Gottschall, Jonathan. 2012. *The Storytelling Animal: How Stories Make Us Human*. Boston: Houghton Mifflin Harcourt.

Hasselberger, William, and Micah Lott. 2023. Where Lies the Grail? AI, Common Sense, and Human Practical Intelligence. *Phenomenology and the Cognitive Sciences* (October). https://doi.org/10.1007/s11097-023-09942-x

Heidegger, Martin. 1966. *Discourse on Thinking*, New York: Harper and Row.

Hilton, Jacob et al. 2024. A Right to Warn about Advanced Artificial Intelligence. Accessed at: https://righttowarn.ai/.

Huxley, Aldous. 2005. *Brave New World and Brave New World Revisited*. New York: Harper Perennials.

James, William. 1890. *The Principles of Psychology*, vol. 1. New York: Henry Holt and Company.

Johnson, Mark. 1993. *Moral Imagination: Implications of Cognitive Science for Ethics*. Chicago: University of Chicago Press.

Kahneman, Daniel. 2011. *Thinking, Fast and Slow*. New York: Farrar, Straus and Giroux.

Koivisto, Mika, and Simone Grassini. 2023. Best Humans Still Outperform Artificial Intelligence in a Creative Divergent Thinking Task. *Scientific Reports* 13, 13601. Accessed at: https://www.nature.com/articles/s41598-023-408 58-3.

Larmore, Charles E. 1987. *Patterns of Moral Complexity*. Cambridge: Cambridge University Press.

Larmore, Charles E. 2001. Moral Judgment. In *Judgment, Imagination, and Politics: Themes from Kant and Arendt*, ed. Ronald Beiner and Jennifer Nedelsky, 47–64. New York: Rowman and Littlefield.

Lee, Kai Fu, and Chen Quifan. 2021. *AI 2041: Ten Visions for our Future*. New York: Currency

MacIntyre, Alasdair. 1981. *After Virtue: A Study in Moral Theory*. Notre Dame: University of Notre Dame Press.

MacIntyre, Alasdair. 1999. *Dependent Rational Animals*. Chicago: Open Court.

Maslej, Nestor et al. 2024. *The AI Index 2024 Annual Report*. AI Index Steering Committee, Institute for Human-Centered AI, Stanford University. Stanford, CA. Accessed at: https://aiindex.stanford.edu/wp-content/uploads/2024/05/HAI_AI-Index-Report-2024.pdf.

McAfee, Andrew, and Erik Brynjolfsson. 2016. Where Computers Defeat Humans, and Where They Can't. New York Times, March 16, 2016. Accessed at: http://www.nytimes.com/2016/03/16/opinion/where-computers-defeat-humans-and-where-they-cant.html?emc=edit_th_20160316&nl=todaysheadlines&nlid=71023177&_r=0.

Oakeshott, Michael. 2001. *The Voice of Liberal Learning*. Indianapolis: Liberty Fund.

Oatley, Keith. 2008. The Mind's Flight Simulator. *In Psychologist* 21: 1030–1032.

Paasonen, Susanna. 2021. *Dependent, Distracted, Bored: Affective Formations in Networked Media*. Cambridge: MIT Press.

Plous, Scott. 1993. *The Psychology of Judgment and Decision Making*. Philadelphia: Temple University Press.

Rolland, Romain. 1920. Un Livre de Raymond Lefebvre: *Le Sacrifice d'Abraham. L'Humanité*, March 19.

Santoro, Erik, and Benoît Monin. 2023. The AI Effect: People Rate Distinctively Human Attributes as More Essential to Being Human After Learning About

Artificial Intelligence Advances. *Journal of Experimental Social Psychology* 107: 104464.

Spice, Byron. 2017. Carnegie Mellon Artificial Intelligence Beats Top Poker Pros. Accessed at: https://www.cmu.edu/news/stories/archives/2017/january/AI-beats-poker-pros.html.

Stanford Digital Economy Lab. *The Digitalist Papers*. 2024. Accessed at: https://www.digitalistpapers.com/essays.

Strickland, Eliza. 2024. 15 Graphs That Explain the State of AI in 2024. *IEEE Spectrum*, April 16. Accessed at: https://spectrum.ieee.org/ai-index-2024.

Suleyman, Mustafa. 2023. *The Coming Wave*. New York: Crown.

Taleb, Nassim Nicholas. 2010. *The Black Swan: The Impact of the Highly Improbable*. New York: Random House.

Tetlock, Philip E. 2005. *Expert Political Judgment*. Princeton: Princeton University Press.

The Nation. 1932. Mahatma Gandhi Meets Romain Rolland. (February 10). Accessed at: https://www.thenation.com/article/archive/mahatma-gandhi-meets-romain-rolland/.

Thiele, Leslie Paul. 2006. *The Heart of Judgment: Practical Wisdom, Neuroscience, and Narrative*. Cambridge: Cambridge University Press.

Thiele, Leslie Paul. 2025. Human Prospects in a World of Smart(er) Machines. In *The Political Economy of Robots: Prospects for Prosperity and Security in the Automated 21st Century*, ed. Ryan Kiggins, 2nd ed. Palgrave Macmillan.

Thiele, Leslie Paul, and Marshall Young. 2016. Practical Judgment, Narrative Experience, and Wicked Problems. *Theoria* 63 (148): 35–52.

Turkle, Sherry. 2015. *Reclaiming Conversation: The Power of Talk in a Digital Age*. New York: Penguin.

Tversky, Amos, and Daniel Kahneman. 1982. Judgments of and by Representativeness. In *Judgment under Uncertainty: Heuristics and biases*, eds. Daniel Kahneman, Paul Slovic and A. Tversky. Cambridge: Cambridge University Press.

Verganti, Roberto, Luca Vendraminelli, and Marco Iansiti. 2020. Innovation and Design in the Age of Artificial Intelligence. *Journal of Product Innovation Management* 37: 212–227.

Weber, Max. 1958. *The Protestant Ethic and the Spirit of Capitalism*. New York: Charles Scirbner's Sons.

Williams, James. 2018. *Stand Out of Our Light: Freedom and Resistance in the Attention Economy*. Cambridge: Cambridge University Press.

CHAPTER 2

Regulation and Self-Governance

This chapter surveys the historical development of the AI-enabled attention economy and outlines the accompanying threats to human agency. It provides a detailed account of the challenges associated with the regulation of AI and digital media. Governmental regulation is of crucial importance. But it cannot actuate the human capacity for reflective self-direction. Self-governance grounded in digital distancing is necessary. Neither governmental regulation nor digital distancing should become a partisan issue.

Today, the most crucial resource is not land or people, armies or navies, gold or silver, coal or petroleum, foundries or factories. It is information. For the first time in human history, the most crucial resource is not something concrete and enduring. It is ephemeral and transient. And while the most crucial resources in times past were always in short supply, information today exists in abundance. This dramatic shift in what we most value changes everything.

In 1948, Claude Shannon, the father of information theory, wrote a paper called "A Mathematical Theory of Communication." Shannon proposed that data could be measured in binary digits or bits—discrete values of zero or one. Shannon's theoretical innovations in the science of digital communication, coupled with the invention of transistors, fiber optic cables, digital circuits, and computers, ushered in the Information

L. P. Thiele, *Human Agency, Artificial Intelligence, and the Attention Economy*, https://doi.org/10.1007/978-3-031-82086-1_2

Age. While former times were characterized by a "fragmenting environment" of separateness, detachment, and nonparticipation, McLuhan observed in 1964, the new world of ubiquitous "electronic feedback" provided an integrative environment defined by "inclusiveness" and "instant involvement." The self of the mechanical age was saddled with "meagerness and poverty" whereas the hyperconnected selves of the electronic age faced "the problem of abundance and superfluity." McLuhan (1966, 108) noted that it would be a "comic irony" if people proved unable to cope with abundance.

When McLuhan penned these words, the Information Age was giving rise to the "Golden Age of Advertising," as the mass marketing of abundance that began after the Second World War came into full swing. In 1962, political philosopher Jürgen Habermas (1989, 171–72) already lamented the "transformation of the public sphere into a medium of advertising." Such a "world fashioned by mass media," Habermas observed, would strip the public sphere of its most important function: serving as a forum for the critical conversations of citizens united by common values and practices, including reasoned debate about political responsibility and authority. Whether the public sphere was ever such a hallowed place of citizen engagement is debatable. Still, Habermas was onto something. Publicly shared spaces were being transformed into venues for advertising. Public forums declined as billboards rose, and marquees rapidly replaced manifestos.

Sudden transformations can be traumatic, Alvin Toffler claimed in his 1970 bestseller, *Future Shock*. When people are subjected to "too much change in too short a time," they experience a "shattering stress and disorientation." The disorientation and "decision stress" suffered by denizens of the twentieth century, Toffler (1970, 2, 355) observed, was largely the product of the "cognitive overstimulation" that occurs as societies experience exponential growth in information and communication technologies (ICTs). As more businesses sought to secure more of the attention of more people, the quantity of circulating information began to exceed people's capacity to process it effectively, reducing the quality of their decision-making while increasing psychological trauma. The irony McLuhan predicted was perhaps not so comic, as many proved psychically unable to cope with information overload.

A year after Toffler published *Future Shock*, Nobel Laureate Herbert Simon (1971, 40–41) presciently observed that "in an information-rich world, the wealth of information means a dearth of something else: a scarcity of whatever it is that information consumes. What information consumes is rather obvious: it consumes the attention of its recipients." An information-rich world, Simon observed, creates a "poverty of attention" wherein the "cost" of abundant information is borne more by its recipients than its producers. McLuhan, Habermas, Toffler, and Simon foresaw the arrival of the attention economy and its inherent risks, as advanced nation-states moved beyond industrialization to ground their societies on data technologies. When the attention economy underwent a "digital transformation" in the ensuing decades, the foreseen dangers were amplified.

The pursuit of attention is older than humankind. Certainly, its pursuit figured prominently in the lives of our forebears. But as Georg Franck (1999) observed in one of the first assessments of the attention economy to employ the term, humanity's age-old penchant for attention has become thoroughly monetized. While there is a long "pre-history" of the economy of attention, the burgeoning of media in the twentieth century, Franck wrote, created a condition of "mental capitalism." The monetization of attention made possible by information and communication technologies ushered in a full-scale "colonisation of virtual space." Capturing attention became the prime directive of the business world, and the recipients of ubiquitous advertising bore most of the cost.

Two years after Franck's seminal article was published, Thomas Davenport and John Beck wrote the first book devoted to the attention economy. "In postindustrial society," they observed, "attention has become a more valuable currency than the kind you store in bank accounts.... Understanding and managing attention is now the simple most important determinant of business success" (Davenport and Beck 2001, 3, 17). This new economy demanded ever more data on how attention gets spent and how it can be secured. The first systematic effort to measure the attention given to media was the metering technology employed for television. Nielson Media Research pioneered these efforts, wherein thousands of TV sets provided data on when they were turned on and what networks and programs their viewers were watching. In 2001, prior to the creation of Friendster, LinkedIn, MySpace, and Facebook, and six years before the arrival of the iPhone, Davenport and Beck (2001, 37, 206) could legitimately claim—notwithstanding available metering technology—that attention measurement remained in "the Dark Ages." But the Dark Ages of the attention economy were short lived. As Davenport and Beck correctly foresaw, "The anonymity once conferred upon individuals by mass society will not last much longer. Your identity and many details of your preferences and purchases will be available to anyone who wants to pay for them." Today, the capacity of ICT corporations to gather and analyze vast amounts of data detailing all the comings and goings of our digital lives grounds and grows the attention economy.

Eric Schmidt, Executive Chairman of Google, and Jared Cohen, Director of Google Ideas, acknowledged that "modern technology platforms, such as Google, Facebook, Amazon and Apple, are even more powerful than most people realize.... Almost nothing short of a biological virus can spread as quickly, efficiently or aggressively as these technology platforms, and this makes the people who build, control and use them powerful too" (Schmidt and Cohen 2014, 9–10). Charting the viral growth of the attention economy over the first decades of the twenty-first century, Shoshana Zuboff (2019) argues that we now live in the era of "surveillance capitalism." In the effort to sell advertising, ICT corporations extract as much data as possible from users. This data is then marketed to businesses as "behavioral futures," predictions of who is amenable to purchasing specific products, and how. Digital platforms whose business models rely on advertising revenues effectively sell forecasts of what their users want to do and buy or can be manipulated to do and buy. Early capitalism was chiefly characterized by industrialists' ownership of the means of production. Surveillance capitalism is characterized by ICT corporations' ownership of "the means of behavioral modification" (Zuboff 2019, 339). Surveillance capitalism has become the "puppet master," Zuboff (2019, 376) warned, and we have become its puppets.

The growing power of surveillance capitalism is best illustrated by following the money. At the end of the Second World War, total advertising sales in the US were under $3 billion. They grew fourfold by the time the Golden Age of Advertising peaked in the 1960s, and they would grow twentyfold in the next half century. The more important statistic, however, involves the transformation of the key outlets of advertising. In 2011, total sales of digital advertising in the US were just one-quarter of sales on non-digital outlets (e.g. newspapers, magazines, direct mail, radio, television, cinema, and billboards). Less than a decade later, digital advertising revenues had grown almost fivefold, while non-digital sales had shrunk to less than 37% of the total (Pew Research Center 2021). In actual figures, US digital advertising grew from just over $8 billion dollars in 2000 to just under $60 billion in 2015 before more than tripling to

$225 billion in 2023 (Statista 2024). Well before the end of the decade, the digital advertising market will account for three-quarters of all media spending in the US and across the globe (eMarketer 2021; Scott-Dawkins 2022).

The growth in digital advertising testifies to the sophistication of the mechanisms employed to keep users online and engaged. Platforms are engineered to be *sticky*—in high use, difficult to turn away from, and frequently returned to. In turn, ever more sophisticated methods of tailoring and targeting advertising are employed. Powered by AI, sticky platforms translate the data of our digital engagements into methods to better harvest our attention and direct our behavior.

ICT corporations at the forefront of the attention economy, such as (US-based) Amazon, Apple, Meta, Google (Alphabet), IBM, Microsoft and (China-based) Alibaba, Baidu, and Tencent, do more than monitor and measure how we spend our time. They systematically create and exploit "opportunities to hijack attention" (Microsoft 2015, 4, 24). The hijacking relies on AI to churn through the massive amounts of data produced by our digital engagements. The scraping and analysis of this data reveals patterns in what we watch, say, do, and buy, and indicates how advertising can be tailored and targeted to better exploit our needs, wants, desires, emotions, and habits. In turn, it reveals how these needs, wants, desires, emotions, and habits can be stimulated and shaped to make us ever more accessible to, and dependent upon, the digital platforms that enable and administer our virtual lives.

The transformations that McLuhan, Habermas, Toffler, Simon, Franck, Davenport, and Beck observed prior to the rise of the Internet and social media and the introduction of AI have since greatly expanded in scope, increased in intensity, and grown in sophistication. And these transformations are not simply going on *around* us. They are happening *within* us. We have become enamored with the power, charms, and conveniences of the attention economy. More importantly, we have become its primary product: attention-seeking and attention-dispersing subjects manipulated by attention-grabbing and attention-shaping digital media.

There are over a billion websites today, and a new one is built every three seconds (Haan 2024). Digital platforms, apps, and devices are proliferating. Echoing McLuhan and Simon, political theorist Danielle Allen (2023) observes that "Before the rise of digital content … humanity's artifacts were precious and rare. Cultural content was rare.… But now, cultural content is abundant. The quantity of cultural creation that has occurred in the last two decades vastly outweighs the content created in the prior millennia of human history.... In a world where cultural content is abundant, what is scarce is attention.... Now, in a world in which attention is scarce, fragmented, and captured, the more important job is to help people rebuild the very muscle of their attention." Growing numbers of parents, teachers, scholars, and organizations are engaging in what has been called *attention activism*. Their core belief is that our "attention is fundamentally endangered," and with it "our ability to think for ourselves and to desire on our own terms." The mission of attention activism is to "resist the non-consensual extraction of our collective attention, to reclaim and deepen this precious faculty, and thereby create a more free, more flourishing, and more compassionate world." This "fundamentally political" work, activists maintain, is the work of freedom, understanding, and world-building (Friends of Attention 2019). We are encouraged to give more "attention to attention ... we must make attention itself the *thing being taught*" (Burnett et al. 2023).

Reclaiming attention will not be easy. The Internet, World Wide Web, email, social media, computers, smartphones, virtual assistants, and AI are powerful, ubiquitous, and increasingly indispensable features of our lives. They provide countless benefits. Their practical applications, and our reliance upon them, steadily increase as their speeds, scales, and functionalities expand. All this digital development, however, may significantly stunt the cultivation of crucial human dispositions, habits, and skills. Human downgrading in the face of digital upgrading has been described as the "cultural equivalent of climate change"; and the effort to combat the erosion of human agency has been defined as a "civilizational moment" (Harris and Raskin 2019; Harris 2019).

The comparison to climate change is revealing. The widely embraced mission of reversing global warming is belied by the stubborn fact that our lives and lifestyles remain enmeshed in economies powered by fossil fuels. Likewise, while fully immersed in an AI-enabled attention economy we are tasked with reversing human downgrading. Just as the fly's desperate struggles to escape the spider's web only entangle it further, our efforts to navigate cyberspace in a way that reclaims our attention threaten to enmesh us more thoroughly in its sticky networks. To ensure that our efforts are not Sisyphean, we must shift social norms and modify individual behavior. We must also transform business practices and craft effective governmental policies.

Digital technologies in general, and AI in particular, have the potential—indeed, the near inevitability—to rapidly advance science, medicine, education, and economic growth. There will be great benefits in health, welfare, and the generation of knowledge. It is an exciting prospect. The problem is that many advances will be *too* rapid. Individuals and societies will not be able to ensure that ever more powerful technologies are deployed wisely and occasionally get prohibited. Innovation will outrun ethical considerations. Social transformations will undermine political capacity and digital power will overwhelm individual competence. "Human autonomy will be increasingly compromised," scholars and ICT professionals assessing the need and means to regulate digital platforms conclude, "if the digital status quo is allowed to continue" (Tambini and Moore 2022, 1). Suleyman (2023) concurs, observing that we will survive the coming waves of technological development, with AI serving as its key accelerant, only if nation-states step up to the plate and regulate.

Regulating the AI-Enabled Attention Economy

Democratic governments are responsible for protecting the health, welfare, and education of youth, such that they can become active and responsible citizens capable of making free and informed choices about their personal and collective lives. In turn, they have a responsibility to protect all citizens from hazards that threaten life or undermine health and welfare. To this end, they protect their citizens from foreign aggression and domestic crime and regulate industrial and commercial practices so people can enjoy foods, merchandise, lands, lakes, rivers, oceans, and the atmosphere without being exposed to toxins and perils. In this section, I argue that governments also have the obligation to regulate AI and ICT platforms to shield youth from noxious digital environments that corrupt or stunt their development and to protect the lives and rights of all citizens. Readers who have little interest in regulatory challenges or prefer first to hear my case for digital distancing might skip this section or return to it at a later time.

The threats of unregulated AI include clandestine discrimination in hiring and commercial practices, heightened electoral manipulation and fraud, increased and more powerful forms of misinformation, massive disruptions to the labor market, the invasion of privacy and undue surveillance, copyright infringements, cyber theft and fraud, cyberterrorism, cyberwarfare, and military weaponization. Accordingly, the dangers posed by artificial intelligence have come under increased scrutiny by legislators, and the total number of AI-related regulations has been growing rapidly in the US and across the globe (Maslej et al. 2024).

The regulation of AI is uniquely challenging. First, most lawmakers do not understand how AI works, and at a fundamental level, the same could be said about industry experts. Second, AI is being deployed in most sectors of the economy, in schools, colleges, and universities, and across diverse arenas of social, political, and cultural life. One-size-fits-all regulation will not work.

Third, there is a "fundamental mismatch" between the speed and unpredictability of AI development and the much slower pace of policy and regulatory processes (Satariano and Kang 2023). In mid-2021, European regulators proposed "future-proof legislation" to address the risks posed by AI, inserting feedback mechanisms that allow "rules to adapt to technological change" (European Commission 2023). Unfortunately, they did not anticipate the speed and scope of technological change. The public release of ChatGPT in 2022, followed quickly by the release of other LLMs, demonstrated that no legislation is future-proof.

Fourth, the growing concentration of corporate power makes for comparatively weakened public oversight. ICT corporations need to be supervised by public institutions to ensure that the common good is served. Efforts in this regard have been spotty at best. And things appear to be going in the wrong direction. As Posner (2024) writes "If AI lives up to its promise and becomes the lifeblood of every sector of the economy—as an input in every industry from law to manufacturing—we can expect a future of economic concentration and corporate political power that dwarfs anything that came before." In short, public oversight of the most powerful digital technologies is undermined by the steadily growing power of the stakeholders deploying them.

Fifth, effective regulation must address the global reach and impacts of AI. Yet sharp competition and hostility between nations, including full-scale AI arms races, thwart the crafting of effective international agreements. Vladimir Putin famously said that AI presents "colossal opportunities" and the state that becomes "the leader in this sphere will be the ruler of the world" (Associated Press 2017). In financial, industrial, commercial, and military arenas, the race is on, and regulation is widely perceived to make victory less likely. The UK, for instance, is one of many countries that wants to become an "AI superpower." To this end, regulation is minimized to ensure that the UK becomes "the most attractive place in the world for AI innovation and support" (United Kingdom Department of Science, Innovation and Technology 2023, 2, 7, 11–12). While acknowledging "potential" risks, including threats to human agency, the UK whitepaper detailing its "pro-innovation approach to AI regulation" does not address the severity of the risks or how they will be mitigated. In late 2024, the US-China Economic and Security Review Commission (2024, 171) recommended that "Congress establish and fund a Manhattan Project-like program dedicated to racing to and acquiring an Artificial General Intelligence (AGI) capability." As the original Manhattan Project produced the atomic bombs that were detonated over Hiroshima and Nagasaki, killing hundreds of thousands of civilians, the nomenclature for the proposed AI arms race is foreboding. Soon after Donald Trump began his second term as President, he signed an executive order removing barriers to AI development to ensure "American technological leadership" (Whitehouse.gov 2025).

While the perils of AI arms races and a wholly unregulated market are intolerable, the means of securing the benefits of AI while avoiding dangers remain elusive. Crafting and enforcing effective regulations that do not undermine economic opportunities and benefits, the pursuit of knowledge, free speech, and other basic rights and liberties for a diverse set of deployers and users across the globe is a monumental challenge. Doing so in the face of accelerating technological development becomes a Herculean task.

Scores of individual nations, as well as many international bodies, are developing principles, recommendations, and binding regulations to guide the development and limit the deployment of AI. In March 2024, for example, the United Nations (2024, 5) adopted its first global resolution on artificial intelligence, which aimed to "promote safe, secure and trustworthy artificial intelligence systems in an inclusive and equitable manner, and for the benefit of all, and foster an enabling environment for such systems to address the world's greatest challenges, including achieving sustainable development in its three dimensions—economic, social and environmental—with specific consideration of developing countries and leaving no one behind." Obviously, all this is much easier said than done.

A voluntary agreement forged by the US federal government in 2023 with the primary purveyors of AI, including Adobe, Amazon, Anthropic, Google, IBM, Inflection, Meta, Microsoft, and OpenAI, is focused on security, including testing, research on bias and privacy, information sharing with governments and other organizations regarding risks, and transparency measures such as watermarks to identify AI-generated content. The agreement, being voluntary and self-policed by corporations, is effectively unenforceable. An executive order signed later that year requires companies deploying advanced AI tools to ensure their systems cannot be used to produce biological or nuclear weapons and to report the findings of their tests to the federal government. The order also recommends, but does not require, the watermarking of photos, videos, and audio developed by AI as a means of combatting deep fake misinformation. In November 2023, eighteen nations, including the US, signed an agreement with the aim of making artificial intelligence "secure by design" while keeping it out of the hands of rogue actors. The agreement is nonbinding.

The European Union's *Artificial Intelligence Act of 2024* has more teeth. It bans AI systems that pose an "unacceptable risk," with exemptions for military and national security purposes, certain forms of police work, and scientific research. Prohibited systems include those that employ subliminal, manipulative, or deceptive techniques to impair informed decision-making and behavior, classify people based on social behavior, personality, or biometric markers, monitor people's emotions in schools or the workplace, or compile facial recognition databases from the Internet or networked cameras. Other "high risk" systems, while not prohibited, require monitoring and human oversight. Deepfake images and text for public consumption must be flagged as AI generated. In turn, AI systems will have to comply with EU copyright law regarding the content employed for their training (European Union 2024).

The terrain of AI regulation is rife with obstacles, and the stakes are high. It will be years before effective regulations are in place in most countries, and many more years before their effectiveness and unintended consequences can be properly assessed. By that time, AI will have greatly advanced in power and scope, likely requiring new approaches to its regulation. With this in mind, agile AI regulation likely will require the formation of governmental departments specifically devoted to the task (Marcus 2024, 149–156). As AI is gaining agency itself, the time is right to have governmental agencies dedicated to its oversight.

Regarding regulation of digital media, the most prevalent concerns are the harms faced by youth and the threat to privacy that ensues from the harvesting of online data. Regarding protection of the general population, data privacy laws have been proposed since the 1970s in Europe, and since 2009 in the US. Over the years, a number of national governments have enacted or are considering regulations to protect the health and welfare of children. One of the earliest efforts was South Korea's "Shutdown Law" of 2011, which prohibited children under 16 from playing online games between midnight and six in the morning. In 2024, Australia legislated a minimum age of 16 years for users of social media, holding platforms responsible for enforcement (Kim 2024).

Some nation's regulations aimed at protecting children have been iterative. In 2019, China established rules to protect the "physical and mental health of minors." Children under the age of 18 were forbidden from playing online games for more than 90 minutes on weekdays and could not play between 10 pm and 8 am (Cuthbertson 2019). In 2021, China strengthened the rules, with children and youth barred from online gaming on schooldays and limited to a single hour a day on weekends and holidays (Buckley 2021). Students were also forbidden to take their smartphones to school and were limited to 40 minutes a day on Douyin, China's version of TikTok (BBC News 2021). In 2023, China reworked the rules yet again, making them broader, more enforceable, and more detailed. A comprehensive "minors' mode" for the mobile Internet was created, requiring app and device developers and providers to comply. Rules vary by age group. Children under eight years of age can only use smart devices 40 minutes a day, and content is limited to general education and interest-area cultivation. Upon turning eight, they get another 20 minutes a day and can view popular science, life skills, age-appropriate news, and "entertainment content with positive guidance." At 12 years of age, academic subject education is added, and at 16 years of age, they can be exposed to "healthy and uplifting information content" that is age appropriate (China Law Translate 2023). While these regulations prevent some palpable harms to youth, parental rights are ignored. Clearly, Big Brother is watching as digital media advance in power and scope and is making up the rules in response.

Children's access to and usage of digital media is still a largely unregulated arena in the US. At a Senate Judiciary Hearing in early 2024, executives from Meta, X, TikTok, Discord, and Snapchat were accused of having blood on their hands, and were asked to take responsibility—and apologize—for their companies' role in harming children. Later that year, Instagram made all new and existing accounts of teenagers private by default and will prevent them from receiving notifications between

10 pm and 7 am. Some platforms have indicated their preference for a unified federal law that sets national standards for children's online safety (Singer 2024). Other ITC corporations argue that proposed federal rules will prove ineffective and will impinge on free speech. One of the more active industry advocates, which has represented Google, Meta, Snapchat, and TikTok, is NetChoice. With a mission "to make the Internet safe for free enterprise and free expression," NetChoice has been quite successful in having legislation aimed at regulating digital platforms struck down by the courts.

Still, restrictions on social media and smartphone use by children are growing. A number of US states have passed legislation restricting minors' access to or use of social media platforms while protecting them from online predators. In turn, federal laws have been proposed, such as the Children and Teens' Online Privacy Protection Act 2.0, the Kids Online Safety Act (2023), and the Protecting Kids on Social Media Act (2023). Most efforts aim to prevent the sexual exploitation and online bullying of minors, ensure parental or guardian consent for social media use by minors, prohibit the use of algorithmic recommendation systems on minors, and, occasionally, deny access to social media for children under a certain age. In turn, an increasing number of US states and school boards have banned smartphones from schools, or severely restricted their use (Hubler 2024). Moreover, there are widespread public efforts to create new social norms, such as preventing children from using social media before 16 years of age, banning smartphones from schools, and restricting their use before high school (Haidt 2024).

Regarding the privacy rights of the general population, European efforts culminated in the 2016 General Data Protection Regulation (GDPR), which was designed to reign in data-collection practices. Fines have been substantial, exceeding one billion euros. Legislation has not yet been passed at the federal level in the US, although a patchwork of state-based regulations is in place, with more in development.

Regarding the health and welfare of the general population, one of the most comprehensive bills introduced to the US Senate, the Media Addiction Reduction Technology Act or SMART Act, required social media platforms to reign in their exploitation of human psychology and brain physiology to heighten user engagement and direct user behavior. The bill required platforms to (1) prohibit the use of auto refill and autoplay functions that encourage incessant engagement; (2) restrict awards (e.g. badges) for higher levels of online engagement; (3) allow self-limitation and limit an individual's time spent on platforms across all devices to a default setting of 30 minutes per day; and (4) conspicuously display pop-up notifications at least twice an hour indicating to users how much time they have spent on the platform that day (U.S. Congress, 2019). The SMART Act never made it out of committee. Less ambitious bills have been introduced, such as the 2022 Platform Accountability and Transparency Act, the objective of which is to support research about the impact of digital communication platforms. It would require platforms to comply when university researchers request data for projects approved by the National Science Foundation.

Section 230 of the Communications Decency Act of 1996 has provided technology companies immunity from liability for 3rd party content hosted on their networks. Effectively, this bars lawsuits against them for defamation, negligence, and violations of state antidiscrimination laws. It also recognizes their right to restrict content they deem inappropriate, obscene, or otherwise objectionable, including child pornography, hate speech, misinformation, and terrorist activity. They are also free to *not* to filter or restrict content. While these provisions are aimed at promoting an open Internet and vibrant online economy, they do not incentivize technology companies to filter or restrict offensive or illegal content and do not prevent their censoring or suppressing legitimate political speech.

In 2023, the US Supreme Court rejected efforts to clarify or limit the sweep of Sect. 230. This will make the reduction of online misinformation and dangerous content more difficult. In 2024, the Supreme Court upheld platforms' free speech rights, which, like newspapers, are protected from government intrusion in determining what to include or exclude from communicating. Effectively, the court argued that the First Amendment protects editorial judgment, regardless of whether the speaker is an individual or a platform.

In late 2024, however, the United States Court of Appeals for the Third Circuit reversed a lower court's ruling that the Communications Decency Act immunized TikTok from responsibility when a 10-year-old girl died of asphyxiation imitating a "Blackout Challenge" provided on her curated *For You Page*. The blackout challenge encourages viewers to record acts of choking themselves with belts, purse strings, or other objects to the point of passing out. The Communications Decency Act absolves TikTok from responsibility for third-party content on its site. However, the algorithm that TikTok employs to generate a *For You Page*, the court ruled, was the platform's own "expressive activity." The algorithm, therefore, is protected as first-party speech. However, the right to free speech is accompanied by responsibility, making it illegal, for example, to scream "Fire!" in a crowded theater, or to incite violence to a mob. In the same manner, the platform remains responsible for the algorithm that determines what content arrives unbidden to a user's *For You Page,* allowing a lawsuit against it by the grieving mother (United States Court of Appeals for the Third Circuit 2024). If the lawsuit proves successful, it will no doubt be appealed. Still, the wheels are in motion to upend or revise the responsibility that digital platforms have for their curated content.

The European Union's 2022 Digital Services Act, in contrast to US standards, compels corporations to police their own platforms with the aim of eliminating hate speech, terrorist propaganda and other illegal postings, reducing misinformation, disclosing how their algorithms amplify divisive content, and allowing users to opt out of recommendation algorithms based on personal data. Substantial fines may be assessed if annual audits demonstrate that "systematic risks" are not reduced. The effectiveness of such legislation depends mostly on its level of enforcement. The Digital Services Act may be contravened by laws in the US and other nations that require social media firms to leave content and speakers unmoderated. As such, platforms may have to choose which laws to uphold, potentially balkanizing the global Internet and expanding the number of national Internets.

Proposals for the further regulation of ICT corporations are diverse and wide-ranging (Bollinger and Stone 2022; Lindsay 2021; Center for Humane Technology 2021b; Aral 2020; Williams 2018, 112–118; Balkin 2016). They include efforts to:

- mandate platform interoperability, with data and identity portability between them, to maximize consumer benefits and users' digital rights while combatting the network effect and the monopoly power it produces
- mandate enhanced transparency of platform's business models, macro designs, and algorithmic persuasion mechanisms
- make platforms limit, disable, or take down botnets and automated sites

- make platforms verify identities of users when opening accounts to decrease activities of (anonymous) trolls, provocateurs, and foreign agents
- hinder or demetricate "like" and "share" counts as a means of reducing virality and increasing individuals' evaluation of content
- mandate ad-blockers as default settings
- ban personalized behavioral advertising and micro-targeting
- make "do not track" and "do not sell" (tracking data) the default settings
- make platforms liable for the content that they promote if they utilize personalized algorithmic amplification to engage their users
- link the provision of immunity from civil suits arising from harmful content to platforms' participation in self-regulatory organizations (SROs) and notice-and-takedown processes for unlawful content
- make programmers and designers take an "oath" to abide by and prioritize public welfare or formally assume a fiduciary obligation, so they are legally required to act with confidentiality and care for their users, just as lawyers, doctors, and financial advisors are legally required to act in the best interests of their clients and patients.

Free speech is a cherished ideal and a right. Virality is neither. We may not want to put many limitations on *what* people can say, and that may include false information, hateful rhetoric, and wild conspiracy theories. Still, there is no inherent right to the technological means of broadcasting and replicating noxious speech. Freedom of speech does not mandate freedom of reach. Content control of social media will always be difficult to achieve if free speech is to be safeguarded. Transmission control, in contrast, could tamper virality while leaving speech largely unrestricted. Options here include restricting or eliminating the creation, replication, and transmission of posts by generative AI and automated bots, and the requirement that extra steps be taken by individual users of social media platforms before content can be shared. Just as speed limits, traffic lights, stop and yield signs, and speed bumps are put on roadways to make them safer, so "friction-by-design" can be integrated into the architecture of digital platforms to induce safer and more civil interactions (Frischmann and Benesch 2023).

In the same vein, while platform users should be able to gain access to all the nooks and crannies of cyberspace, their psychologies need not be exploited such that they are enticed into chronic engagement and lured into echo chambers. So, for example, posts, feeds, and sites that are related to each other could be viewed when explicitly selected but would not appear by default in an "infinite scroll." This would reduce the number of users that stay online endlessly and aimlessly, only to get sucked into rabbit holes. Likewise, default settings could have notifications coming in batches, rather than in an endless stream throughout the day, so that users would not be chronically interrupted. In turn, social media sites could require paid subscriptions that are free of advertising, rather than being costless to the user but designed to maximize engagement.

Each of these measures has limits and downsides. A platform moving to a subscription business model, for example, would exclude less well-to-do users, and would likely find itself losing large numbers of clients to "free" platforms. Likewise, the effort to combat inauthentic postings, bots, and trolls might prevent citizens from voicing their opinions anonymously, leaving them open to retaliation by authoritarian regimes, extremist groups, and violent individuals.

In the needed effort to protect the public realm from organized and automated misinformation, scholars observe that the "unintended consequences from over-regulation may be as damaging to democratic systems as the threats themselves" (Woolley and Howard 2019, 245). The same is true for regulatory efforts aimed at protecting individual agency. Many regulations will be gamed, undermining their positive impacts. Others will lack teeth or prove too challenging to effectively enforce. Some proffered cures may prove worse than the disease. To be sure, it would be a disastrous misstep if the cherished values of free speech, privacy, security, and the dissemination of knowledge got thrown out with the bathwaters of the attention economy. Still, regulatory remedies are needed, notwithstanding the impossibility of perfect solutions. The best, as Aristotle observed long ago, should not become the enemy of the good.

Unfortunately, not all good things go together. Free speech, privacy, security, and the dissemination of knowledge cannot all be maximized simultaneously. Free speech and the dissemination of knowledge, for example, can undermine security and privacy. This occurs, for example, when people are doxed, with private or identifying information such as home addresses published online, typically with malicious intent. And, of course, the question of *whose* free speech, privacy, and security comes into question. Some US courts, for example, have allowed legislation that bans platforms from "censoring" users' posts by way of removal or algorithmic demotion. Here, the protection of users' free speech is of concern. Other courts have deemed the "curation" of content by platforms—by way of algorithmic ranking, deprioritizing content in feeds or search results, amplification, or other means—to be "inherently expressive." Here, the protections of the First Amendment are secured for digital platforms, just as they are for newspapers and magazines. As one Supreme Court Justice said in reviewing these matters, "It is not at all obvious how our existing precedents, which predate the age of the Internet, should apply to large social media companies" (Oremus 2022; Liptak 2022). The political and legal battles surrounding the regulation of digital media will be challenging, to say the least.

H. L. Mencken (1921, 158) quipped that "there is always a well-known solution to every human problem—neat, plausible, and wrong." Simple solutions to address the dangers of an AI-enabled attention economy do not exist. But the claim that we must choose between free speech, privacy, security, and the dissemination of knowledge *or* the governmental regulation of digital technologies is simplistic and misleading. Rather, effective policy will include "a spectrum of flexible, calibrated options while avoiding reductionist or binary thinking and false dichotomies" (Center for Humane Technology 2021a). Effective regulation of AI and the attention economy is crucial, notwithstanding inherent challenges. False dichotomies dangerously muddy the waters.

Digital platforms' business models, macro designs, and algorithmic mechanisms should preserve the privacy, free speech, and security of users while securing the benefits of open and expansive digital communications. The right sort of regulation could push platforms to focus more on the value they provide to users than the value they extract from them (Aral 2020, 320). As always, the devil will be in the details. While regulations and legislation cannot be future-proofed, they can and should be made *robust*. Rather than solely addressing current problems, robust regulations anticipate a wide range of plausible futures and are designed to be adaptable (Thiele 2020).

Self-Governance and the Problem with Technological Solutions

The fundamental concept of economics is the law of supply and demand. Consumer demand stimulates the supply of goods and services. In turn, the availability of goods and services, when advertised and marketed, creates demand. To focus on the regulation of ICT corporations is to address the supply side of the attention economy. To focus on personal governance—individuals' ability and responsibility to creatively resist the digital hijacking of their attention and the downgrading of their aptitudes—addresses the demand side.

This book does not further address governmental intervention and public policy. Its focus is self-governance. This goal is complementary to, but distinct from, public oversight of ICT corporations and AI. We can illuminate this complementary relationship by comparing the crisis of human downgrading to the crisis of climate change, with attention to the law of supply and demand.

If we want to effectively address climate change, it would be a grave mistake to focus solely on the demand side. Of course, it is a good thing for individuals to reduce their carbon footprints by limiting the amount of goods and services they purchase and by consuming goods and services that have a low-carbon impact. If enough individuals did so, the supply of high-carbon goods and services would drastically shrink, and global warming would be mitigated. But this one-way approach is insufficient. We simply cannot count on enough people "doing the right thing" if such actions are difficult or costly. Moreover, the efforts of those individuals (or nations) who do the right thing will be partially if not wholly offset by free-riders who gain the benefits of others' sacrifices without having to alter their own behavior.

To effectively combat climate change, we must also address the supply side. A key mechanism for doing so, and likely a necessary one, is a tax on carbon. Such a tax would induce the market to decrease the production of high-carbon goods and services. In turn, it would induce consumers to live low-carbon lives by avoiding highly taxed, and hence more expensive, goods and services. A carbon tax allows consumers to do the right thing by doing the economically rational thing.

To be effective, a carbon tax must be large enough to incentivize the development and provision of low-carbon goods and services. At the same time, it must be small enough to prevent massive economic disruptions. To avoid unnecessary burdens on individuals, a carbon tax should also partially or wholly replace other taxes, such as the income tax, making it "revenue neutral" for governments. If revenues are generated by a carbon tax, they should be used both to mitigate climate change and to alleviate the hardships that the tax imposes on the poor. In turn, a carbon tax must be globally administered so that cheap, untaxed imports do not undercut the market for domestically produced low-carbon goods and services.

Climate change has rightfully been deemed the greatest market failure of all time. Economies powered by fossil fuels have not been structured to account for their chief externality or byproduct—carbon pollution and the global warming it produces. Carbon intensive economic activity generates upstream benefits, namely inexpensive goods and services. But downstream costs are not figured into the prices that consumers pay for these goods and services. Instead, future generations get saddled with the environmental bill (Thiele 2024). Carbon taxes reduce or eliminate this generational injustice by internalizing the externalities of fossil fuel-based economies.

What is the equivalent of a carbon tax for the digital economy, one that will reduce its toxic externalities in the same way that a carbon tax reduces greenhouse gases? Unfortunately, there is no straightforward means of taxing the supply side of the digital economy in a way that effectively and fairly reduces or eliminates the erosion of human agency. Still, the rationale for focusing on the demand side is not simply that regulating the supply side is difficult. Rather, we need to focus on the demand side because the crisis of human downgrading is fundamentally different from the climate crisis.

Individuals who benefit from fossil fuel economies may be reluctant to endorse a carbon tax, or any other structural change. However, their cognitive capacities are not eroded by their carbon intensive lifestyles, such that they find themselves incapable of understanding the nature of the problem or the necessity of a structural solution. In contrast, when agency is eroded by an AI-enabled attention economy, people are hampered in understanding and effecting its remedy.

In the face of this dilemma, the idea that governments should come to the rescue has strong appeal. The problem is that focusing solely on regulating the supply side may weaken efforts to address the demand side. In effect, we would be adopting a technocratic solution, and in doing so we would undermine the fuller development and exercise of self-leadership and service.

The notion that we can solve all the problems that technologies create with more technology is representative of an "unspoken orthodoxy," the widespread assumption that every problem has a technological solution (Noble 1998, 207; and see Morozov 2013). Government regulation is not typically understood as a form of technology. But technology is not limited to tools, machines, and devices. Technology includes anything—simple or complex, concrete or conceptual—that is designed and constructed to fulfill a specific purpose (Thiele 2019). Tools, machines, and devices are technologies. So are economic artifacts such as money and contracts, and political institutions such as legislatures, constitutions, public policies, and regulations.

Technocratic solutionism is consistently optimistic. This optimism is unfounded when it is misdirected. Consider the current enthusiasm for blockchain technologies and their Web 3.0 possibilities. With Web 1.0, the public gained access to the Internet and became consumers of its content. Web 2.0, via social media, turned Internet consumers into the producers of online content. The vast majority of those producing content, and thereby contributing to the fortunes of Facebook, Instagram, YouTube, TikTok, and other network platforms, did not get paid for their efforts. By utilizing blockchain technology, Web 3.0 promises to turn the Internet's consumer-producers into co-owners, distributing the profits generated from digital interactions. As such, Web 3.0 is touted as the antidote to the inequities of surveillance capitalism.

Web 3.0 may not be any more effective at achieving equity in the digital world than share-holder capitalism was in stemming gross inequalities in wealth. In turn, Web 3.0 may further undermine the development of civic skills. It promises to release us from centralized and mostly corporate or, in some nations, governmental control of cyberspace by providing peer-to-peer management of digital information sharing. In many respects, Web 3.0 is driven by the distrust of centralized data management. That distrust is warranted. Rather than developing strong relationships among social peers, it promises a cyber solution. The result may be the wholesale replacement of trust in people with a new, and misguided trust in blockchain technology. Massive frauds and malfeasance carried out in the realm of cryptocurrencies, which are grounded on blockchain technology, suggest that technology is not a good substitute for social capital and civic trust.

"Give a small boy a hammer," the philosopher Abraham Kaplan observed in the early 1960s, "and he will find that everything he encounters needs pounding." Kaplan's so-called *law of the instrument* was aimed at social scientific methodologies that were inappropriately deployed simply because they were available. The law of the instrument was reiterated a few years later by psychologist Abraham Maslow in the form that it most often takes today: "If all you have is a hammer, everything looks like a nail." Bringing governmental regulations to bear on the problem of human downgrading in the attention economy amounts to swinging away with a hammer. Much good work can be done with such tools, but there are limits to what can be accomplished with any blunt instrument.

Human downgrading is not a nail that can be pounded away by technocratic means. There is no alternative to self-governance, which includes our development of healthy dispositions, habits, skills, and relationships. And the hard work of this development might not be initiated or sustained if it is believed that technologies—whether of a mechanical, digital, or governmental variety—will solve the problem for us.

There is a story about Gandhi, possibly apocryphal, that illustrates the point. A woman walked a great distance with her young son to seek council from the Mahatma. Upon gaining an audience, she announced that her boy ate large amounts of sugar and would not stop. The mother pleaded for Gandhi to intervene. After some reflection, he told the woman to return in two weeks. She duly complied, again presenting her sweet-toothed son a fortnight later. Gandhi looked sternly yet lovingly at the young lad and solemnly said: "Stop eating sugar. It isn't good for you." The child, with widened eyes, immediately agreed. The mother was both pleased and irked. She thanked the Mahatma, but asked why he did not command the boy two weeks ago, saving them another grueling journey. "Two weeks ago," Gandhi simply replied, "I was still eating sugar."

A spiritual leader's efforts to better a child's nutrition, like government interventions to better our digital diets, can be of great benefit. At the same time, there is no substitute for personal self-governance, as exhibited by Gandhi when he changed his own behavior before deeming himself fit to recommend abstemiousness to the child. If the boy kept his promise to Gandhi by exercising self-governance, likely it was because he knew that the great man had made a sacrifice on his behalf. Gandhi demonstrated the synergistic power of self-leadership and service.

Effective regulation of the AI-enabled attention economy is desperately needed. But it is no substitute for self-governance. This point is often lost on prominent critics of the attention economy. Zuboff, for example, decries behavioral assaults on human agency, and strongly argues the case for legislation to severely curtail if not outlaw surveillance capitalism. Beyond extolling the human capacity for promise-making, however, she does not address how individuals can or should resist the attention economy or regain their agency. Making promises is a laudable human capacity and an important means of expressing agency. Still, empty promises are commonplace—including those we make to others and, in the fashion of quickly abandoned New Year's resolutions, those we make to ourselves. That should alert us to the insufficiency of promise-making as a bulwark against the attention economy.

Zuboff (2019, 332) rightly notes that our "freedom flourishes only as we steadily will to close the gap between making promises and keeping them." But she does not address the means to close this gap, short of legislation. When Zuboff and her family built their new house, she reminisces, they used old stone for its massive walls and filled the abode with antique furniture. Still, surveillance capitalism breaches these thick walls, such that there are "no corners in which to curl up and taste the pleasures of solitary inwardness" (Zuboff 2019, 478). No doubt digital technology can make our lives busier and more transparent. Whether we allow smartphones at our dining tables or in our bedrooms, however, is still up to us.

Nobel Laureate economist Amartya Sen rightly insists that "A division of responsibility that places the burden of looking after a person's interest on another person can lead to the loss of many important things in the form of motivation, involvement and self-knowledge that the person herself may be in a unique position to have. Any affirmation of social responsibility that *replaces* individual responsibility cannot but be, to varying extents, counterproductive. There is no substitute for individual responsibility" (Sen 1999, 283). To be sure, social structures, political institutions, and environmental conditions greatly impact the extent to which individual responsibility can be cultivated and embraced. To advocate for social and political mechanisms to protect and advance individuals' capabilities, Sen therefore insists, *is* to advocate for individual responsibility. A key justification for regulating the AI-enabled attention economy is that it safeguards human capacities for self-governance.

Still, safeguarding the capacity for self-governance does not guarantee its cultivation. Regulation cannot generate self-leadership. Were strong, effective legislation enacted tomorrow, we would still inhabit a world in which vast numbers of people have diminished capacities to focus attention and reflectively direct their lives. Individuals must work to develop the needed dispositions, habits, and skills. Cutting the puppet master's strings is all well and good. Absent the cultivation and exercise of agency, however, it will not produce responsible acts of freedom.

Practicing self-leadership, at the end of the day, is an individual obligation. To fulfill it, we must regain our capacity for attention and reflective self-direction. Much of the literature that addresses this challenge falls into the category of *digital detoxing*. A digital detox is an exercise in abstinence. The goal is to wean oneself from digital media, breaking compulsive habits. Available resources include self-help manuals, monographs, memoirs, illustrated workbooks, card decks, websites, and apps, as well as therapeutic counseling, training, coaching, consultancy, unplugged weekends and summer camps, screen-free vacations, and full-scale residential retreats.

Some view digital detoxing, along with deleting one's social media accounts and other forms of digital abstinence, as complicit in a neoliberal ideology wherein individuals are expected to take responsibility for their own lives and pull themselves up with their own bootstraps (Syvertsen 2020; Syvertsen and Enli 2020). To be sure, digital abstinence may align with neoliberal ideology when citizens limit themselves to "lifestyle politics" and spurn collective action. Such efforts may even deteriorate into digital attention seeking, a form of branding in which individuals advertise their "conspicuous non-consumption" (Portwood-Stacer 2012).

While digital detoxing need not be aligned with neoliberalism or spurn collective action, it is of limited value. In *Stolen Focus,* Johann Hari maintains that we should neither blame the victims of surveillance capitalism, whose ability to focus has been stolen, nor give them false hopes by suggesting they can remedy the problem by exercising greater self-control. Instead, Hari recommends collective action that will force governments to "ban" surveillance capitalism. New laws and regulations can be modeled, Hari believes, on the laws and regulations that forced manufacturers to stop putting lead in paint and gasoline.

Lead is toxic if ingested or inhaled in sufficient quantities, so we should not be putting it into the air we breathe, into the water we drink, or onto the walls of our homes. Making it illegal for manufacturers to add lead to paint or gasoline, or anything else that will cause it to be widely ingested or inhaled, is a straightforward solution. Hari suggests that controlling the complex forces that diminish our capacity for focused attention in the digital world is as straightforward as the task of eliminating toxins from a few commodities that function perfectly well in their absence.

The primary toxins of the attention economy can be identified: they are distraction, dependence, deskilling, dogmatism, and despair. Unfortunately, there is no easy way to measure their levels or effects. In turn, there is no straightforward mechanism to remove them from people's hearts and minds, or from the body politic. The extraction of lead from paint and gasoline had uniformly positive impacts on human health and only temporarily hurt the bottom lines of a few industries. The attempt to regulate digital life, in contrast, could have wide-ranging repercussions and might endanger many cherished goods, including free speech, privacy, security, social connectivity, and the dissemination of knowledge.

Hari repeatedly encourages his readers to organize the struggle against surveillance capitalism. To be sure, collective action is needed. But it will not occur if individuals cannot focus their attention sufficiently to understand the problem at hand and develop workable solutions. Chanting the mantra "Organize!" does not help distracted, dependent, deskilled, dogmatic, or despairing people collectively craft and steadfastly implement remedies. Organization demands the collective focusing of attention. If people cannot focus their attention individually, how will they do so collectively?

To wean himself from his own digital distractions and dependencies, Hari arranged a three-month digital detox at a Cape Cod beach house: no smartphone, no email, no social media, no Internet. It had the desired effect. Just as highly restrictive food diets often get abandoned after brief victories, soon after quitting Cape Cod Hari found himself enthralled by the same digital habits that originally sent him packing for three months of cold turkey. At the end of his abstinence, Hari (2022, 264) acknowledges that his attention "has never been worse." The honesty is refreshing. One senses, however, that the exuberance of Hari's call for governmental bans and regulatory controls stems from the frustration and disappointment of seeing his self-governance evaporate outside the supportive environment of an unplugged beach house.

Hari concludes his wide-ranging and insightful book with the claim that the climate crisis is the greatest threat that our species has ever faced. He rightly observes that we will not be able to avert climate catastrophe if we do not figure out how to reclaim our capacity for attention. *Stolen Focus* is based on interviews that Hari conducted with dozens of people across the globe. That is one of its strengths. Still, the greenhouse gases released by Hari's air travel would surpass the carbon footprint of entire villages in many parts of the world. I say this not to relish a "Gotcha" moment. If I am casting the first stone in this matter, it is certainly not because I am without carbon sins myself. None of us are. For all I know, Hari purchased carbon offsets for the tens of thousands of air miles he flew to conduct his research, getting a few trees planted in compensation. The merits of offset programs are debatable, given that air travel immediately injects massive amounts of carbon into the atmosphere, and the newly planted trees would take three decades to soak up the equivalent amount. But that is not the issue at hand. The point is that preaching is not practice, any more than the making of promises ensures their keeping. Whatever vows are made, and whatever legislation is passed, there is no substitute for individual responsibility. Governmental regulation may complement but cannot replace self-leadership. The reverse is also true.

While it is indispensable that individuals take responsibility for their attention, the solution is not abstinence. Unplugging for long periods of time is impractical and often ineffective. One study had 600 individuals either wholly abstain from using their smartphones for a week, or reduce use by an hour a day—an average reduction of about one-third of daily use. All participants recorded their efforts in a "Daily Compliance Diary," which effectively also served as a prompt. Four months after the intervention, both abstainers and reducers reported lower levels of smartphone use, less problematic use, fewer symptoms of depression and anxiety, and higher levels of physical activity and life satisfaction. But these effects were stronger and remained more stable in the reduction group. In turn, the reduction group, but not the abstinence group, reported a decrease in smoking. The researchers concluded that "a moderate and controlled reduction" of smartphone use leads to greater well-being and healthier lifestyles, and does so more effectively than bouts of abstinence (Brailovskaia et al. 2022). Self-restraint is more effective than self-denial, and its benefits extend across a range of behaviors.

The fundamental problem with digital detoxing, as will be addressed in Chapter 7, is that bad habits are never vanquished by abstinence alone. They must be replaced by good habits. For full-scale addictions, cold turkey or graduated abstinence may help reset hyperstimulated nervous systems (Kardaras 2016, 238–239). Abstinence also has the benefit of freeing up time and neural space so that new behaviors can be initiated. If firm habits are not developed by the end of the detox period, however, they will be overwhelmed by old habits upon return to the real world. As detoxing sojourns typically occur in unusual settings, like an unplugged beach house on Cape Cod, they do not present the conditions in which new behaviors can take root. So, upon return to workaday life, they wither.

Our withdrawal of attention from the digital economy will only be sustainable if that attention is well invested in other places. As Jenny Odell (2019, 93) writes: "What is needed, then, is not a 'once-and-for-all' type of quitting but ongoing training: the ability not just to withdraw attention, but to invest it somewhere else, to enlarge and proliferate it, to improve its acuity." This reinvestment receives the greatest returns if it is directed to the development of the dispositions, habits, and skills of self-leadership and service.

The Political but Non-Partisan Nature of Digital Distancing

Advocates of regulation often argue that we need to shift from "focusing on personal responsibility to making technology companies accountable for their addictive design choices… [T]o regain what we have lost, we need to stop blaming ourselves and engage in concerted legal action to exert pressure on governments and the real choice-maker—the technology industry—to redesign products and spaces in a way that will put us back in the driver's seat" (Bernstein 2023, xii–xiii). To be sure, effective regulation is required. Addressing the need for and benefits of self-governance, however, is not blaming the victim. The point is to empower ourselves, both as individuals and in our collective ventures.

To advocate self-governance is not to endorse neoliberalism or "rugged individualism," wherein heroic selves display unmitigated autonomy, crafting their visions, cultivating their strengths, pursuing their bliss, and determining their fates regardless socio-economic circumstances, technological conditions, and all obstacles in their way. Healthy dispositions, good habits, and effective skills are only ever developed and sustained in the context of supportive environments. Of that there can be no doubt. Developing social structures, laws, and regulations that empower individuals and enable collective action is crucial. In any case, neoliberalism is less a tribute to rugged individualism than the banner under which the "haves" of society organize themselves, employing the mechanisms of the state to protect and increase their wealth, generally at the expense of the "have nots," democracy, and the natural environment.

Across the social sciences, there is an old but still vibrant debate about whether structure or agency most shapes human behavior. Structures include genetic dispositions, cultural norms and customs, social mores, political institutions, and legal arrangements. Agency refers to the capacity of individuals to act autonomously and independently, freely making choices and taking initiative. Structure *and* agency are always interactive. To cultivate and exercise agency today, however, is to swim against an ever-strengthening current. Most will make little headway in the absence of external means of support, including governmental regulations. Still, there is no substitute for individual efforts to develop the dispositions, habits, and skills that constitute our characters and competencies.

There are limits to what we can do as individuals to address complex issues. Systemic problems require structural solutions. Governments should regulate digital platforms and technologies to protect people's health, secure their rights, and safeguard their polities. Still, the limits of structural solutions become apparent when the problem, in large part, occurs inside hearts and minds. Just as the false choice of *either* free speech, privacy, security, and the dissemination of knowledge *or* regulation should be rejected, so we need not choose between governmental oversight *or* personal responsibility. Both are required, and both are possible. Indeed, they are complementary. Trying to sustain human agency absent governmental oversight of the AI-enabled attention economy is unrealistic. In turn, effective regulation cannot succeed in a world of downgraded people with degraded agency.

Digital platforms extracting behavioral data have the power to micro-target individuals with personalized stimuli, keeping them online, engaged, and in docile service to the attention economy. Decades before the AI-enabled attention economy developed these powers, Marshall McLuhan (1964, 73) wrote: "Once we have surrendered our senses and nervous systems to the private manipulation of those who would try to benefit from taking a lease on our eyes and ears and nerves, we don't really have any rights left." The violation of rights remains a vital concern. When the attention economy takes lease on our senses and synapses, however, an equally if not more pressing issue is the erosion of agency. Safeguarding agency is more complex than protecting rights. Rights may persist in statute form, and be upheld *de jure* if not de facto, long after the agency required for their meaningful exercise has disintegrated. Governments can protect our rights absent much effort on our part. To safeguard agency, however, we must regularly exercise it. Governments cannot do that for us.

Notwithstanding these caveats, the reader may still be inclined to view my focus on self-governance as an endorsement of a classically liberal political philosophy. To be sure, I do believe that adults should have rights and liberties that allow them to live out their lives as they see fit, assuming their activities do not negatively impact the rights or liberties of others. Governments, by and large, should not be paternalistic. Individuals' efforts to flourish ought not to be gainsaid or restricted by laws or regulations unless such efforts prevent others from the equal opportunity to pursue flourishing lives.

In the highly connected, globally interdependent, and ecologically precarious world that we now live in, however, there are few actions an individual can take that do not impact the welfare of others. Even the effort to satisfy the basic needs of feeding, clothing, and sheltering ourselves on a planet undergoing anthropogenic climate change has the potential to impinge the rights and liberties of others by way of the carbon produced and the ecologies disrupted. The classic liberal principle also begs important questions about how individuals come to judge what is in their interest and under what conditions free and informed choices can be made. Making good choices is an unrealizable goal if agency is absent.

In any case, liberal principles are not practical policies. They bespeak the aspiration of human flourishing gained through the exercise of agency. We can only meaningfully approximate the liberal ideal of a free society, therefore, if we devote ourselves to the social and political task of cultivating and sustaining people's capacity for self-leadership and service. Too often "liberal agnosticism" about the contents of a good life conveniently lines up with the free market ideology of unconstrained consumer choices bathed in the "softly egalitarian, flattering light of autonomy" (Crawford 2015, 18). This vision of egalitarian autonomy does not well represent a condition wherein individual preferences have become the object of engineering by ICT corporations. In an AI-enabled attention economy that systematically undermines agency, those who celebrate liberal principles should forgo any faith in the panacea of wholly unregulated markets.

This should not be a partisan issue. Daniel Patrick Moynihan (1996, 63), the four-term Democratic Senator from New York, maintained that "The central conservative truth is that it is culture, not politics, that determines the success of a society. The central liberal truth is that politics can change a culture and save it from itself." Helping individuals develop and exercise their agency, both sides of the aisle should agree, should be a priority for governments and an avenue of innovation for the business community.

David Brooks was a traditional conservative fighting statism and "sclerotic, dynamism-sapping bureaucracies" before becoming a staunch anti-Trumpist "on the rightward edge of the leftward tendency." He observes that the core threat today is social decay. Brooks (2022, 103) writes that "The danger we should be most concerned with lies in family and community breakdown, which leaves teenagers adrift and depressed, adults addicted and isolated. It lies in poisonous levels of social distrust, in deepening economic and persisting racial disparities that undermine the very goodness of America—in political tribalism that makes government impossible." By no means is digital culture—the norms, values, and practices associated with online activities—the singular cause of these ominous trends. Still, the AI-enabled attention economy has exacerbated and accelerated social decay and the erosion of human agency. A viable solution must be both cultural and political.

In 1996, John Perry Barlow, a lyricist, journalist, and co-founder of the Electronic Frontier Foundation, posted "A Declaration of the Independence of Cyberspace." Barlow wrote from Davos, Switzerland, where he was observing the proceedings of the World Economic Forum. The Telecommunications Act had just been passed by the US Congress, the first significant overhaul of telecommunications policy in 60 years. Importantly, the law now included the Internet within its jurisdiction. Governments, Barlow (1996) wrote in protest, were not welcome in cyberspace, whose "great and gathering conversation" was "an act of nature" that grows by way of a people's "collective actions." In cyberspace, Barlow lyricized, "all may enter without privilege or prejudice" and everyone may express his or her beliefs "without fear of being coerced into silence or conformity." The state should back off. "Where there are real conflicts, where there are wrongs," Barlow added, "we will identify them and address them by our means. We are forming our own Social Contract… our governance will emerge." Things did not turn out as Barlow hoped. Self-governance within cyberspace did not develop to an extent nearly sufficient to manage the conflicts and wrongs. Barlow was naïve to believe that the online world would secure the health and welfare of its denizens absent governmental regulation. But he was right to demand that self-governance play a crucial role.

Politically speaking, our liberties are only ever secured by the responsibilities we accept. Witnessing the counterculture movement of the 1960s, and having survived the Nazi's death camps two decades earlier, Viktor Frankl recommended that a Statue of Responsibility be erected on the west coast of the United States to complement the Statue of Liberty that adorns its east coast (see statueofresponsibility.org and Hạnh 2006: 137).

To retain and be worthy of liberties we must accept the responsibility of cultivating and exercising our agency. Zuboff and Hari decry the attention economy's erosion of agency. But their plea for collective action to foster legislation, like Barlow's plea for collective action to supplant legislation, does not address how agency can be cultivated and strengthened for this purpose.

For the benefits of cyberspace to outweigh its harms, we need both well-designed governmental policy *and* robust self-governance. Which comes first? You can't hatch chickens without eggs, and you can't get eggs without chickens. Likewise, you cannot cultivate human agency in the absence of supportive social and political structures, and you cannot create these structures without human agency. There is no good answer to the chicken-or-egg question. Still, there is an effective means to address the dangers of digital technologies while securing their benefits. We need structural reform of the AI-enabled attention economy by way of effective governmental regulation, *and* we need individuals to cultivate and exercise self-leadership and service.

I have chosen to focus on the egg of human agency, with the hope that this will abet collective efforts to establish effective public oversight of digital technologies. Before hatching this egg, however, we must first account for its systematic scrambling in the AI-enabled attention economy. In the penultimate chapter, I explain how we can safeguard and strengthen human agency by way of digital distancing. To prepare the ground, we investigate how a pandemic of distraction and dependence, deskilling, dogmatism, and despair has been loosed on the world.

References

Allen, Danielle. 2023. The Humanities and the Rise of the Terabytes. *American Academy of Arts and Sciences*. Accessed at: https://www.amacad.org/news/humanities-and-rise-terabytes.

Aral, Sinan. 2020. *The Hype Machine: How Social Media Disrupts Our Elections, Our Economy, and Our Health—and How We Must Adapt*. New York: Currrency.

Associated Press. 2017. *Putin: Leader in Artificial Intelligence Will Rule World*. September 1. Accessed at: https://apnews.com/article/bb5628f2a7424a10b3e38b07f4eb90d4

Balkin, Jack M. 2016. Information Fiduciarierts and the First Amendment. *UC Davis Law Review* 49 (4): 1185–1234.

Barlow, John Perry. 1996. A Declaration of the Independence of Cyberspace. Accessed at: https://www.eff.org/cyberspace-independence.

BBC News. 2021. China: Children Given Daily Time Limit on Douyin—Its Version of TikTok. Accessed at: https://www.bbc.com/news/technology-58625934.

Bernstein, Gaia. 2023. *Unwired: Gaining Control over Addictive Technologies*. Cambridge: Cambridge University Press.

Bollinger, Lee, and Geoffrey Stone. 2022. *Social Media, Freedom of Speech and the Future of Our Democarcy*. New York: Oxford University Press.

Brailovskaia, Julia, Jasmin Delveaux, Julia John, Vanessa Wicker, Alina Noveski, Seokyoung S. Kim, Holger Schillack, and Jürgen Margraf. 2022. Finding the "Sweet Spot" of Smartphone Use: Reduction or Abstinence to Increase Well-Being and Healthy Lifestyle?! An Experimental Intervention study. *Journal of Experimental Psychology*. Accessed at: https://psycnet.apa.org/record/2022-50916-001?doi=1.

Brooks, David. 2022. I Remember Conservativsm. *The* Atlantic (January/February): 94–103.

Buckley, Chris. 2021. China Tightens Limits for Young Online Gamers and Bans School Night Play. *New York Times*, August 30. Accessed at: https://www.nytimes.com/2021/08/30/business/media/china-online-games.html.

Burnett, D. Graham, Alyssa Loh, and Peter Schmidt. 2023. Powerful Forces Are Fracking Our Attention. We Can Fight Back. *New York Times*, November 24. Accessed at: https://www.nytimes.com/2023/11/24/opinion/attention-economy-education.html.

Center for Humane Technology. 2021a. Accessed at: https://humanetech.com/.

Center for Humane Technology. 2021b. Policy Reforms Toolkit. Accessed at: https://www.humanetech.com/policy-reforms.

China Law Translate. 2023. Guidelines for the Establishment of Minors' Modes for the Mobile Internet. Accessed at: https://www.chinalawtranslate.com/en/kid-mode-guidelines/.

Crawford, Matthew B. 2015. *The World Beyond Your Head: On Becoming an Individual in an Age of Distraction*. New York: Farrar, Straus and Giroux.

Cuthbertson, Anthony. 2019. China Bans Children Playing Video Games from More Than 90 Minutes a Day or at Night. *Independent* (November 7). Accessed at: https://www.independent.co.uk/life-style/gadgets-and-tech/gaming/china-gaming-ban-video-game-addiction-a9188806.html.

Davenport, Thomas H., and John C. Beck. 2001. *The Attention Economy: Understanding the New Currency of Business*. Boston: Harvard Business School Press.

eMarketer. 2021. Insider Intelligence's Digital Advertising Trends to Watch in 2022. Accessed at: https://www.emarketer.com/content/insider-intelligence-digital-advertising-trends-watch-2022.

European Commission. 2023. Regulatory Framework Proposal on Artificial Intelligence. Accessed at: https://digital-strategy.ec.europa.eu/en/policies/regulatory-framework-ai.

European Union. 2024. The EU Artificial Intelligence Act. Accessed at: https://artificialintelligenceact.eu/.

Franck, Georg. 1999. The Economy of Attention. *Telepolis* (December 7). Accessed at: https://www.heise.de/tp/features/The-Economy-of-Attention-3444929.html.

Friends of Attention, 2019. Accessed at: https://www.schoolofattention.org/about-1 and https://friendsofattention.net/sites/default/files/2020-05/TWELVE-THESES-ON-ATTENTION-2019.pdf

Frischmann, Brett, and Susan Benesch. 2023. Friction-In-Design Regulation as 21st Century Time, Place, and Manner Restriction. *Yale Journal of Law & Technology* 25: 376.

Haan, Katherine. 2024. Top Website Statistics for 2024. *Forbes Advisor*, April 2. Accessed at: www.forbes.com/advisor/business/software/website-statistics/.

Habermas, Jürgen. 1989. *The Structural Transformation of the Public Sphere*. Cambridge: MIT Press.

Haidt, Jonathan. 2024. *The Anxious Generation: How the Great Rewiring of Childhood Is Causing an Epidemic of Mental Illness*. New York: Penguin.

Hanh, T. N. 2006. We Have the Compassion and Understanding Necessary to Heal the World. In *Mindful Politics*, ed. M. McLeod. Boston: Wisdom Publications.

Hari, Johann. 2022. *Stolen Focus: Why You Can't Pay Attention—and How to Think Deeply Again*. New York: Crown.

Harris, Tristan, and Aza Raskin. 2019. Your Undivided Attention. (April 16). Accessed at: https://your-undivided-attention.simplecast.com/episodes/coming-soon-your-undivided-attention-qK0Fq1Al.

Harris, Tristan. 2019. A New Agenda for Tech. *Center for Human Technology*. Accessed at: https://www.humanetech.com/news/newagenda.

Hubler, Shawn. 2024. California Joins Growing National Effort to Ban Smartphone Use in Schools. *New York Times*, June 18. Accessed at: https://www.nytimes.com/2024/06/18/us/california-newsom-smartphone-ban.html.

Kardaras, Nicholas. 2016. Glow Kids: How Screen Addiction is Hijacking our Kids—and How to Break the Trance. New York: St. Martin's Press.

Kim, Victoria. 2024. Australia Has Barred Everyone Under 16 From Social Media. Will It Work? *New York Times*, November 28. Accessed at: https://www.nytimes.com/2024/11/28/world/asia/australia-social-media-ban-law.html.

Lindsay, Roddy. 2021. I Designed Algorithms at Facebook. Here's How to Regulate Them. *New York Times*, October 6. Accessed at: https://www.nytimes.com/2021/10/06/opinion/facebook-whistleblower-section-230.html.

Liptak, Adam. 2022. Supreme Court Blocks Texas Law Regulating Social Media Platforms. *New York Times,* May 31. Accessed at: https://www.nytimes.com/2022/05/31/us/politics/supreme-court-social-media-texas.html?campaign_id=60&emc=edit_na_20220531&instance_id=0&nl=breaking-news&ref=cta&regi_id=71023177&segment_id=93831&user_id=1b8a7a6c418ba0b24e53c8b7ae5ab191.

Marcus, Gary. 2024. *Taming Silicon Valley: How We Can Ensure that AI Works for Us*. Cambridge: MIT Press.

Maslej, Nestor et al. 2024. *The AI Index 2024 Annual Report*. AI Index Steering Committee, Institute for Human-Centered AI, Stanford University. Stanford, CA. Accessed at: https://aiindex.stanford.edu/wp-content/uploads/2024/05/HAI_AI-Index-Report-2024.pdf.

McLuhan, Marshall. 1964. *Understanding Media: The Extensions of Man*. New York: McGraw Hill.

McLuhan, Marshall. 1966. Cybernation and Culture. In *The Social Impact of Cybernetic*, ed. Charles R. Dechert, 95–108. Notre Dame: University of Notre Dame Press.

Mencken, H. L. 1921. *Prejudices, Second Series*. London: Gape.

Microsoft (Consumer Insights, Canada). 2015. Attention Spans. Accessed at: https://www.scribd.com/document/265348695/Microsoft-Attention-Spans-Research-Report.

Morozov, Evgeny. 2013. *To Save Everything, Click Here: The Folly of Technological Solutionism*. New York: Public Affairs.

Moynihan, Daniel Patrick. 1996. *Miles to Go*. Cambridge, MA: Harvard University Press.

Noble, David F. 1998. *The Religion of Technology*. New York: Alfred A. Knopf.

Odell, Jenny. 2019. *How to Do Nothing: Resisting the Attention Economy*. Brooklyn: Melville House.

Oremus, Will. 2022. Want to Regulate Social Media? The First Amendment May Stand in the Way. *The Washington Post,* May 30. Accessed at: https://www.washingtonpost.com/technology/2022/05/30/first-amendment-social-media-regulation/?mc_cid=c6290965dd&mc_eid=8710c3458f.

Pew Research Center. 2021. *Digital News Fact Sheet*. Washington, DC: Pew Research Center.

Portwood-Stacer, L. 2012. Media Refusal and Conspicuous Non-consumption: The Performative and Political Dimensions of Facebook Abstention. *New Media and Society* 15 (7): 1041–1057.

Posner, Eric. 2024. The AI Octopus. *Project Syndicate,* January 8. Accessed at: https://www.project-syndicate.org/commentary/ai-will-strengthen-big-tech-oligopoly-market-concentration-and-corporate-political-power-by-eric-posner-2024-01.

Satariano, Adam, and Cecilia Kang. 2023. How Nations Are Losing a Global Race to Tackle A.I.'s Harms. *New York Times*, December 6. Accessed at: https://www.nytimes.com/2023/12/06/technology/ai-regulation-policies.html.

Schmidt, Eric, and Jared Cohen. 2014. *The New Digital Age: Reshaping the Future of People, Nations and Business*. New York: Vintage.

Scott-Dawkins, Kate. 2022. This Year Next Year: Global 2022 End-of-the-Year Forecast. Group M. Accessed at: https://www.groupm.com/longform/this-year-next-year-global-2022-end-of-year-forecast/.

Sen, Amartya. 1999. *Development as Freedom*. New York: Random House.

Simon, Herbert. 1971. Designing Organizations for an Information-Rich World. In *Computers, Communications, and the Public Interest*, ed. M. Greenberger, 38–52. Baltimore: Johns Hopkins Press.

Singer, Natasha. 2024. Silicon Valley Battles States Over New Online Safety Laws for Children. *New York Times,* January 31. Accessed at: https://www.nytimes.com/2024/01/31/technology/social-media-free-speech-netchoice.html.

Statista. 2024. Online Advertising Revenue in the United States from 2000 to 2023. Accessed at: https://www.statista.com/statistics/183816/us-online-advertising-revenue-since-2000/.

Suleyman, Mustafa. 2023. *The Coming Wave*. New York: Crown.

Syvertsen, Trine, and Gunn Enli. 2020. Digital Detox: Media Resistance and the Promise of Authenticity. *Convergence: The International Journal of Research into New Media Technologies* 26: 1269–1283.

Syvertsen, Trine. 2020. *Digital Detox: The Politics of Disconnecting*. UK: Emerald.

Tambini, Damian, and Martin Moore, eds. 2022. *Regulating Big Tech*. New York: Oxford University Press.

Thiele, Leslie Paul. 2019. *The Art and Craft of Political Theory*. New York: Routledge.

Thiele, Leslie Paul. 2020. Integrating Political and Technological Uncertainty into Robust Climate Policy. *Climatic Change* 163 (November 2020): 521–538.

Thiele, Leslie Paul. 2024. *Sustainability*, 3rd ed. Cambridge: Polity Press.

Toffler, Alvin. 1970. *Future Shock*. New York: Bantam Books.

U.S. Congress. 2019. S.2314—SMART Act. Accessed at: https://www.congress.gov/bill/116th-congress/senate-bill/2314.

U.S.-China Economic and Security Review Commission. 2024. 2024 Report to Congress. Accessed at: https://www.uscc.gov/sites/default/files/2024-11/2024_Annual_Report_to_Congress.pdf.

United Kingdom Department of Science, Innovation and Technology. 2023. *AI Regulation: A Pro-innovation Approach*. Accessed at: https://www.gov.uk/government/publications/ai-regulation-a-pro-innovation-approach.

United Nations. 2024. Resolution A/78/L.49. Accessed at: https://daccess-ods.un.org/access.nsf/Get?OpenAgent&DS=A/78/L.49&Lang=E.

United States Court of Appeals for the Third Circuit. 2024. Tawainna Anderson, Individually and as Administratix of the Estate of n.a., a Deceased Minor, Appellant v. Tiktok, Inc.; Bytedance, Inc. Accessed at: https://www2.ca3.uscourts.gov/opinarch/223061p.pdf.

Whitehouse.gov. 2025. *Fact Sheet: Executive Order to Establish United States Leadership in Digital Financial Technology*. Accessed at: https://www.whitehouse.gov/fact-sheets/2025/01/fact-sheet-president-donald-j-trumptakes-action-to-enhance-americas-ai-leadership/

Williams, James. 2018. *Stand Out of Our Light: Freedom and Resistance in the Attention Economy*. Cambridge: Cambridge University Press.

Woolley, Samuel C., and Philip N. Howard, eds. 2019. *Computational Propaganda*. New York: Oxford University Press.

Zuboff, Shoshana. 2019. *Surveillance Capitalism: The Fight for a Human Future at the New Frontier of Power*. New York: PublicAffairs.

CHAPTER 3

Distraction and Dependence: The Loss of Reflective Self-Direction

Before the widespread use of the Internet, or the existence of social media and smartphones, George Ainslie, a psychiatrist who worked with people suffering from addictions in his clinical work, flagged the impact of the attention economy on desire and the capacity for self-direction. Ainslie (1992, 330, 366) observed that "mass media impoverish a society in the same way as drugs and other addictions, although more subtly, by draining away more attention than they return.... When people discover the consequences of their consumption patterns and try to give them up, they often find that controls adequate to overcome the motivation for these patterns are not in their repertoire." What Ainslie says about earlier forms of mass media applies with much greater force in today's digital culture.

Many of the most prominent digital devices, platforms, and apps are engineered to distract us, to create habits of ever-increasing use, and to transform these habits into full-scale dependence (McNamee 2019, 43; Doshi 2014; Center for Humane Technology 2019). With education, jobs, careers, social engagements, and "pleasure cycles" inextricably bound up with ubiquitous and incessant digital engagement, we have become "terminally distracted" (Anderson 2009). We have also become utterly dependent. Before we can muster effective resistance to agency-eroding forms of distraction and dependence, we need to understand their mechanisms. This chapter addresses the AI-enabled attention

L. P. Thiele, *Human Agency, Artificial Intelligence, and the Attention Economy*, https://doi.org/10.1007/978-3-031-82086-1_3

economy from a neuroscientific and psychological perspective, illustrating how digital platforms exploit prominent human proclivities to maximize engagement and hijack attention.

Nowhere is digital distraction and dependence more apparent than in people's engagement with social media. These Internet-based platforms, with over five billion users worldwide, facilitate the creation and exchange of user-generated content employing mobile or web-based technologies. They are the most popular feature of the Internet. The majority of social media users employ a wide array of social messaging, networking, and content-sharing sites. Globally, some of the most popular are Douyin, Facebook, Kuaishaou, Instagram, Messenger, Pinterest, QQ, Reddit, Sina Weibo, Snapchat, Telegram, Threads, TikTok, YouTube, WeChat, WhatsApp, and X (formerly Twitter). Inevitably, new platforms will arise and gain prominence.

What Nicholas Carr (2011, 157) said of Google applies to many social media platforms: they are in "the business of distraction." They coax users to incessantly browse, search, surf, post, share, like, play, click, and buy. This elicited behavior crowds out reflection. The ensuing scarcity of "slow, concentrated thought," as Carr observes, heightens the likelihood that users will develop compulsive online habits. Distraction fosters dependence.

Nearly half of teens are now on the Internet "almost constantly," which is double the amount reporting this much usage a decade ago (Andersonet al. 2023). In fact, teens spend about five hours a day on social media (Rothwell 2023), and almost a third of this use can be accounted for by habit formation compounded by users' lack of self-control (Allcott et al. 2022, 2458). Designers of these digital platforms intentionally generate and exploit "a stable, consistent and thoroughly addicted user base" (Doshi 2014).

Many executives and designers of digital media platforms are well aware of the dangers, and no small number now rue their contributions. The engineer who helped build Facebook's "like" button and group chat, Justin Rosenstein, regrets his contributions to psychologically destructive technology, observing that distraction has become ubiquitous and continual. Rosenstein now refuses to use most social media platforms, strictly limits his time to others, and has installed a parental-control feature on his iPhone to prevent the downloading of new apps (Lewis 2017). Likewise, Aza Raskin bemoans his invention of the infinite scroll, a mechanism that enables users of social media sites to scroll through content without having to click a button or refresh their pages. Currently, Raskin heads the *Center for Humane Technology*, an organization that addresses the psychological, social, and political dangers of digital culture.

Sean Parker, founding president of Facebook, helped transform the site from a small college project into a highly lucrative global corporation. When Facebook was being developed, he explained, the objective was to "consume as much of your time and conscious attention as possible." That goal led to the creation of the "like" button and other features that stimulate engagement. Parker observes that "It's a social-validation feedback loop ... exactly the kind of thing that a hacker like myself would come up with, because you're exploiting a vulnerability in human psychology" (Solon 2017). Testifying at a Congressional hearing about the impact of social media, Tim Kendall, Facebook's first Director of Monetization and, subsequently, the President of Pinterest, described how distraction and dependence are achieved: "At Facebook I believe we sought to mine as much human attention as possible and turn it into historically unprecedented profits. To do this ... we took a page from Big Tobacco's playbook, working to make our offering addictive at the outset.... So we added updates, photo tagging, likes, which made status and reputation primary, and laid the groundwork for the growing mental health issues we see resulting from these services" (Kendall 2020). Current Facebook executives do not acknowledge the validity of Parker's and Kendall's accusations. However, in an internal memo based on its own research, the company revealed that about one in three teenage girls reports "problematic" (overuse) of its Instagram platform (Wells et al. 2021; Facebook 2021). Of course, today's parents did not need to get a memo from Facebook to know that their high school, middle school, and often elementary school children suffer from digital distraction and dependence.

Indeed, parents who occupy a position of power on the supply side of the attention economy are aware of the dangers. Steve Jobs, who rolled out the iPad and iPhone, Evan Williams, a founder of Twitter and Blogger, Chamath Palihapitiya, former vice president of user growth at Facebook, and other luminaries of digital culture have severely restricted or forbidden their children's use of electronic devices and social media. William Siu (2022), a prominent video game developer, did not allow his children to use his products. "Thinking about my games in my daughters' hands," Siu writes, "I had to confront what these products really were and what they could do. Knowing all the techniques with which we tried to bring about addiction, I realized I didn't want my children exposed to that risk." Chris Anderson, the former editor of *Wired* and chief executive of a robotics and drone company began severely limiting his children's access to digital media after observing symptoms and consequences. Anderson acknowledged that "On the scale between candy and crack cocaine, it's closer to crack cocaine….We thought we could control it. And this is beyond our power to control. This is going straight to the pleasure centers of the developing brain" (Bowles 2018). Inventors, producers, and promoters of digital platforms, apps, games, and devices, have often adopted the primary precept of drug dealing for their own families: Never get high on your own supply (Alter 2017, 2).

Social media platforms made their fortunes by creating habits in users and then transforming these habits into full-scale dependence. They did so, early Facebook investor Roger McNamee (2019, 43) writes, by focusing their technology on the "weakest elements of human psychology." Many medical professionals concur, observing that digital technology, no less than drugs, pave "the road to hell" of addiction (Lustig 2017, 147). Indeed, up to 40% of the general population may suffer some form of Internet-based addiction (Alter 2017, 26–28).

Whether the compulsive use of social media sites, apps, and games constitutes a bona fide addiction is much debated. To be sure, digital media use can be as compulsive as behavioral addictions to gambling, sex, or shopping (Tullett-Prado et al. 2023). The use of digital media often occupies most of people's time, interferes with the fulfillment of obligations at work, school, or home, is difficult to stop or curtail, undermines social relations and physical health, and continues unabated even in the face of its recognized harm. When deprived of their digital devices, many people display psychological symptoms, such as depression or irritability, consistent with the effects of withdrawal from addictive substances. In turn, high digital media use in children produces some of the same structural changes in the brain as those associated with the early use of alcohol (Zhao et al. 2022).

To be sure, there are growing numbers of organizations and health providers devoted to the assessment and treatment of digital addictions. Still, it is unnecessary to insist, and difficult to substantiate, that digital media are addictive to the same extent as alcohol, nicotine, opioids, or other illicit or prescription drugs. Consequently, I generally forego the terminology of addiction in what follows and speak instead of digital dependence.

Psychologists distinguish between egodystonic and egosyntonic disorders. The latter are not recognized to be unhealthy and destructive by the people suffering from them. Many personality disorders, such as narcissism, are egosyntonic. Those afflicted typically view their attitudes and behavior as normal and healthy. Egosyntonic disorders may be particularly difficult to treat, as their noxiousness is denied by those suffering from them. Other personality disorders, such as gambling addictions or obsessive–compulsive disorders, may be egosyntonic or egodystonic, depending upon the self-perceptions of the afflicted. Digital dependence falls into this category.

Many compulsive users of digital technology recognize their behaviors to be unhealthy, and some believe themselves to be addicted. Others deny the problem. Transforming an egosyntonic state of digital dependence into an egodystonic state of self-awareness may be the first step toward recovery. Psychologists maintain that "radical honesty" about the condition of one's mental health is a (pre)condition for controlling or ending compulsive behaviors (Lembke 2021). Regardless of the level of self-awareness, digital dependence can be considered a disorder to the extent that it undermines the development of crucial skills, opportunities for flourishing, and life satisfaction. For vast numbers of people, it does just that.

Digital Perils for Young and Old

Nicholas Negroponte of MIT's Media Lab wrote a book entitled *Being Digital* a dozen years before the arrival of the iPhone. "Being digital," Negroponte (1995, 231) states, "is here. It is now. It is almost genetic in its nature, in that each generation will become more digital than the preceding one." Negroponte was prophetic, for better or worse. While digital distraction and dependence occur across all demographics, it is particularly salient for youth, the digital natives who never knew a world without the Internet and social media. Generation Z, or Gen Z, born from the mid-1990s into the 2010s, have used smartphones, tablets, computers, and social media from a very young age. Some scholars label Gen Z as the iGen, the first fully digital generation whose formative teen years coincided with the arrival of the smartphone (the first iPhone was released in 2007, and the first android smartphone a year later).

Most web traffic comes from smartphones (Haan 2024). While they are employed for multiple purposes, youth predominantly use their smartphones to "pass time" (Schaeffer 2019). Gen Z spends over 6 hours each day on their phones (Harmony Healthcare 2024). The last thing that more than seven out of ten do before going to bed is check their phones. It is also the first thing most do upon arising. And the time between the last online exposure of the evening and the first of the morning is decreasing. Indeed, the use of digital technology, and social media specifically, is a major contributor to the chronic shortage of sleep suffered by large segments of the population, including children and youth (Gull and Sravani 2024).

Insufficient sleep negatively impacts physical and mental health along with a wide range of cognitive faculties, including memory. Lack of sleep also causes a deterioration of the capacities for focused attention and self-control (Schmeichel and Baumeister, 2004, 95). Heavy use of social media has the same effects. If time on social media is replacing sleep, as it does for much of the population including growing numbers of children and youth, then the capacity for focused attention and self-control suffers a double whammy. To make matters worse, sleep deprivation also heightens reward-seeking and risk-taking behavior. We should not be surprised, therefore, that those who sacrifice sleep for social media find themselves impaired in their offline lives.

For every hour that digital technologies are utilized for learning purposes, youth spend more than ten hours employing their devices for entertainment (Hovrath 2024). In turn,when youth are on phones or computers, they are typically multitasking. The word *multitasking* is mostly a misnomer. People can walk and chew gum at the same time. When we are engaged in cognitive endeavors, however, we do not really multitask. Rather, we engage in "attention switching," serially devoting attention to different tasks, one after the other, often in cycles. Multitasking has become the norm for users of digital technologies. It contributes significantly to diminished cognitive abilities and performance, reducing both the time and quality of effort devoted to specific tasks.

The majority of youth employ multiple electronic devices at the same time, and more than three-quarters believe that multitasking is the only way they can function effectively (Microsoft 2015, 7, 37). Over half of young media users are online while doing homework, and they perform more poorly in school as a result (Kaiser Family Foundation 2010). About one-third of middle school teachers consider students distracted by their phones in the classroom to be a "major problem." By high school, that figure rises to 72%. Meanwhile, 58% of high school teachers identify the second most problematic behavior to be students showing little interest in learning (Lin et al. 2024). The former problem exacerbates the latter.

One study demonstrated that students who use their smartphones in college classrooms record less than half as much information in their notes, have poorer recall, and score a full letter grade lower on examinations (Kuznekoff and Titsworth 2013). Other studies demonstrated a similar drop in final exam scores in college classrooms where electronic devices of any sort are allowed (Glass and Kang 2019; Carter et al. 2016). Importantly, multitasking in the classroom decreases comprehension and retention for both the users of digital devices and their offline neighbors (Sana et al. 2013; Glass and Kang 2019). Distraction is contagious.

In leisure hours, young people switch tasks every two minutes; and students who are studying only manage to extend focused attention for one to three minutes longer (Gazzaley and Rosen 2016, 110–111, 115). Undergraduates on computers switch tasks, on average, every 65 seconds. One-fifth of all content is viewed for five seconds or less, and three-quarters of content is viewed for less than a minute (Yeykelis et al. 2014).

When online and multitasking, performance is degraded and susceptibility to distraction increases. Children engaging in more than two hours of screen time each day demonstrate noticeable losses in overall cognitive capacity, or global cognition (Walsh et al. 2018). Increased use of social media by students correlates with greater fatigue in class, a reduction in study hours, education decline, and poorer mental health (Sangani and Razavi 2024). Indeed, time spent on social media and smartphones serves as a reliable predictor of lower grade point averages (GPAs) in students from middle school through college (Kuss and Griffiths 2011; Lepp et al. 2015; Gazzaley and Rosen 2016, 124).

As online activities increase, face-to-face interactions decrease. In the late 1970s, over half of 12th graders said they got together with their friends almost every day for social activities. Four decades later, that figure was down to 28%, the equivalent of a drop of an hour a day of direct encounters. The decline in face-to-face interactions cannot be explained by increased curricular or extra-curricular obligations, as today's adolescents devote less time to paid employment and homework than earlier generations (Twenge et al. 2019). Rather, youth are now more likely to message friends and colleagues rather than communicate directly, and the vast majority employ their devices purposefully to avoid social interactions (Kardaras 2016; Turkle 2015; Rainie and Zickuhr 2015; Schaeffer 2019). This behavior often starts in the home. Children currently spend about twice the number of hours each week on their digital devices as they do interacting with parents (Donnelly 2019). Of course, it is often a two-way street, as parents are increasingly preoccupied with their own devices.

When parents and children are *paying attention* to digital media, a former Google employee observed, they "pay" with "the heart-to-heart talk you could have had with your anxious child, [or] the sleep you didn't get and the fresh feeling you didn't have the next morning" (Burkeman 2019). Today's households—especially relatively well-to-do ones—have multiple televisions, computers, tablets, smartphones, and video game consoles. Unsurprisingly, screen time is extensive and rising, while face-to-face communication and interaction are in decline. This starts at an early age, and it has a demonstrably negative effect. Higher screen time for one-year-old children produces a greater probability of developmental delays the following year in communication, problem-solving, and personal and social skills. Most of these delays continue over the subsequent two years (Takahashi et al. 2023).

Aptitudes are skills we are genetically disposed to learn. One distinctly human aptitude is the ability to discern the cognitive and emotive states of others and understand how these mental states impact behavior. Humans are proficient mind readers. Developmental psychologists call this aptitude a *Theory of Mind*, though mind reading is a naturally occurring aptitude that develops well before we engage in any sort of theorizing. Like the aptitude for verbal communication, the aptitude for mind reading is not functional at birth. Infants and toddlers cannot read minds. The ability to discern the mental states of others and predict their behavior based on their ascertained beliefs, intentions, motivations, and goals is developed through experience.

A child less than four years old will typically fail to predict the behavior of individuals who hold false beliefs. In a "false belief test," young children might observe a cookie being moved from a jar to a cupboard. After all but one of the children have left the room, the cookie is moved back to the jar. The rest of the children then return. When asked where the returning children will look for the cookie, the child who witnessed both switches indicates that they will look in the jar (because that is where the cookie is). The effect of the false belief held by the returning children, who did not see the cookie returned to the jar, is not taken into account. Very young children have not yet learned how to get inside the minds of others to accurately predict their behavior. They remain "mind-blind."

The aptitude for mind reading distinguishes human beings from all other animals. Young children perform it at about the same level as adult chimpanzees. Older children, with the exception of the severely autistic or otherwise mentally disabled, become adept, easily interpreting and accurately predicting others' behavior in light of their beliefs, intentions, motivations, and goals (Dunbar 2000; Tomasello, Call, and Hare 2003; Premack and Premack 2003, 139–157; Lillard and Skibbe 2005). Mind reading is a common feature of everyday life. For human beings, it is second nature.

Like the human aptitude for verbal communication, our aptitude for mind reading requires regular exposure to a social environment that is rich in interactions to develop. In turn, social interactions are greatly facilitated by the aptitude for mind reading. Anything that undermines the development of mind reading is likely to erode a broad array of social skills.

The term "virtual autism" has been coined to describe a syndrome in which autistic tendencies are displayed in young children who spend a great deal of time with electronic devices, and whose symptoms significantly decline when screen time is reduced (Cytowic 2017; and see Wedge 2017). While the linkage between digital culture and the rise of (virtual) autism appears plausible, it is too soon to draw firm conclusions. Still, we know that social and emotional intelligence declines with too much time spent on screens, as such aptitudes primarily develop by way of face-to-face encounters (Small et al. 2020). And we know that such declines are reversible. After only five days at an outdoor camp without screen-based media, for example, preteens significantly improved their ability to infer the emotional states of people in photographs and videos compared to a control group (Uhls et al. 2014).

The capacity for and display of compassion and sympathetic understanding appears to decrease when people are absorbed in fast-paced, multitasking activity online (Carr 2011, 221). When more screen time substitutes for familial and playmate interactions, social and emotional aptitudes—and perhaps even basic mind-reading skills—do not well develop. Harms are compounded when online youth, who lose the benefits of social interactions at dining tables and playgrounds, are also exposed to high levels of age-inappropriate content in cyberspace (Haidt 2024).

When more screen time takes the place of play with friends in unstructured circumstances, children are also deprived of the opportunity to exercise their independence and creativity. A mother who provided an "analog" (screen-free) home for her family suggested a simple fix for parents struggling to wean their children from digital devices: "Kids – Digital Media + Independence = Creative Free Play" (Martiniko 2023, 9). The exercise and expression of creativity is not the only benefit of an analog lifestyle. Independent free play also provides children with opportunities to learn how to cooperate and resolve disputes—crucial civic skills that ground democratic life (Haidt and Lukianoff 2018, 191–194).

If face-to-face exchanges become less effective, more awkward, and increasingly unpleasant for youth owing to their lack of mind-reading and social aptitudes, they are likely to retreat further into virtual life, exacerbating a vicious cycle. This process of online enclosure accounts for the growth of counseling services and therapy centers for digitally distracted and dependent youth. Some parents whose children opt for social media and screen time in lieu of homework, conversation, play, showering, and eating have taken digital media providers to court, claiming that they are producing and marketing an addictive substance.

Parents are right to be concerned. In his Advisory Report of 2023, US Surgeon General Vivek Murthy (2023a, 4) observed that social media present "a profound risk of harm to the mental health and well-being of children and adolescents." The following year, Murthy (2024) advised Congress to place Warning Labels on social media, like those on tobacco and alcohol products, that list the significant mental health harms for adolescents who use them.

The evidence is compelling, with scores of studies validating the Surgeon General's assessment (Messena and Everri 2023). Adolescents who spend more than three hours per day on social media have a significantly heightened risk of mental health problems (Riehm et al. 2019). Heavy digital technology use is associated with decreased emotional and social intelligence, impaired brain development, and disrupted sleep (Small et al. 2020). Adolescents with high digital media use are also more than twice as likely to demonstrate attention deficits (Ra et al. 2018). Young adults engaging in high social media use, in turn, are two to three times more likely to experience social isolation than those with low social media use (Primack et al. 2017). Youth experiencing emotional distress who reduce social media use by even one hour a day demonstrate significant decreases in symptoms of depression and anxiety, and get more sleep (Davis and Goldfield 2024).

Appeals to the courts for redress have begun. In 2023, Meta, which owns Facebook, Instagram, WhatsApp, and Messenger, was sued by more than three dozen US states for designing "psychologically manipulative product features to induce young users' compulsive and extended use," thereby violating consumer protection laws by unfairly ensnaring children and deceiving users about the safety of its platforms (King and Singer 2023). In 2024, attorneys general of thirteen US states and the District of Columbia sued TikTok for intentionally addicting children and teenagers while falsely claiming its use to be safe and knowingly contributing to a mental health crisis among American youth to maximize its advertising revenue.

Distraction and dependence and the accompanying psychological and social impacts for young people are particularly problematic because these digital natives will largely determine the trajectory of the attention economy over the coming decades. A paternalistic approach toward the protection of children and youth seems warranted. This protection should shield children from induced behavior that undermines the acquisition and development of cognitive and social aptitudes and skills. It should also shield them from online exposure to age-inappropriate content and sexual abuse (Thorn 2021).

How adult denizens of cyberspace can and should be protected, given the dangers and pitfalls of governmental paternalism, is a more complex challenge. Clearly something needs to be done. The harmful impacts of digital media, while most pronounced among youths, cross all demographics. Including time spent on tablets, laptops, and desk computers, waking hours for many adults have become a continuous monitoring and exchanging of texts, notifications, updates, emails, photos, and videos. Adults today spend over ten hours each day connected to media, with almost nine of those hours devoted to screen time (Nielson 2018). Daily time online is approaching seven hours a day globally, with the residents of some countries logging in well over nine hours a day (Datareportal 2023). The vast majority of that time, 70%, is devoted to social media, including video and photo apps such as TikTok and YouTube. In the top markets around the globe, users across all ages average about a third of their waking hours on mobile apps, and time online is increasing rapidly—as much as 30% in a two-year span (App Annie 2022). One-third of adults say they are now "almost constantly" online (Perrin and Atske 2021).

In the mid-2010s, people accessed their smartphones about 80 times each day, logging over 2600 taps, types, swipes, and clicks, with the heaviest users more than doubling that number of touches (Winnick 2016). Almost half of those surveyed admitted they could not bear to live without their devices; even more said they have become dependent on social media (Ryan et al. 2014). By the early 2020s, people were accessing their smartphones over 140 times a day; the majority also slept with their phones at night, never went without them for a full day, and felt addicted to them (Kerai 2023).

In turn, office workers check emails over 70 times a day, and multitasking has become "a way of life." With the exception of conversations, office workers only spend about two minutes focused on any single activity (e.g. attending to email, working on the computer, examining documents) before switching tasks (Mark 2015, 24, 26, 56, 79). The rates of digital technology use and multitasking have been rising steadily in the 2020s across all demographics. While multitasking is more prevalent among youth, it is particularly baleful for older adults, as they are less adept at filtering out distractions (Gazzaley and Rosen 2016, 86).

At work, home, and play, we are incessantly distracted and have become chronically distractable. Many of the heaviest users display symptoms that mimic those of Attention Deficit Hyperactivity Disorder (ADHD), including "distraction, fidgeting, having trouble sitting still, difficulty doing quiet tasks and activities, restlessness, and difficulty focusing and getting bored easily when trying to focus" (Kushlev et al. 2016). A study by Microsoft (2015, 4, 23) concluded that "digital lifestyles deplete the ability to remain focused on a single task, particularly in non-digital environments.... Overall, digital lifestyles have a negative impact on prolonged focus." The study revealed that more than two-thirds of heavy social media users acknowledge that they find it difficult to stay focused on specific tasks. The impact is most profound for those with a high volume of media consumption, high social media usage, extensive multi-screen activity, and early adoption of new technologies. In short, the more digital activities one engages in, the less adept one becomes at concentrating the mind and marshaling thought (Microsoft 2015, 17). The upshot, some scholars conclude, is that digital culture is "dumbing down" the world by "creating people who are unable to think well and clearly" (Nass 2009; and see Gazzaley and Rosen 2016, 78–80).

While most prominent for youth, digital distraction and dependence is suffered by all. As will be further detailed in Chapter 7, it is highly correlated with decreases in well-being and life satisfaction. Still, it would be a mistake to throw out the baby of a digitally enabled life with the bathwater of the attention economy. To understand how the former can be maintained while the latter is curtailed requires an investigation of the mechanisms employed to engineer distraction and dependence.

The Neuroscience of Distraction and Dependence

In his second epistle to the Corinthians (3:18), Paul the Apostle marveled at the transformation of the early Christians who, he believed, were directly witnessing the work of God. What we behold, Paul observed, we become. Such has always been the power of attention. Today, what we behold are multiple, frequent, short, ever-changing stimuli from digital media. We have become increasingly unresponsive to anything else.

Greater speed and briefer, more rapid-fire stimuli are a culture-wide phenomenon. The shot length of film, television, and most other visual media has been declining steadily over the years, going from 12 seconds for film in 1930 and 13 seconds for television in 1950 to less than four seconds for both by 2010 (Mark 2023, 225). Current music and social media videos have even shorter shot lengths. "It's a cultural indoctrination," Mark observes, "to grow used to and even prefer these kinds of highly stimulating fast scene shifts, and as a result, it's growing even harder for young people to pay attention to something that's more slow-moving" (Barnett 2023). Viewers of digital media today are chronically exposed to rapid-fire stimuli. A quarter of TikTok's highest-performing videos are less than 30 seconds long, half of the platform's users find videos that are longer than one minute to be "stressful," and a third of users watch videos at double speed (Stokel-Walker 2022). When multitasking, the pace is effectively increased. So-called *dual videos* on TikTok feature side-by-side clips of colorful, quickly shifting content. Such rapid-fire stimuli demand multitasking, and their popularity portends the erosion of the capacity for slow, concentrated thought. Such is the power of digital media to shape its beholders. We have become habituated to fast-paced variety, and quickly tire of anything requiring prolonged attention.

The term *brain hacking* is employed for the intentional engineering of compulsive digital habits. Brain hacking is not just behavior manipulation. Digital media are changing the way the brain functions and altering its structures. Neuroimaging studies of people who are digitally dependent have demonstrated decreases in white matter (myelinated axons) in brain regions that play critical roles in cognitive control, emotional processing, and decision-making (Lin et al. 2012). Corporations at the forefront of the attention economy are aware of these neurological impacts. The "extremely useful" techniques and mechanisms employed by digital marketers to produce "rapid-fire attention shifting," Microsoft industry representatives acknowledge, have "rewired" the brains of clients and customers (Microsoft 2015, 46). Marketing efforts are developed with the full understanding that "digital lifestyles are changing the brain, decreasing the ability for prolonged focus and increasing their appetite for more stimuli" (Microsoft 2015, 4).

By rewiring our brains, digital media have reduced our capacities for the exercise of intentional, reflective reasoning in decision-making, leaving only impulses to drive much of our behavior. With this in mind, Williams (2018, 68, xii) states that "We therefore have an obligation to rewire this system of intelligent, adversarial persuasion before it rewires us." Williams rightly highlights the imperative of retaining our capacity for prolonged, focused attention. But he underestimates the urgency, as the neural rewiring is already underway.

The psychological and bio-chemical mechanisms for neurological rewiring are well known. Humans instinctively crave food, yearn for sex, heed fast-moving objects, are disposed to learn useful information and acquire skills, and engage in many other behaviors that, over the ages, have increased their evolutionary fitness. One of the most prominent mechanisms for the development and maintenance of such traits is dopamine, a neurotransmitter (also known as a neuromodulator). The brain's dopamine cycle facilitates "seeking behaviors." It is crucial to the pursuit of rewards, pushing us to strive and work. As such, dopamine is the key neurotransmitter of motivation.

When an individual is presented with opportunities that might produce rewards, such as food, sex, or learning, dopamine is transported by cerebral neurons (in the ventral tegmental area or VTA) to the nucleus accumbens, known as the brain's pleasure center, which is the neural interface between motivation and action. Here other neurochemicals called endogenous opioid peptides (EOPs) are released. These opioid peptides have a similar effect on the brain as morphine. They make you feel good. Dopamine also interacts with the prefrontal cortex, which allows for the design and execution of plans, and the hippocampus, which enables memory. All of these neurological activities facilitate reward-seeking behavior. Dopamine motivates us to engage in a wide range of behaviors that are essential to human survival.

EOP receptors in the brain that are stimulated as part of this cycle normally shut down further dopamine transmission over time, causing seeking behaviors to decrease or terminate upon the gaining of a reward. That is why eating ends hunger, having sex lowers arousal, and intense exercise leaves one prone to rest. However, EOP receptors can also be "downregulated" by chronic exposure. An overstimulated reward pathway may increase the tolerance of these receptors, making greater amounts of stimulation necessary to achieve the same level of satisfaction. Owing to this process, medical professionals deem the dopamine cycle "the dark underbelly of our consumer culture" (Lustig 2017, 93).

Addictive drugs cause the quick release of large amounts of dopamine and EOPs, bringing about a feeling of euphoria, a "high." The brain's homeostatic mechanisms respond by reducing the production of dopamine and its receptors. Consequently, drug addicts, whose brains have reduced levels of dopamine and fewer dopamine receptors, often require heavier and more frequent doses to achieve the sought for feelings of pleasure (or absence of pain). When downregulation occurs, increased stimulation is required to produce the same satisfaction. In turn, other rewards, such as food and social interaction, may steadily lose their appeal. People who incessantly spike their dopamine can be pitched into a state of anhedonia, wherein the capacity to feel normal pleasures without pharmacological stimulation is severely degraded. In the parlance of previous eras, they suffer from *decadence,* a decline in well-being that follows from excessive indulgence.

Synthetic opioids and stimulants release the dopamine that induces reward-seeking behavior. This causes addicts to be afflicted with an overpowering yearning for the instant, intense pleasure associated with drug use, and often brings them to take life-threatening risks to secure it. The current opioid crisis illustrates the danger, with the number of fatal drug overdoses in the US increasing dramatically in the wake of widely available opioid pharmaceuticals (National Institute of Health 2019). Between 2015 and 2022, fatal overdoses doubled, hitting the grim milestone of well over 100,000 deaths in 2021, more than gun-related and automobile fatalities combined (Rabin 2021). The opioid epidemic has become a public health crisis, much exacerbated by the widespread availability of fentanyl, a powerful synthetic opioid. Drug treatment typically entails a period of abstinence or detoxification, which is aimed at recalibrating the brain's reward system, allowing the dopamine receptors to calm down.

Digital distraction and dependence have analogous neural mechanisms, as do the means for their reversal. Developers of social media platforms use various mechanisms to provide users with "a little dopamine hit" on a regular basis (Solon 2017). As Anna Lembke (2021, 1) observes, "The smartphone is the modern-day hypodermic needle, delivering digital dopamine 24/7 for a wired generation." The cosponsor of new legislation in Florida to restrict the platforms that children can access characterized the "dopamine hits" users get from social media as "digital fentanyl" (Atterbury 2024). That is hyperbole. The levels of dopamine released by synthetic opioids and stimulants are much higher—by a factor of ten for methamphetamine, for example—than that produced by working your smartphone or engaging in social media. Still, online activities that spike dopamine can become so compulsive that normal self-care is neglected in favor of chronic digital stimulation.

People glued to their smartphones, so-called *smombies* (smartphone zombies), have been killed, or have killed others, walking absent-mindedly into traffic or crashing their cars (Gazzaley and Rosen 2016, 129, 132). Cricked-neck owners of these attention-hungry devices require governance mechanisms to ensure public safety. Texting while driving is now widely prohibited. In turn, many cities have created dedicated pedestrian lanes for people using smartphones to prevent bump-ups. Prominent *smombie* warning signs have also been erected on roadsides, prompting drivers—assuming they are not distracted by their smartphones—to be on the lookout for digitally absorbed pedestrians. In turn, municipalities have embedded traffic signals in the pavement where smartphone users are more likely to notice them, preventing their stepping off curbs and into oncoming traffic. As large numbers of pedestrians employ their smartphones at street crossings, heightening injuries and fatalities, protective measures are warranted (Noake 2015). And that is to say nothing of sharply rising levels of "text neck," the repetitive strain injury that comes to those whose heads are regularly in the down-forward position required for phone use.

Compulsive video gaming provides a dramatic illustration of the physical dangers presented by compulsive digital behaviors. Between 2002 and 2021, at least twenty-three cases of "sudden gamer death" have occurred. The deaths followed continuous playing of action-rich multiplayer games, mostly lasting from one to several days. The cause of death is typically pulmonary embolism, cerebral hemorrhage, or fatal cardiac arrhythmia brought on by long sedentary positions, dehydration, sleep deprivation, exhaustion, stress, and heightened blood pressure (Kuperczko et al. 2022). While fatal digital overdoses remain relatively rare, about 60% of US gamers neglect sleep to keep playing, and about 40% miss meals. A study of players of one of the more popular online games found that 44% of players believed themselves to be addicted (Oggins and Sammis, 2012).

Gaming stands at the forefront of concerns for health professionals grappling with compulsive online behavior in youth. Gamers often exhibit the same neurological pathologies as people suffering from drug addiction, including degraded memory, impaired regulation of emotion, disrupted electrochemical activity in their reward circuits, and inhibited prefrontal cortex functioning (e.g. decision-making, self-regulation, long-term goal setting, impulse control, and predicting the consequences of behavior). Medical professionals have taken notice. The latest edition of the Diagnostic and Statistical Manual of Mental Disorders (DSMD), recognizes "Internet gaming disorder" as a condition warranting more research. In 2019, the World Health Organization included "gaming disorder" in the substance use and addictive behaviors section of its International Classification of Diseases.

While addictive gaming is a primary concern for many health professionals, it is widely acknowledged that "nearly all the major uses of the internet—gaming, surfing, social media, texting, sexting, porn—are dopamine drivers" (Diamandis and Kotler 2020, 247). In turn, emerging digital media that provide heightened sensory stimulation, such as virtual reality and mixed reality, will take the exploitation of the dopamine cycle to new heights—or perhaps better said, new depths.

For a time it was rather trendy, at least in Silicon Valley, to engage in *dopamine fasting* as a means of fighting digital addictions (Kamphuis 2017; Sepah 2019; Lanese 2019). The effort is poorly labeled, as dopamine is required for learning, working, and the development of healthy dispositions, habits, and skills. Dopamine motivates us to make the future a more rewarding place. It pushes us to strive. In short, "no dopamine, no effort" (Lieberman and Long 2018, 68). Still, the recommended activities for a dopamine fast are generally well targeted. For the most part, dopamine fasters focus on stimulus control. The idea is to recalibrate the brain's reward system by abstaining from compulsive behavior and promoting alternative activities. Dopamine fasting is a digital detox. Like any detox, dopamine fasting can help recalibrate the biochemistry of the brain. Just as portion control is a more effective means of weight loss than outright fasting, however, so digital distancing is more effective than online abstinence.

The exploitation of the dopamine cycle in the attention economy can be usefully compared to the biochemistry of overeating in the industrial food system. Worldwide, obesity rates have nearly tripled since 1975 (World Health Organization 2019). Today, four out of every ten Americans are obese, and by 2030, one in two adults in the US will be considered obese, with half of those falling under the category of severely obese (Centers for Disease Control and Prevention 2018; Ward et al. 2019). Many other countries demonstrate similar levels of obesity. In large part, this dangerous trend is a product of the industrial food system, which exploits our natural desire to consume fatty, sweet, and salty foods. This gustatory yearning served Pleistocene humans well enough, as fatty, sweet, and salty edibles were both crucial for survival and in relatively short supply. Today, however, these foodstuffs are ubiquitous and cheap. Consequently, entire populations are subject to the "conditioned hypereating" of fat, salt, and sugar (Kessler 2010). The results are widespread and growing levels of obesity, diabetes, and coronary artery disease.

With processed "hyperpalatable foods" overstimulating opioid centers, both our brains and our palates get rejiggered—one might say, downregulated—such that healthier options taste bland or obnoxious. The rejiggering begins at an early age, with hyperpalatable foods providing the mainstay of "kids' menus" at restaurants while dominating the processed food options for children available in grocery stores and supermarkets. Consequently, childhood obesity in the US has more than tripled since the 1970s, with one in every five school children now clinically obese (Centers for Disease Control and Prevention 2019). As David Kessler (2010, 38, 244) observes: "In a cyclical process, eating highly palatable food activates the opioid circuits, and activating these circuits increases consumption of highly palatable foods.... With its ability to create superstimuli, coupled with its marketing prowess, the [food] industry has cracked the code of conditioned hypereating and learned exactly how to manipulate our eating behavior." As Kessler (2010, 48) observes, an advertisement for the chain restaurant T.G.I. Friday's captures the phenomenon with remarkable frankness: "This isn't about grabbing a bite. It's about a bite grabbing you. 'Cause when Friday's gets hold of 'your appetite, we're not letting go.'" In a similar fashion, digital technologies have gotten hold of our natural appetite for attention and the tendency to distribute it widely. And they're not letting go.

Human beings have always craved attention and approval from those they depend upon for their survival and welfare. Crying and smiling are innate capacities for babies, and mothers instinctively respond (hormonally and physiologically) to their infant's cries and smiles. Gaining the attention of mothers, and in later life, of others, was crucial to individual survival throughout the evolution of our species. Human beings are born attention seekers. This natural hankering for attention often took the form of a yearning for social acceptance and validation. Attention-seeking abetted fitness.

Distributing attention widely also served our species well. Had early hominins been prone to narrowly focus on a specific task, say searching for berries, while blocking out all other awareness of their environment, such as the muted sounds of large carnivores creeping through the underbrush, their survival would have been imperiled. Rather than finding lunch, they would have become it. Sensory multitasking kept them alive. Distributing attention widely abetted fitness.

Here's the problem. The attention economy exploits the natural human propensities to seek attention and distribute it widely in the same way the industrial food system exploits our natural craving for fat, salt, and sugar. Notifications, tags, shares, posts, likes, positive emojis, and views on social media platforms tap into our innate drive for social acceptance and validation (McNamee 2019, 86). They also tap into our tendency to widely distribute our attention. Turkle (2017, 227) observes the impact on brains: "Our neurochemical response to every ping and ring tone seems to be the one elicited by the 'seeking' drive, a deep motivation of the human psyche. Connectivity becomes a craving; when we receive a text or an e-mail, our nervous system responds by giving us a shot of dopamine. We are stimulated by connectivity itself. We learn to require it, even as it depletes us." Like the dogs that Pavlov habituated to salivate at the clang of a bell, we live in anticipation of digital notifications, which bespeak not only the possibility of increased social recognition and approval but also the arrival of information that alerts us to potential benefits, harms, or threats.

Just as evolution hardwired us to seek attention and distribute it widely, it also hardwired us to crave intensity. The latter instinct helped us escape threats to life and limb. Natural opioids are produced by the body as part of the dopamine cycle to reward intense activities, such as the endorphins generated by vigorous exercise that are responsible for the "runner's high." Endorphins (a contraction of *endogenous morphine*) keep us poised for fights, flights, and fornication. We naturally yearn for them.

Our brain chemistry and neurological hardwiring for sensory stimuli and intense activity increased fitness in a Pleistocene environment. In the contemporary context, they become a liability that is easily hijacked (Hendlin 2019). Such is the case with the "multitasking high" that Internet and social media users experience as they surf the web while monitoring social media for the latest notifications (Turkle 2017, 163). For multitaskers, approval, novelty, and danger ever beckons and they instinctively heed the call by spreading themselves thinly across cyberspace in an intense search for stimuli. Dopamine is involved in the process, serving as a neuromodulator that prompts the wide distribution of attention (Whitten 2022). That is why people multitasking on digital devices demonstrate physical indicators of arousal many seconds before they switch tasks (Yeykelis et al. 2014). They are being bio-chemically encouraged to maximize attention distribution and reward seeking in cyberspace.

Instincts that abetted survival over our evolutionary history now push us into compulsive and often destructive behaviors. These maladaptive habits have brought us to a point where "normal" drives—the sort that secured our evolutionary survival—get waylaid or suppressed. Many forego regular meals and socialization with friends and families to satisfy their digital cravings. And one out of every three Americans today would prefer to give up sex rather than live without a smartphone (Foroohar 2019, 21). Digital distraction and dependence have become the new normal, upending activities and relationships that over the eons defined the human condition. This is our brave new world.

Supernormal Stimuli and System 1 Tendencies

How could such dramatic shifts to age-old human drives occur? Just as the industrial food system exploits our proclivities for gustatory stimuli, so digital media exploit our yearnings for electronic stimuli. People who incessantly spike their dopamine by way of digital distractions have downregulated their neural receptors. This leaves them chronically yearning for more, and often incapable of feeling much pleasure absent online engagements. For the digitally overstimulated, like the gustatorily overstimulated, activities and options that are healthy and nourishing may be experienced as tasteless and boring. The "supernormal stimuli" of digital media—fast-moving, ever-changing, and emotionally powerful—have become irresistible (Barrett **2010**). Approval, novelty, danger, intensity, and instant gratification beckon, and we react instinctively. Like moths to a flame, we incessantly hover over devices, our inherited predispositions dangerously outmatched in a digital world.

The debilitating impact of digital technologies is best understood in terms of the systematic exploitation of what the Nobel-prize-winning psychologist and economist Daniel Kahneman (**2011**) called our System 1. System 1 is a set of innate capacities and psychological dispositions. It is intuitive, largely automatic, emotional, and impulsive, producing gut reactions and snap judgments. It helps us react quickly to threatening situations, to bond with family and friends, and to be wary of strangers. System 1 has been crucial to our species' survival, helping us gain sustenance, reproduce, and preserve life and limb in situations where speedy reactions and solidarity were of the essence. It allowed the primeval forest dweller to react instantly to the sound of a snapping twig, and just as quickly to flee the tiger leaping out from behind the bush. It fostered enduring loyalties and tribal cohesion in the face of external threats. Thoughtful reflection and critical analysis were not required for it to function. System 1 provides us with a crucial set of mechanisms to satisfy basic needs and to react quickly when the options are fight or flight and there is no time to think.

In contrast to the instinctive drives of System 1, human beings distinguish themselves among the planet's animals by also having System 2. System 2 is a product of what is often called the "higher" brain functions. It is the home of reflective thought and the self-direction it enables. System 2 does not stimulate speedy reactions to threats. But it can help us avoid dangerous situations in the first place, and aid in the development of beneficial practices and technologies.

Today, System 1 leaves us prey to supernormal stimuli that our Pleistocene forebears never had to face. The means to its systematic exploitation are well known. All you need is a trigger, an easy response, and variable rewards. The trigger jump-starts the dopamine cycle. It can be the ping of an arriving text, notification, or email, or the sights and sounds of a video game. This trigger is followed by an undemanding response, such as checking an incoming text, notification, or email, looping into exchanges on social media, or working the aptly named joystick. This nearly costless action produces rewards that vary in both their rate (how often they are gained) and their magnitude (how big they are). Digital rewards span the gamut from junk mail to crucial messages, from the flaming responses of trolls and ideological foes to the likes and raves of online acquaintances and friends, from virtual death at the hands of well-armed supervillains to shiny coins, medals, and badges that testify to online prowess. The trigger, (easy) response, and (variable) reward sequence keep us coming back for more. Our dopamine cycle motivates us to seek the goodies and figure out how to secure them with ever greater frequency and efficiency.

Triggers, easy actions, and variable rewards provide the scaffold for the engineering of digital dependence. Casinos deploy this scaffold most prominently in the way they arrange and program their slot machines. The gambler is triggered by the raucous sounds and vivid images; the easy action of inserting coins, barcoded tickets, or credit cards; the pulling of levers or pushing of buttons; and the variable rewards of coming up empty, getting a small payout, or winning the jackpot. While risking money is a means to winning more, the odds greatly favor the house's machines (by as much as 25%). Playing the slots is a very ineffective method of monetary gain. Yet slot machines are the most popular form of gambling, with casinos deriving about 70% of their revenues from them. The reason is that the experience of playing the slots is highly rewarding neurologically. Gamblers are drawn to participate in the full warp and woof of the dopamine cycle. While the trigger, response, and reward sequence of working the slots are three distinct events, they are typically experienced as an integrated process. That is to say, the cycle as a whole—rather than a specific payout—draws gamblers in and keeps them coming back for more. It is "addiction by design" (Schüll 2012).

The same can be said of digital media. No one wants junk mail, flames from trolls and ideological foes, or death at the hands of virtual villains. We would prefer important messages, likes and raves, and shiny trophies. Still, we are drawn to the digital drama as a whole. So we find ourselves navigating the variable terrain in chronic pursuit of occasional rewards.

In the 1950s, B.F. Skinner, a Harvard psychologist dubbed the father of behaviorism, first observed the impact of variable rewards on animals in his laboratory. Pigeons and mice that received either large, small, or no food treats for pressing a lever would exhibit much more compulsive behavior than those receiving the same reward each time. Variable rewards produced the highest rates of lever pressing and the highest levels of persistence. The predilection for variable rewards in animals is likely grounded in the fitness benefit of diligence in foraging. Here the persistent investment of time and energy, notwithstanding many fruitless efforts, eventually pays off in the discovery of life-sustaining food.

For our Pleistocene ancestors, likewise, the hunting of prey and foraging for food was often an exercise in frustration. If intermittent rewards were experienced intensely and disappointments spurred on rather than dissuaded further efforts, however, hunting and foraging would be dutifully—even eagerly—pursued. The penchant for variable rewards would pay off in the long term, providing an adaptive benefit. The persistence of people waving magnetic wands over beaches for hours in search of coins or jewelry buried in the sand is grounded on the same psychological propensity. Likewise, the frenzied searches associated with annual spring (Easter) egg hunts generate a level of pleasure that far surpasses what would be produced simply by providing children an equivalent amount of treats. In the same manner, variable rewards within the attention economy exploit our evolved neurological wiring and brain chemistry, keeping us persistently and often compulsively engaged in the face of uncertain gains.

If your brain is predicting that a certain behavior or event will produce a small pleasure, but you unexpectedly receive a great pleasure, then your dopamine will spike owing to the unanticipated bonanza. This is known as "reward prediction error" by neuroscientists. The experience of the unexpected produces a pleasant surge in dopamine. Our predilection for variable rewards prompts us to invest attention in opportunities to be surprised.

We enjoy the reward of pleasant revelations and are saddened, angered, or frustrated by unpleasant ones. Still, unpleasant shocks may heighten anticipation of forthcoming pleasures. That is why gambling addicts are prone to "chasing losses." For pathological gamblers, the dopamine release following a loss is much higher than that exhibited by most people, while the dopamine release following a win is the same for both (Linnet et al. 2010). That is to say, gambling addictions are grounded in dopamine surges that are associated with both losing and winning. While wins are obviously gratifying, they remain relatively rare events. In contrast, losses are quite frequent. As every loss can be interpreted as a prelude to future success, the pathological gambler gets continuously stimulated by hits of dopamine, win or lose. The success of Tinder, the dating app, is grounded in an analogous mechanism. Users never know which swipe will bring a match. While there are relatively few successes compared to the number of tries, each swipe portends a potential victory. Consequently, "serial swiping" is widespread among Tinder users and bears all the characteristics of a behavioral addiction (Purvis 2017).

The downsides of compulsive digital activity are obvious. Like many gambling addicts, the digitally distracted and dependent often acknowledge, and rue, the emotional, social, and physical costs associated with their activities. Yet they persist nonetheless, impelled by triggers, easy actions, and variable rewards. The level of compulsiveness varies greatly, from the mildly dysfunctional to the fatal. Still, it is fair to say that digital culture is turning millions of people into dopamine junkies.

Wanting More and Becoming Bored with Ourselves

More is better. That was a stable truth throughout much of our species' evolution. Faced with a relative scarcity of food, shelter, sexual partners, and friends, our earliest ancestors experienced the availability of multiple options as a rare luxury. An abundance of choices was much to be desired.

Circumstances that allow for selection and choice have clear benefits. Abundance is preferable to scarcity. In turn, being self-directed rather than (oppressively) managed is psychologically and physically good for human beings, improving health and moods (Iyengar 2010, 15–18). Having many choices, and being able to select from them, is universally perceived as a good. Of course, there are cultural differences. Members of communitarian societies, for example, are more likely to respond favorably to the limitation of individual self-direction by authorities. Still, most people like to make their own decisions and prefer choice abundance.

Indeed, many other animals also demonstrate a bias for abundance. But human beings are particularly disposed to prefer a wide range of options. We are wired with a "buffet mentality" (Iyengar 2010, 9–10, 203). More is better, and much more is generally perceived to be much better. The common belief is that more freedom means more well-being, and more choices mean more freedom, so more choices mean more well-being (Schwartz (2016, ix). The logic is straightforward. Still, as ancient wisdom suggests and recent scholarship confirms, less can be more.

Studies demonstrate that "when people are given a moderate number of options (4 to 6) rather than a large number (20 to 30), they are more likely to make a choice, are more confident in their decision, and are happier with what they choose" (Iyengar 2010, 190; see also Chernev et al. 2015). With a surfeit of options, the investment of time and cognitive resources required to make good decisions increases dramatically, as does the sense of uncertainty regarding final selections and the regret of missing out on potentially superior outcomes. As a result, people faced with choice overload often become exhausted, confused, and regretful. Generally speaking, they would be better off with fewer alternatives. Given the natural predilection for choice abundance, however, people are loath to limit their options, even when doing so would increase overall levels of satisfaction (Iyengar 2010, 202).

When people are faced with practically limitless choices, as often occurs in cyberspace, dissatisfaction is common. With an endless bonanza of options before them, settling for anything less than a perfect match is perceived as intolerable. Yet faced with a plethora of posts, pics, clips, music, movies, goods, services, dates, and mates to consider, the chance of making the optimal choice is slim. Consequently, a surplus of opportunities produces "not happiness, but ... the haunting fear of a suboptimal choice" (Vallor 2016, 247). Choice abundance produces anxiety. This negative inner state gets interpreted as an indicator of a world that is not providing good options. So, we mistakenly come to believe that increasing the number of options would improve matters. Interpreting excess as scarcity, we seek ever greater abundance, creating a vicious cycle of discontent.

Facing an endless array of online activities, people choose to distribute their attention widely, effectively hedging their bets against missing out on something worthwhile. In the college classroom of yesteryear, for instance, options for students were quite limited: they could listen to the professor and take notes or doodle, or they could daydream or snooze.

For today's students equipped with laptop computers and smartphones, the options are endless. The tendency is to multitask—one ear to the professor, another listening to bud-conveyed music or podcasts; two eyes alternating between PowerPoint slides and an endless array of online offerings including shopping, games, and social media; fingers tapping, scrolling, and texting cyberspace co-travelers or, less frequently, typing out class notes. Unsurprisingly, nine out of every ten students today use their digital devices for non-academic purposes in the classroom (Gazzaley and Rosen 2016, 115).

"The most important function of attention," Davenport and Beck (2001, 58) recognized at the onset of the attention economy, "isn't taking in information, but screening it out." To direct and devote our attention to any specific task requires us to filter out the rest of the clamoring world. Multitaskers screen out fewer distractions. So they struggle to direct and devote attention to specific tasks. Performance and learning suffer accordingly.

Effective memory is the retention of relevant information. It is grounded in the ability to ignore what is irrelevant. You can only hear the signal if you block out the noise. The inability to ignore distractions impedes memory and the goal-directed behavior that depends upon it (Gazzaley and Rosen 2016, 66). Heavy media multitaskers demonstrate both poorer working memory and diminished long-term memory. Consequently, they have a "reduced ability to draw on the past–be it very recent or more remote–to inform present behavior" (Uncapher et al. 2016). People who engage in heavy media multitasking also perform more poorly in cognitive tasks. Ironically, they also do relatively worse in activities that entail task-switching. This unexpected deficit appears to be the product of their heightened distractibility, that is, their inability to ignore environmental stimuli (Firth et al. 2019). In short, heavy multitasking increases attentional impulsivity, making people more distractible. And the more distractible they become, the poorer they perform across various cognitive domains, including multitasking.

Clifford Nass (2009), the Director of the Communication Between Humans and Interactive Media Lab at Stanford University, confirms that people who multitask are "terrible at every aspect of multitasking. They're terrible at ignoring irrelevant information; they're terrible at keeping information in their head nicely and neatly organized; and they're terrible at switching from one task to another.... One would think that if people were bad at multitasking, they would stop. However ... they seem to think they're great at it...virtually all multitaskers think they are brilliant at multitasking." Notwithstanding its inefficacy, multitasking produces the illusion of efficiency by feeding off our natural penchant for choice abundance.

As Turkle observes, "Those who multitask most frequently don't get better at it; they just want more of it.... [M]ultitasking feels good because the body rewards it with neurochemicals that induce a multitasking 'high.' The high deceives multitaskers into thinking they are being especially productive. In search of the high, they want to do even more" (Turkle 2017, 43, 163). To be clear, multitaskers do not *accomplish* more. They are simply being *aroused* more. Multitaskers are being bio-chemically rewarded for their distractedness, and this deceives them into thinking that they are skilled acrobats expertly navigating a fast-paced world. Effectively, they are being manipulated by technologies that encourage multitasking, and they are being transformed into dopamine-driven samplers of digital wares who cannot concentrate, focus attention, perform optimally, or delay gratification.

Most of us partake in this self-deception and suffer the consequences. Young people are particularly vulnerable. Those who frequently engage in multitasking on digital media have smaller gray matter density in their anterior cingulate cortexes (Loh and Kanai 2014). This area of the brain is involved in impulse control, emotional regulation, attention allocation, and decision-making. The capacity for what is called "executive function," which allows self-direction and impulse control, deteriorates. Developing the capacity for delayed gratification in pursuit of worthy goals is one of the more important lessons we learn as children. It is a chief vehicle for and indicator of maturation, as virtually all worthwhile aspirations entail the delay of gratification. Youth who demonstrate greater self-control, and hence are capable of delaying gratification, fare better in life: they have better family and interpersonal relationships, better scholastic performance, better mental, emotional, and physical health, lower levels of delinquency, greater prosperity, and more fulfillment (Moffitt et al. 2011; Twenge and Baumeister 2002; Mischel et al. 1988). To the extent that life online deprives youth of opportunities and rewards for delaying gratification, it blocks pathways to maturity.

Youth today are at the forefront of the cultural drift toward instantaneity. The smartphone is an instant gratification device par excellence. When tapped into social media, online retailers, and the Internet, smartphones allow users to get what they want, whenever they want it, wherever they are. Young multitaskers engaged in countless hours of surfing, sampling, and sharing on the smartphones are chipping away at the psychological foundations of responsible adulthood by foregoing the challenge of focusing their attention for extended periods. Almost half of all Internet users will not wait longer than two seconds for a website to load, and they spend less than six seconds examining the written content of sites (Haan 2024). When young multitaskers are presented with a post or blog that cannot be digested at a glance, they are inclined to respond, in a predictably abbreviated fashion: TLDR—Too Long, Didn't Read. The more accurate response might be TDCR—Too Distracted, Couldn't Read.

Youth who revel in choice abundance by way of digital multitasking do not earn the fruits of focused attention and delayed gratification. They do not become "sophisticated decision makers" who deploy "self-imposed constraints" to avoid "self-defeating" actions (Duckworth et al. 2018, 107). That may explain why today's youth, at least in some countries, are taking longer to grow up (Twenge 2017). The span between the late teens and the late twenties has become known as "emerging adulthood." Thirty is the new twenty. A "slow life" strategy is a reasonable way to prepare for adulthood if the extended time is well spent. The problem is that preparing for adulthood requires crucial learning experiences, including the experience of dedication, focused attention, and delayed gratification producing worthwhile achievements. Yet today's youth have fewer such life experiences, as increasing amounts of their lives are consumed by an ever-expanding array of digital distractions (Twenge 2017). As a purveyor of instantaneity, the attention economy is a force for infantilization.

The term *adulting* is often employed by emerging adults to designate the tasks and responsibilities associated with maturity. It has become, in many respects, a pejorative term, as newly assumed tasks and responsibilities are seen as boring and tiresome in comparison to the fast-paced gratifications available online. Increasingly, the primary metric for evaluating activities, relationships, employment, and even professions is whether they are "fun." Now there is nothing wrong with having fun. Indeed, it is a crucial part of the good life. Beyond the world of toddlers, however, entertainment should not ground a value system.

The drive for instantaneity heightens trend chasing, with an accompanying deterioration of reflective, critical thought. It has impacts well beyond popular culture. With academic journals and volumes available online and increasingly open-access, for example, the scholarly books and articles that get referenced by researchers tend to be fewer, more recent, and restricted to a smaller number of journals and outlets. In earlier days, the browsing of print archives in brick-and-mortar libraries encouraged academics to anchor their work upon a relatively broad and deep foundation of scholarship. The ready availability of online, hyperlinked sources today, in contrast, encourages scholars to follow trending articles and opinions, narrowing the range of referenced research and potentially generating premature consensus (Evans 2008). Most institutions and sectors of the economy, and virtually every facet of our public and private lives, are impacted by the drive for instantaneity.

Over the course of human history, survival and flourishing depended on choice-expanding learning. Consequently, the penchant for exploration became a defining feature of our species. But in modern times, novelty has been transformed into a fetish. With the development of consumer societies, the novel stopped being subversive. Instead, constant innovation is what allowed the system to stay the same (Vattimo 1985, 7). The new became the norm. The attention economy intensifies the routinization of novelty that began in early modern times, detaching it further from ultimate ends and moral standards while expanding its range by way of surplus choice. Today, we are not only neophiles, lovers of novelty. We are hyper-neophilic.

Microsoft apprises its advertising clients how to take full advantage of hyper-neophilia, observing that people in the digital age are "suckers for novelty" but "lose interest, FAST." Recognizing that "today's marketing is about instant gratification," Microsoft recommends the full exploitation of our "rapid-fire attention shifting" propensities. Deploying such tactics will allow businesses to "hook consumers right off the bat" and keep them on the line (Microsoft 2015, 25, 46).

Novelty is a key dopamine driver in the attention economy, and it is strategically exploited. We are bombarded by an endless stream of advertisements, emails, texts, posts, newsfeeds, graphics, pics, and videos. There is always something new to be consumed, be it gossip, gizmos, or GIFs. Even as we gorge ourselves at the all-you-can-eat digital buffet, however, we feel uneasy about overlooking the most recently proffered dish. The mere thought of missing out on trending news, events, or gadgets leaves us anxious. This existential distress bears the acronym FoMO—fear of missing out.

FoMO is an aversion to overlooked information, entertainment, interactions, or opportunities. But FoMO is misnamed. The emotion in question is not fear, but anxiety. Fear relates to specific things, such as snakes, spiders, hypodermic needles, and public speaking. In contrast, anxiety does not have a clearly identifiable source. When anxious one fears nothing in particular. Yet there remains a nervous unease, a foreboding. FoMO is anxiety in the face of missing out on as-yet unspecified information, entertainment, interactions, or opportunities. Digital devices, platforms, and apps are constantly titillating us with stimuli, foremost of which are undiscovered prospects. The urge to scratch these chronic digital itches makes us anxious.

The bio-chemical fingerprint of FoMO is the spike in cortisol that occurs every time our phones, computers, or watches ping or vibrate, signaling an incoming notification, text, or email. It also rises steadily albeit more slowly in the absence of notifications. Indeed, we check our phones without any external prompting about half the time. Our cortisol-induced anxiety is temporarily assuaged when we monitor our devices for the latest posts. Still, compulsive monitoring only reinforces the bio-chemical reaction and ensuing behavior. The digital leash that keeps us emotionally tethered is quite short. Younger populations feel moderately to highly anxious if they cannot check their digital devices at least once every fifteen minutes (Gazzaley and Rosen 2016, 171). Older generations are not much behind. And the leash is getting shorter.

FoMO has a partner: boredom. Boredom would seem to be the antipode of FoMO. When bored we are listless and apathetic. When anxious we are nervously alert. Yet these emotional states prove symbiotic. The philosopher Martin Heidegger first systematically addressed how the anxious yearning for technological innovation feeds off boredom while simultaneously intensifying it. The pursuit of endless novelty is both a response to and exacerbates boredom, which Heidegger identified as the "basic mood" of the technological age.

Most fundamentally, Heidegger argued, anxiety is the dread of confronting our limitations and, ultimately, our mortality. As fallible creatures destined to die, we cannot escape existential anxiety. Accordingly, Heidegger counsels, we should learn to live well with it. That entails avoiding the systematic repression of anxiety. If we allow much of our cognitive and emotional energy to be spent fleeing anxiety or shielding ourselves from it, we will become enervated. In turn, numbing ourselves from the harsher truths of the human condition pushes us into a more generalized indifference. In this devil's bargain, the repression of anxiety leaves us anesthetized and bored (Thiele 1997, 1998).

The capacity for boredom might be considered a defining feature of our species. Having a reach that exceeds their grasp, human beings are haunted by the anxiety of failure and finitude. The capacity for boredom allows us to dull the pain of unmet aspirations and powerlessness. The debilitating boredom of Shakespeare's Hamlet comes to mind. "How weary, stale, flat and unprofitable/ Seem to me all the uses of this world," Hamlet broods. The Danish prince is bored because he cannot remedy villainy close at hand. Incapable of righting the wrong of a murdered father and a complicit mother—or even squarely facing the implications—the sulking, immobilized Hamlet becomes psychically numb. This is the etiology of deep boredom. Unwilling to confront anxieties, we envelop ourselves in the mute fog of indifference.

Digital distractions promise an escape from boredom. However, digital overstimulation coupled with anxiety actually heightens boredom. The dynamic is relatively straightforward. The experience of FoMO generates negative emotions. These negative emotions decrease the pleasure gained from online activities, and this lack of fulfillment increases boredom (Wu-Ouyang 2023, 299). Our boredom, in turn, leads us to believe that we are understimulated, that we are missing out on something that might truly satisfy us. So we seek more frequent and intense digital distractions. FoMo boosts boredom, and boredom feeds FoMO.

Boredom can be temporarily alleviated by novelties, but the anxiety of missing out always oozes through the cracks, stimulating the pursuit of more frequent and intense distractions. Incessantly pursuing the latest stimulations, impulsiveness becomes a chronic state, a conditioned response to boredom and the sense of meaninglessness that accompanies it (Moynihan et al. 2017). Distraction effectively downregulates us. We become, simultaneously, bored to distraction and chronically anxious about missing out. Digital technology drives this positive feedback loop (Eyal 2019b, 13), a downward spiral around shimmering screens.

The attention economy may be psychologically propelled less by FoMO than EBA—Evasion of Boredom and Anxiety. Digital distractions ensure that every moment of the day is spoken for. We are guaranteed never to be alone with our thoughts, thoughts that might bring to mind unmet aspirations, elicit negative feelings, or precipitate a debilitating ennui. As one college student observed, an incessantly monitored smartphone provides an "insurance policy" against boredom (Turkle 2015, 74). But the policy is a sham or rather a Ponzi scheme.

The college students who spend the most time on smartphones report being more susceptible to boredom and more likely to use their devices to escape it (Lepp et al. 2015). The same is true for adolescents. Bored teenagers employ social media excessively, and excessive use of social media increases their experience of boredom (Donati et al. 2022). People who are bored often engage in *digital switching,* a kind of multitasking where users switch between and within media content. People who engage in digital switching become even more bored, however, experiencing less satisfaction, lowered meaning, and reduced attention (Tam and Inzlicht 2024). In short, the incessant dopamine spikes associated with intense digital activities downregulates us into boredom.

As surfing, swiping, and switching across, between, and within digital media exacerbate rather than alleviates boredom, it follows that the antidote is not heightened distraction. Rather, the antidote for boredom is the focused investment of attention that allows our immersion in worthy content. Yet the focused investment of attention is precisely what the digitally distracted and dependent find too demanding to muster or sustain.

In her popular book *How to Break up with your Phone,* Catherine Price observes that people become addicted to their devices, in large part, because they provide an ever-handy means to escape boredom. Indeed, almost 8 out of every 10 youth acknowledge that the moment something is not occupying their attention, they reach for their phones (Microsoft 2015, 7). Accordingly, Price (2018, 2) observes, to effectively combat phone addiction we must embrace opportunities "to stop and think." For those deeply dependent upon digital distractions, however, an opportunity to stop and think is hardly welcome. Indeed, it is experienced with dread. That is why the abrupt absence of ready distractions—as experienced by those who lose or break their smartphones, for example—elicits greatly heightened levels of anxiety. This deep and chronic nervousness reflects a truth that the digitally dependent would rather not face: they have become bored with themselves.

When 1000 students in ten countries abstained from using all media for 24 hours, the results were revealing. Across the board, students acknowledged how "desperately bored" they were when they went unplugged (International Center for Media & the Public Agenda 2019b).

Absent digital distractions, they literally did not know what to do with themselves. To be sure, time became available for reflection. But that was experienced as a problem, not an opportunity. It was a problem because those unpracticed in art of reflection seldom find it rewarding. Indeed, a broad swath of intellectual opportunities became insufferable. Conditioned by digital hyperstimulation, many students could not tolerate the boredom of being in the classroom without the opportunity to multitask on their devices. Many of the youth who were deprived of these devices for 24 hours found the best experience of a very long and tedious day to be the act of closing their eyes at night. Lying in bed, having done their time, they could anticipate waking up to a world of digital distractions, a world that would keep boredom at bay.

Taken together, anxiety and boredom provide the emotional foundation for the attention economy. "When people say they're 'addicted' to their phones," Turkle (2015, 37–38) observes, "they are not only saying that they want what their phones provide. They are also saying that they don't want what their phones allow them to avoid. The thing I hear most is that going to your phone makes it easier to avoid boredom or anxiety." The ultimate seductiveness of digital media, Turkle (2015, 357) concludes, is that "they respond to our fantasies, our wishes, that we will always be heard, that we can put our attention wherever we want it to be, and that we will never have to be alone. And, of course, they respond to an implied fourth fantasy: that we will never have to be bored." Digital distractions assuage anxieties of missing out with dopamine-rich activities and supernormal stimuli that stave off boredom. Digital dependence deepens when this neuro-chemically induced behavior is misunderstood—as it generally is by multitaskers—to be a functional coping mechanism.

Prominent among the ADHD symptoms displayed by youth in digital culture is "getting bored easily when trying to focus" (Kushlev et al. 2016). That may account, in part, for the higher levels of boredom experienced by those who engage in social interaction while using their phones (Dwyer et al. 2018). Face-to-face experiences cannot deliver the speed and novelty available through online engagements. So inveterate multitaskers often find themselves chronically bored by RL (real life). Digital distractions relieve boredom while reinforcing the tendency to become bored in their absence, heightening the dependence on digital salves.

Given the nature of the dopamine cycle and downregulation, keeping boredom at bay by way of digital stimulation is a Sisyphean task, or, to employ Turkle's terminology, a fantasy. It should come as no surprise, then, that many of today's youth increasingly find themselves "phone bored." The affliction occurs when "you're technically 'on your phone,' but you're still bored out of your mind. It's that feeling when you're mindlessly clicking around, opening and closing apps, looking for something to do digitally and finding the options uninteresting" (Lorenz 2018). For the digitally wired and emotionally tired, the sensible thing would be to ponder their situation and take appropriate action to escape the downward spiral. With the capacity for reflective self-direction severely eroded, however, agency seldom rises to the occasion.

One of the Ten Commandments of "Screen Hygiene," according to Susan Maushart (2010, 267), is "Thou shalt not fear boredom." When embraced, the experience of boredom can serve as a prelude to thoughtful creativity. When feared and shunned, boredom becomes the greatest ally of distraction. Incessantly fleeing boredom by way of digital media ensures that every moment will be spoken for and all opportunities for reflection will be forgone. Scholars who investigate the cognitive impacts of digital distractions and multitasking conclude that "we appear to have lost the ability to simply be alone with our thoughts" (Gazzaley and Rosen 2016, 112). The problem is not anxiety or boredom per se, as such "moods" are inevitable (Thiele 1995). The problem is the digital transformation of these challenging yet potentially fecund burdens of the human condition into chronic states of dysfunction. In short, the attention economy is downregulating our capacity to tolerate boredom, work through anxieties, and profitably enjoy the exercise of agency.

The consequences are psychologically, morally, and politically perilous. Anxiety abets our distraction and dependence. And it is out of boredom, as Fyodor Dostoevsky (1972, 33) remarked in his *Notes from Underground*, that people are brought to take delight in sticking pins in others simply to observe the reaction. "Boredom," he wryly observes, "leads to every possible kind of ingenuity." To be sure, bored people often become cruel, even to themselves. In one experiment, participants were asked to sit quietly without distractions for fifteen minutes. They also had the option of giving themselves painful electric shocks. None believed they would shock themselves at the start of the session, but after just six minutes of inactivity, two-thirds of the men and a quarter of the women were torturing themselves with electricity to escape their boredom (Wilson et al. 2014). The profoundly bored are a menace to themselves. In turn, they often make others pay dearly for their existential malaise.

Reinhard Kuhn (1976, 93) observes that "Violent and indiscriminate action is as natural a consequence of ennui as is total inaction." So, as Charles Baudelaire suggested in *Les Fleurs du Mal*, boredom may turn the earth into a shambles and swallow the world with a yawn.

The downregulation of the capacity to tolerate boredom in the attention economy likely contributes to the prominence of digital aggression and trolling. Trolls infest online communities with inflammatory messages and misinformation. Their flamboyant aggression may serve as a temporary salve to a debilitating cycle of anxiety and ennui. Bored with themselves, they take delight in manufacturing shock value online, sticking digital pins into others.

Arguably, the rise of authoritarianism is also facilitated by widespread ennui. Discussing an upcoming visit with former President Donald Trump in 2024, Viktor Orbán, Hungary's semi-autocratic prime minister, observed that "One of the most boring elements of my profession is that the leaders are getting more and more boring." Trump, in contrast, was a "maverick" who Orbán found "refreshing" (Harris 2024). To be sure, Trump has often been embraced by foreign admirers and domestic followers alike because he "shakes things up," refusing to abide by protocol, norms, rules, or laws. In this respect, he is attractive to people who otherwise would be bored. These issues will be further addressed in Chapter 5. Before assessing the repercussions of digital distraction and dependence on public life, however, we need to explore the cognitive and social deskilling that contributes to the peril.

References

Ainslie, George. 1992. *Picoeconomics: The Strategic Interaction of Successive Motivational States with the Person*. Cambridge, UK: Cambridge University Press.

Allcott, Hunt, Matthew Gentzkow, and Lena Song. 2022. Digital Addiction. *American Economic Review*. 112 (7): 2424–2463.

Alter, Adam. 2017. *Irresistible: The Rise of Addictive Technology and the Business of Keeping Us Hooked*. New York: Penguin.

Anderson, Monica, Michelle Faverio, and Jeffrey Gottfried. 2023. *Teens, Social Media and Technology 2023*. Pew Research Center. Accessed at: https://www.pewresearch.org/internet/2023/12/11/teens-social-media-and-technology-2023/.

Anderson, Sam. 2009. In Defense of Distraction. *New York Magazine* (May 15). Accessed at: https://nymag.com/news/features/56793/.

App Annie. 2022. *State of Mobile 2022*. Accessed at: https://www.appannie.com/en/go/state-of-mobile-2022.

Atterbury, Andrew. 2024. Florida's GOP-controlled House passes strict social media restrictions for minors. *Politico*. (January 24). Accessed at: https://www.politico.com/news/2024/01/24/florida-social-media-restrictions-00137670

Barnett, Sofia. 2023. The Newest Threat to Your Attention Span? TikTok 'Dual' Videos. *Wired* (August 10). Accessed at: https://www.wired.com/story/tiktok-dual-videos-attention-spans/.

Barrett, Deidre. 2010. *Supernormal Stimuli*. New York: W.W. Norton.

Bowles, Nellie. 2018. A Dark Consensus About Screens and Kids Begins to Emerge in Silicon Valley. *New York Times*. Accessed at: https://www.nytimes.com/2018/10/26/style/phones-children-silicon-valley.html.

Burkeman, Oliver. 2019. How the News Took Over Reality. *The Guardian* (May 3). Accessed at: https://www.theguardian.com/news/2019/may/03/how-the-news-took-over-reality.

Carr, Nicholas. 2011. *The Shallows: What the Internet is Doing to Our Brains*. New York: Norton.

Carter, Susan Payne, Kyle Greenberg, and Michael Walker. 2016. The Impact of Computer Usage on Academic Performance: Evidence from a Randomized Trial at the United States Military Academy. *Economics of Education Review* 56: 118–132.

Center for Humane Technology. 2019. Accessed at: https://humanetech.com/.

Centers for Disease Control and Prevention. 2018. Adult Obesity Facts. Accessed at: https://www.cdc.gov/obesity/data/adult.html.

Centers for Disease Control and Prevention. 2019. Childhood Obesity Facts. Accessed at: https://www.cdc.gov/healthyschools/obesity/facts.htm.

Chernev, Alexander, Ulf Böckenholt, and Joseph Goodman. 2015. Choice Overload: A Conceptual Review and Meta-Analysis. *Journal of Consumer Psychology* 25 (2): 333–358.

Cytowic, Richard. 2017. There Is a New Link Between Screen-Time and Autism. *Psychology Today*. Accessed at: https://www.psychologytoday.com/us/blog/the-fallible-mind/201706/there-is-new-link-between-screen-time-and-autism.

Datareportal. 2023. *Digital 2022 Global Overview Report*. Accessed at: https://datareportal.com/reports/digital-2023-global-overview-report.

Davenport, Thomas H., and John C. Beck. 2001. *The Attention Economy: Understanding the New Currency of Business*. Boston: Harvard Business School Press.

Davis, Christopher, and Gary Goldfield. 2024. Limiting Social Media Use Decreases Depression, Anxiety, and Fear of Missing Out in Youth with

Emotional Distress: A Randomized Controlled Trial. *Psychology of Popular Media*. Accessed at: https://psycnet.apa.org/record/2024-76138-001.

Diamandis, Peter, and Steven Kotler. 2020. *The Future is Faster Than You Think: How Converging Technologies are Transforming Business, Industries, and Our Lives*. New York: Simon and Schuster.

Donati, Maria Anna, Carola Beccari, and Caterina Primi. 2022. Boredom and Problematic Facebook Use in Adolescents: What Is the Relationship Considering Trait or State Boredom? *Addictive Behaviors* 125 (February): 107132.

Doshi, Suhail. 2014. Mixpanel: How Addictive Is Your App? *Vox Recode*. Accessed at: https://www.vox.com/2014/3/6/11624264/mixpanel-how-addictive-is-your-app.

Dostoevsky, Fyodor. 1972. *Notes from Underground*. New York: Penguin.

Duckworth, Angela L., Katherine L. Milkman, and David Laibson. 2018. Beyond Willpower: Strategies for Reducing Failures of Self-Control. *Psychological Science in the Public Interest*. 19 (3): 102–129.

Dunbar, Robin. 2000. On the Origin of the Human Mind. In *Peter Carruthers and Andrew Chamberlain*, ed. Evolution and the Human Mind: Modularity, Language and Meta-Cognition, 238–253. Cambridge: Cambridge University Press.

Dwyer, Ryan J., Kostadin Kushlev, and Elizabeth W. Dunn. 2018. Smartphone Use Undermines Enjoyment of Face-to-Face Social Interactions. *Journal of Experimental Social Psychology* 78: 233–239.

Donnelly, Laura. 2019. Children Spend Twice as Long on Smartphones as Talking to Parents. *The Telegraph*. Accessed at: https://www.telegraph.co.uk/news/2019/02/07/children-spend-twice-long-smartphones-talking-parents/.

Evans, James A. 2008. Electronic Publication and the Narrowing of Science and Scholarship. *Science* 321 (5887): 395–99.

Eyal, Nir. 2019b. *Indistractable: Supplemental Workbook*. N.p. Accessed at: https://www.nirandfar.com/.

Firth, Joseph, et al. 2019. The “Online Brain”: How the Internet May be Changing Our Cognition. *World Psychiatry* 18 (2): 119–129.

Facebook. 2021. Hard Life Moments: Mental Health Deep Dive. Accessed at: https://about.fb.com/wp-content/uploads/2021/09/Instagram-Teen-Annotated-Research-Deck-1.pdf.

Foroohar, Rana. 2019. *Don’t be Evil: How Big Tech Betrayed is Founding Principles—and All of Us*. New York: Currency.

Gazzaley, Adam, and Larry. D. Rosen. 2016. *The Distracted Mind: Ancient Brains in a High-Tech World*. Cambridge: MIT Press.

Glass, Arnold, and Mengxue Kang. 2019. Dividing Attention In the Classroom Reduces Exam Performance. *Educational Psychology* 39 (3): 395–408.

Gull, Mubashir, and Bulle Ruth Sravani. 2024. Do Screen Time and Social Media use Affect Sleep Patterns, Psychological Health, and Academic Performance Among Adolescents?. *Children and Youth Services Review* 164: 1–9.

Haan, Katherine. 2024. Top Website Statistics for 2024. *Forbes Advisor*, April 2. Accessed at: www.forbes.com/advisor/business/software/website-statistics/.

Haidt, Jonathan. 2024. *The Anxious Generation: How the Great Rewiring of Childhood Is Causing an Epidemic of Mental Illness*. New York: Penguin.

Haidt, Jonathan, and Greg Lukianoff. 2018. *The Coddling of the American Mind*. New York: Penguin.

Harmony Healthcare. 2024. Black Mirror or Black Hole? American Phone Screen Time Statistics. Accessed at: https://www.harmonyhit.com/phone-screen-time-statistics/.

Hendlin, Yogi Hale. 2019. I Am a Fake Loop: The Effects of Advertising-Based Artificial Selection. *Biosemiotics* 12 (1): 131–156.

Harris, Tim. 2024. Hungarian PM Orban: The Only Serious Chance For Peace Is A Trump Win. *RealClear Politics*, March 3. Accessed at: https://www.realclearpolitics.com/video/2024/03/03/hungarian_pm_orban_trump_comeback_is_the_only_serious_chance_for_peace_in_europe.html.

Hovrath, Jared Cooney. 2024. The EdTech Revolution Has Failed: The Case Against Student Use of Computers, Tablets, and Smartphones in the Classroom. Accessed at: https://www.afterbabel.com/p/the-edtech-revolution-has-failed.

International Center for Media & the Public Agenda. 2019a. *The World UNPLUGGED*. Accessed at: https://theworldunplugged.wordpress.com/emotion/confusion/.

Iyengar, Sheena. 2010. *The Art of Choosing*. New York: Twelve.

Kahneman, Daniel. 2011. *Thinking, Fast and Slow*. New York: Farrar, Straus and Giroux.

Kaiser Family Foundation. 2010. Generation M^2: Media in the Lives of 8- to 18-Year-Olds. Accessed at: https://www.kff.org/other/event/generation-m2-media-in-the-lives-of/.

Kamphuis, Greg. 2017. *A 40 Day Dopamine Fast*. NP: NP.

Kardaras, Nicholas. 2016. Glow Kids: How Screen Addiction is Hijacking our Kids—and How to Break the Trance. New York: St. Martin's Press.

Kendall, Tim. 2020. Testimony at House Committee on Energy and Commerce Hearing on "Mainstreaming Extremism: Social Media's Role in Radicalizing America." Thursday, September 24, 2020. Accessed at: https://energycommerce.house.gov/committee-activity/hearings/hearing-on-mainstreaming-extremism-social-media-s-role-in-radicalizing.

Kerai, Alex. 2023. Cell Phone Usage Statistics. *Reviews.Org*. Accessed at: https://www.reviews.org/mobile/cell-phone-addiction/.

Kessler, David A. 2010. *The End of Overeating: Taking Control of the Insatiable American Appetite*. New York: Rodale Books.

King, Cecilia, and Natasha Singer. 2023. Meta Accused by States of Using Features to Lure Children to Instagram and Facebook. *New York Times*, October 24. Accessed at: https://www.nytimes.com/2023/10/24/technology/states-lawsuit-children-instagram-facebook.html.

Kuhn, Reinhard. 1976. *The Demon at Noontide: Ennui in Western Literature*. Princeton: Princeton University Press.

Kuperczko, Diana, Peter Kenyeres, Gergely Darnai, Norbert Kovacs, and Jozsef Janszky. 2022. Sudden Gamer Death: Non-violent Death Cases Linked to Playing Video Games. *BMC Psychiatry* 22: 824. Acccessed at: https://bmcpsychiatry.biomedcentral.com/articles/10.1186/s12888-022-04373-5

Kushlev, Kostadin, Jason Proulx, and Elizabeth W. Dunn. 2016. 'Silence Your Phones': Smartphone Notifications Increase Inattention and Hyperactivity Symptoms. *Proceedings of the 2016 CHI Conference on Human Factors in Computing Systems*, 1011–1020. Accessed at: http://dl.acm.org/citation.cfm?d=2858036.2858359.

Kuss, Dara J., and Mark D. Griffiths. 2011. Online Social Networking and Addiction: A Review of the Psychological Literature. *International Journal of Environmental Research and Public Health*. 8 (9): 3528–3552.

Kuznekoff, Jeffrey, and Scott Titsworth. 2013. The Impact of Mobile Phone Usage on Student Learning. *Communication Education* 62 (3): 233–252.

Lanese, Nicoletta. 2019. Is There Actually Science Behind 'Dopamine Fasting'? *LIVESCIENCE* (November 19). Accessed at: https://www.livescience.com/is-there-science-behind-dopamine-fasting-trend.html.

Lembke, Anna. 2021. *Dopamine Nation: Finding Balance in the Age of Indulgence*. New York: Dutton.

Lepp, A., J. Li, J.E. Barkley, and S. Salehi-Esfahani. 2015. Exploring the Relationships between College Students' Cell Phone Use, Personality and Leisure. *Computers in Human Behavior* 43: 210–219.

Lewis, Cecil Day. 1951. *The Poet's Task*. Oxford: Clarendon Press.

Lewis, Paul. 2017. 'Our Minds Can Be Hijacked': The Tech Insiders Who Fear a Smartphone Dystopia. *The Guardian*, October 6. Accessed at: https://www.theguardian.com/technology/2017/oct/05/smartphone-addiction-silicon-valley-dystopia.

Lieberman, Daniel, and Long Michael. 2018. *The Molecule of More*. Dallas: BenBella Books.

Lillard, Angeline, and Lori Skibbe. 2005. Theory of Mind: Conscious Attribution and Spontaneious Trait Inference. In *Ran Hassin, James Uleman and John Bargh*, ed. The New Unconscious, 277–305. Oxford: Oxford University Press.

Lin, Fuchun, et al. 2012. Abnormal White Matter Integrity in Adolescents with Internet Addiction Disorder. *PLoS ONE* 7 (1): e30253.

Linnet, Jakob, E. Peterson, D. J. Doudet, A. Gjedde, and A. Møller. 2010. Dopamine Release in Ventral Striatum of Pathological Gamblers Losing Money. *Acta Psychiatrica Scandinavica* 122 (4): 326–333.

Loh, Kep Kee, and Ryota Kanai. 2014. Higher Media Multi-Tasking Activity is Associated with Smaller Gray-Matter Density in the Anterior Cingulate Cortex. PLoS One 9 (9). Accessed at: https://doi.org/10.1371/journal.pone.0106698.

Lorenz, Taylor. 2018. Generation Z Is Already Bored by the Internet. *Daily Beast* (April 3). Accessed at: https://www.thedailybeast.com/generation-z-is-already-bored-by-the-internet?ref=scroll.

Lustig, Robert H. 2017. *The Hacking of the American Mind: The Science Behind the Corporate Takeover of Our Bodies and Brains*. New York: Avery.

Mark, Gloria. 2015. *Multitasking Int the Digital Age*. NP: Morgan and Claypool.

Mark, Gloria. 2023. *Attention Span: A Groundbreaking Way to Restore Balance, Happiness and Productivity*. New York: Hanover Square Press.

Martiniko, Katherine Johnson. 2023. *Childhood Unplugged*. Gabriola Island, BC: New Society Publishers.

Maushart, Susan. 2010. *The Winter of our Disconnect: How Three Totally Wired Teenagers (and a Mother Who Slept with Her iPhone) Pulled the Plug on Their Technology and Lived to Tell the Tale*. New York: Penguin.

McNamee, Roger. 2019. *Zucked: Waking Up to the Facebook Catastrophe*. New York: Penguin.

Messena, Mattia, and Marina Everri. 2023. Unpacking the Relation Between Children's Use of Digital Technologies and Children's Well-being: A Scoping Review. *Clinical Child Psychology and Psychiatry* 28 (1): 161–198.

Microsoft (Consumer Insights, Canada). 2015. Attention Spans. Accessed at: https://www.scribd.com/document/265348695/Microsoft-Attention-Spans-Research-Report.

Mischel, Walter, Y. Shoda, and P. K. Peake. (1988). The Nature of Adolescent Competencies Predicted by Preschool Delay of Gratification. *Journal of Personality and Social Psychology* 54: 687–696.

Moffitt, T. E., L. Arseneault, D. Belsky, N. Dickson, R. J. Hancox, H. L. Harrington, and A. Caspi. 2011. A Gradient of Childhood Self-Control Predicts Health, Wealth, and Public Safety. *Proceedings of the National Academy of Sciences* 108: 2693–2698.

Moynihan, Andrew B., Eric R. Igou, and Wijnand A. P. van Tilburg. 2017. Boredom Increases Impulsiveness: A Meaning-Regulation Perspective. *Social Psychology* 48 (5): 293–309.

Murthy, Vivek. 2023a. *Social Media and Youth Mental Health: The U.S. Surgeon General's Advisory*. Accessed at: https://www.hhs.gov/sites/default/files/sg-youth-mental-health-social-media-advisory.pdf.

Murthy, Vivek. 2024. Surgeon General: Why I'm Calling for a Warning Label on Social Media Platforms. *New York Times,* June 17. Accessed at: https://www.nytimes.com/2024/06/17/opinion/social-media-health-warning.html.

Nass, Clifford. 2009. Interview for *Frontline: Digital Nation.* Accessed at: https://www.pbs.org/wgbh/pages/frontline/digitalnation/interviews/nass.html.

National Institute of Health. 2019. Overdose Death Rates. Accessed at: https://www.drugabuse.gov/related-topics/trends-statistics/overdose-death-rates.

Negroponte, Nicholas. 1995. *Being Digital.* New York: Alfred A. Knopf.

Nielson. 2018. The Nielsen Total Audience Report: Q3 2108. Accessed at: https://www.nielsen.com/us/en/insights/report/2019/q3-2018-total-audience-report/

Noake, Rick. 2015. This City Embedded Traffic Lights in the Sidewalks so that Smartphone Users Don't Have to Look Up. *The Washington Post*, April 25. Accessed at: https://www.washingtonpost.com/news/worldviews/wp/2016/04/25/this-city-embedded-traffic-lights-in-the-sidewalks-so-that-sma rtphone-users-dont-have-to-look-up/.

Oggins, Jean, and Jeffrey Sammis. 2012. Notions of Video Game Addiction and Their Relation to Self-Reported Addiction Among Players of World of Warcraft. *International Journal of Mental Health Addiction* 10: 210–230.

Perrin, Andrew, and Sara Atske. 2021. About Three-In-Ten U.S. Adults Say they are 'Almost Constantly' Online. Pew Research Center. Accessed at: https://www.pewresearch.org/fact-tank/2021/03/26/about-three-in-ten-u-s-adults-say-they-are-almost-constantly-online/.

Premack, David, and Ann Premack. 2003. *Original Intelligence.* New York: McGraw Hill.

Price, Catherine. 2018. *How to Break Up with Your Phone: The 30-Day Plan to Take Back Your Life.* California: Ten Speed Press.

Primack, Brian, et al. 2017. Social Media Use and Perceived Social Isolation Among Young Adults in the U.S. *American Journal of Preventative Medicine* 53 (1): 1–8.

Purvis, Jeanette. 2017. Finding Love in a Hopeless Place: Why Tinder Is So 'Evilly Satisfying.' *Salon.* Accessed at: https://www.salon.com/2017/02/12/finding-love-in-a-hopeless-place-why-tinder-is-so-evilly-satisfying/.

Ra, Chaelin, et al. 2018. Association of Digital Media Use With Subsequent Symptoms of Attention-Deficit/Hyperactivity Disorder Among Adolescents. *JAMA* 320 (3): 255–263.

Rabin, Roni Caryn. 2021. Overdoes Deaths Reach Record High as the Pandemic Spread. *New York Times*, November 17. Accessed at: https://www.nytimes.com/2021/11/17/health/drug-overdoses-fentanyl-deaths.html?campaign_id=9&emc=edit_nn_20220213&instance_id=53054&nl=the-morning&regi_id=71023177&segment_id=82588&te=1&user_id=1b8a7a6c418ba0b24e53c8b7ae5ab191.

Rainie, Lee, and Kathryn Zickuhr. 2015. Americans' Views on Mobile Etiquette. Pew Research Center, August 26.

Riehm, Kira et al. 2019. Associations Between Time Spent Using Social Media and Internalizing and Externalizing Problems Among US Youth. *JAMA Psychiatry* 76 (12): 1266–1273.

Rothwell, Jonathan. 2023. Teens Spend Average of 4.8 Hours on Social Media Per Day. *Gallup*. Accessed at: https://news.gallup.com/poll/512576/teens-spend-average-hours-social-media-per-day.aspx.

Tracii, Ryan, et al. 2014. The Uses and Abuses of Facebook: A Review of Facebook Addiction. *Journal of Behavioral Addiction* 3 (3): 133–148.

Sana, Faria, Tina Weston, and Nicholas Cepeda. 2013. Laptop Multitasking Hinders Classroom Learning for Both Users and Nearby Peers. *Computers and Education* 62: 24–31.

Sangani, N. Janji, and M. R. Razavi. 2024. Investigating the Relationship Between the Use of Social Media, Educational Decline, and Student Mental Health. *Journal of Practice in Clinical Psychology* 12 (2): 123–136.

Schaeffer, Katherine. 2019. Most U.S. Teens Who Use Cellphones Do It to Pass Time, Connect with Others, Learn New Things. Pew Research Center. Accessed at: https://www.pewresearch.org/fact-tank/2019/08/23/most-u-s-teens-who-use-cellphones-do-it-to-pass-time-connect-with-others-learn-new-things/.

Schmeichel, B. J., and R. F. Baumeister. 2004. Affect Regulation. In *Handbook of Self-Regulation: Research, Theory, and Applications*, ed. R.F. Baumeister and K.D. Vohs, 84–98. New York: Gilford Press.

Schüll, Natasha Dow. 2012. *Addiction by Design: Machine Gambling in Las Vegas*. Princeton: Princeton University Press.

Schwartz, Barry. 2016. *The Paradox of Choice: Why More is Less*. New York: EccoPress.

Sepah, Cameron. 2019. The Definitive Guide to Dopamine Fasting 2.0: The Hot Silicon Valley Trend. *Medium*, October 26. Accessed at: https://medium.com/swlh/dopamine-fasting-2-0-the-hot-silicon-valley-trend-7c4dc3ba2213.

Siu, William. 2022. I Make Video Games. I Won't Let My Daughters Play Them. *New York Times*, October 2. Accessed at: https://www.nytimes.com/2022/10/02/opinion/video-game-addiction.html?campaign_id=9&emc=edit_nn_20221003&instance_id=73583&nl=the-morning&regi_id=71023177&segment_id=108807&te=1&user_id=1b8a7a6c418ba0b24e53c8b7ae5ab191.

Small, Gary W., et al. 2020. Brain Health Consequences of Digital Technology Use. *Dialogues in Clinical Neuroscience* 22 (2): 179–187.

Solon, Olivia. 2017. Ex-Facebook President Sean Parker: Site Made to Exploit Human 'Vulnerability.' *The Guardian*, November 9. Accessed at: https://www.theguardian.com/technology/2017/nov/09/facebook-sean-parker-vulnerability-brain-psychology.

Stokel-Walker, Chris. 2022. TikTok Wants Longer Videos—Whether You Like It or Not. *Wired* (February 21). Accessed at: https://www.wired.com/story/tiktok-wants-longer-videos-like-not/.

Takahashi, Ippei, et al. 2023. Screen Time at Age 1 Year and Communication and Problem-Solving Developmental Delay at 2 and 4 Years. *JAMA Pediatrics*. Accessed at: https://jamanetwork.com/journals/jamapediatrics/fullarticle/2808593.

Tam, K. Y. Y., and M. Inzlicht. 2024. Fast-forward to Boredom: How Switching Behavior on Digital Media Makes People more Bored. *Journal of Experimental Psychology: General*. Accessed at: https://psycnet.apa.org/fulltext/2025-11974-001.pdf?auth_token=dbe63f30544cc6199b3c2defe2a8e923e30a8e26.

Thiele, Leslie Paul. 1995. *Timely Meditations: Martin Heidegger and Postmodern Politics*. Princeton: Princeton University Press.

Thiele, Leslie Paul. 1997. Postmodernity and the Routinization of Novelty: Heidegger on Boredom and Technology. *Polity* 29: 489–517.

Thiele, Leslie Paul. 1998. *Timely Meditations: Martin Heidegger and Postmodern Politics*. Princeton: Princeton University Press.

Thorn. 2021. Responding to Online Threats: Minors' Perspectives on Disclosing, Reporting, and Blocking: Findings from 2020 Quantitative Research Among 9–17 year olds. Accessed at: https://info.thorn.org/hubfs/Research/Responding%20to%20Online%20Threats_2021-Full-Report.pdf.

Tomasello, Michael, Josep Call, and Brian Hare. 2003. Chimpanzees Understand Psychological States—The Question is Which Ones and to What Extent. *Trends in Cognitive Sciences* 7: 153–156.

Tullett-Prado, Deon, Jo R. Doley, Daniel Zarate, Rapson Gomez, and Vasileios Stavropoulos. 2023. Conceptualising Social Media Addiction: A Longitudinal Network Analysis of Social Media Addiction Symptoms and Their Relationships with Psychological Distress in a Community Sample of Adults. *BMC Psychiatry* 23: 509.

Turkle, Sherry. 2015. *Reclaiming Conversation: The Power of Talk in a Digital Age*. New York: Penguin.

Turkle, Sherry. 2017. *Alone Together: Why We Expect More from Technology and Less from Each Other*, 3rd ed. New York: Basic Books.

Twenge, Jean M. 2017. *IGen: Why Today's Super-Connected Kids are Growing Up Less Rebellious, More Tolerant, Less Happy—And Completely Unprepared for Adulthood*. New York: Atria Books.

Twenge, Jean M., and Roy F. Baumeister. 2002. Self-Control: A Limited Yet Renewable Resource. In *Self and Identity: Personal, Social, and Symbolic*, ed. Yoshihisa Kashima, Margaret Foddy, and Michael Platow, 57–70. Mahwah, NJ: Lawrence Erlbaum Associates.

Twenge, Jean, Brian Spitzberg, and W. Keith. Campbell. 2019. Less In-person Social Interaction with Peers Among U.S. Adolescents in the 21st Century and Links to Loneliness. *Journal of Social and Personal Relationships* 36 (6): 1892–1913.

Uhls, Yalda T., et al. 2014. Five Days at Outdoor Education Camp Without Screens Improves Preteen Skills with Nonverbal Emotion Cues. *Computers in Human Behavior* 39: 387–392.

Melina, Uncapher, Monica Thieu, and Anthony Wagner. 2016. Media Multitasking and Memory: Differences in Working Memory and Long-Term Memory. *Psychonomic Bulletin and Review* 23 (2): 483–490.

Vallor, Shannon. 2016. *Technology and the Virtues: A Philosophical Guide to a Future Worth Wanting*. New York: Oxford University Press.

Vattimo, Gianni. 1985. *The End of Modernity*. Baltimore: Johns Hopkins University Press.

Walsh, Jeremy, Joel D. Barnes, Jameason D. Cameron, Gary S. Goldfield, Jean-Philippe, Chaput, Katie E. Gunnell, Andrée-Anne, Ledoux, Roger Zemek, and Mark S. Tremblay. 2018. Associations between 24-Hour Movement Behaviours and Global Cognition in a Cross-Sectional Sample of American Children. *The Lancet* 2 (11): 783–791.

Ward, Zachary J., Sara N. Bleich, Angie L. Cradock, Jessica L. Barrett, Catherine M. Giles, Chasmine Flax, Michael W. Long, Steven L. Gortmaker. 2019. Projected U.S. State-Level Prevalence of Adult Obesity and Severe Obesity. *New England Journal of Medicine* 381: 2440–2450.

Wedge, Marilyn. 2017. 'Virtual Autism' May Explain Explosive Rise in ASD Diagnoses. Accessed at: https://www.madinamerica.com/2017/08/virtual-autism-explain-rising-asd-diagnoses/.

Wells, Georgia, Jeff Horwitz, and Deepa Seetharaman. 2021. Facebook Knows Instagram is Toxic for Teen Girls, Company Documents Show. *Wall Street Journal*, September 14. Accessed at: https://www.wsj.com/articles/facebook-knows-instagram-is-toxic-for-teen-girls-company-documents-show.

Whitten, Allison. 2022. Brain Chemical Helps Signal to Neurons When to Start a Movement. *Quanta Magazine* (March 22). Accessed at: https://www.quantamagazine.org/brain-chemical-helps-signal-to-neurons-when-to-start-a-movement-20220322/#comments.

Wilson, T., D. Reinhard, E. Westgate, D. Gilbert, N. Ellerbeck, C. Hahn, C. Brown, and A. Shaked. 2014. Just Think: The Challenges of the Disengaged Mind. *Science* 345 (6192): 75–77.

Williams, James. 2018. *Stand Out of Our Light: Freedom and Resistance in the Attention Economy*. Cambridge: Cambridge University Press.

Winnick, Michael. 2016. Putting a Finger on Our Phone Obsession: Mobile Touches: A Study on How Humans use Technology. *Dscout*. Accessed at: https://pages.dscout.com/hubfs/downloads/dscout_mobile_touches_study_2016.pdf.

World Health Organization. 2019. Obesity and Overweight. Accessed at: https://www.cdc.gov/obesity/data/adult.html.

Wu-Ouyang, Biying. 2023. More Mobile Connectedness, Less Well-Being? *Journal of Media Psychology* 35 (5): 291–302.

Yeykelis, Leo, James J. Cummings, and Byron Reeves. 2014. Multitasking on a Single Device: Arousal and the Frequency, Anticipation, and Prediction of Switching Between Media Content on a Computer. *Journal of Communication* 64 (1): 167–192.

Zhao, Yihong, et al. 2022. Brain Structural Covariation Linked to Screen Media Activity and Externalizing Behaviors in Children. *Journal of Behavioral Addictions* 11 (2): 417–426.

CHAPTER 4

Deskilling: The Atrophy of Cognitive and Social Aptitudes

Digital distraction and dependence degrade our capacity for reflective self-direction. The AI-enabled attention economy also undermines the development of diverse cognitive and social skills that enhance agency. There can be no doubt that digital technologies greatly augment many human faculties. But they also produce widespread deskilling. Resisting and reversing this trend is crucial for human flourishing. It is mistaken to believe that technological fixes can do the job for us.

Deskilling is not unique to our time or the attention economy. At the onset of the industrial age in the mid-eighteenth century, unskilled workers and farmers were drawn to cities to work in manufacturing. Sites of small-scale production, formerly manned by artisans with years of training, were displaced by factories where laborers carried out the routinized procedures that enabled mass production. As machines dispatched tasks that formerly were the responsibilities of skilled workers, artisanry declined. Such deskilling became a prominent concern in the nineteenth century, and it produced widespread social disruptions. Protests and sabotage by Luddites, who destroyed mechanized weaving looms and other labor-replacing technologies, were not uncommon. Reprisals by state officials were harsh, including both police and military actions. Capital punishment was meted out to some saboteurs.

Adam Smith, the eighteenth-century Scottish economist and philosopher, was an early observer of factory production and deskilling. Smith,

L. P. Thiele, *Human Agency, Artificial Intelligence, and the Attention Economy*, https://doi.org/10.1007/978-3-031-82086-1_4

the so-called father of capitalism, famously begins *The Wealth of Nations* with a description of a pin factory where "One man draws out the wire, another straights it, a third cuts it, a fourth points it, a fifth grinds it at the top for receiving the head" (Smith 2003, 11). Making a pin, Smith observed, might entail eighteen separate operations, each of which was carried out by a worker restricted to that task. If an individual was to fashion a pin from start to finish, the daily production of pins at the factory would at best be in the double digits. The division of labor greatly increased efficiency and productivity, Smith marveled, allowing the production of tens of thousands of pins a day.

Smith was eager to celebrate the wonders of capitalism, but he was not oblivious to its costs. Sharp increases in efficiency and productivity were accompanied by equally significant decreases in the manual skill sets of workers. There were cognitive losses as well. Smith (2003, 987) writes:

> In the progress of the division of labour, the employment of the far greater part of those who live by labour, that is, of the great body of people, comes to be confined to a few very simple operations, frequently one or two. But the understandings of the greater part of men are necessarily formed by their ordinary employments. The man whose whole life is spent in performing a few simple operations, of which the effects too are, perhaps, always the same, or very nearly the same, has no occasion to exert his understanding, or to exercise his invention in finding out expedients for removing difficulties which never occur. He naturally loses, therefore, the habit of such exertion, and generally becomes as stupid and ignorant as it is possible for a human creature to become. The torpor of his mind renders him, not only incapable of relishing or bearing a part in any rational conversation, but of conceiving any generous, noble, or tender sentiment, and consequently of forming any just judgement concerning many even of the ordinary duties of private life.... But in every improved and civilized society this is the state into which the labouring poor, that is, the great body of the people, must necessarily fall, unless the government takes some pains to prevent it.

We become what we do, Smith was saying. That presents an unhappy fate for deskilled factory workers. Of course, those who pursue power and wealth are subject to the same truth, which explains why the elite in Smith's time did not consistently demonstrate generous, noble, or tender sentiments and just judgments when ruling over the laboring class and colonized peoples. Still, the fact remains that cognitive and social skills are gained through experience. To the extent that experiences are degraded, skill acquisition is diminished.

Smith's dour assessment was shared by the French political theorist Alexis de Tocqueville, who traveled to the US at the start of the nineteenth century. In *Democracy in America,* de Tocqueville (1958, 52–53) addressed the "democratic principle" that animated the new world. This cultural force, he observed, had a powerful economic impact, stimulating the production "with great rapidity of many imperfect commodities" that are "abundant and cheap." Like Smith, de Tocqueville bemoaned the deskilling of the industrial laborers producing these cheap, abundant commodities. What is lost to the individual, de Tocqueville (1958, 168–169) writes, is "the general faculty of applying his mind.... The man is degraded.... [H]is thoughts are forever set upon the object of his daily toil; his body has contracted certain fixed habits, which it can never shake off; in a word, he no longer belongs to himself.... The workman becomes more weak, more narrow minded, and more dependent." The unelevated character of the laboring producers, de Tocqueville observed, was paralleled by the low standards of the purchasing consumers.

Shortly after the publication of *Democracy in America*, Karl Marx, the primary theorist of communism, penned *The Economic and Philosophic Manuscripts.* The term that Marx employed for the mind-narrowing dependency undergone by industrial workers was *alienation*, and in particular, alienation from the "act of production." In the tedious repetition of factory production, Marx (1978, 74, 95) wrote, the industrial laborer "does not develop freely his physical and mental energy but mortifies his body and ruins his mind. The worker therefore only feels himself outside his work, and in his work feels outside himself.... Machine labour is simplified in order to make a worker out of the human being .... The machine accommodates itself to the *weakness* of the human being in order to make the *weak* human being into a machine." Like de Tocqueville, Marx read Smith and had directly observed the degradations of industrialized labor. His conclusions were similar. In the eighteenth and nineteenth centuries, the seminal theorists of capitalism, democracy, and communism all deplored the deskilling suffered by a rising proportion of the population.

The degraded state of early industrial laborers found its parallel in the fate of "knowledge workers" over the next two centuries. Here the worker's days were spent in the monotonous and relatively mindless fingering of typewriters, and later keyboards. As industrialism and automation progressed, machines took over an increasing proportion of the simple, repetitive components of factory production. With the development of computers in the twentieth century and AI in the twenty-first century, machines are taking over an increasing proportion of knowledge production. In both cases, the use of machines first reduced the skill levels demanded of many workers, and subsequently came to displace them altogether.

Experts in the field of machine learning give even odds that all currently performed human tasks will be better performed by smart machines by the early 2060s (Grace et al. 2016). Estimations of levels of unemployment range widely. In the late 2010s, it was estimated that half the activities that people do across all sectors of most economies were susceptible to replacement by machines, as they involve physical work in highly structured and predictable environments or data collection and processing (Manyika and Bughin 2018, 4–5; Lee 2018, 199; McAfee and Brynjolfsson 2017, 332). With the arrival of generative AI in the early 2020s, it is now estimated that two-thirds of all jobs in developed economies will become automated (Hatzius et al. 2023).

The problem is not solely, or perhaps even primarily, about the loss of jobs. Much human skill development, as Smith observed, occurs in the workplace. So, widespread automation will necessarily result in deskilling. This is a significant loss, as Smith, de Tocqueville, and Marx observed. But there will also be gains. As Charles Babbage, the English polymath already observed in the early 1800s, deskilling has historically been accompanied by reskilling and upskilling. The repetitive, relatively low-skill tasks taken on by machines create opportunities for workers to transition into high-skill careers (Attewell 1987; Vallor 2015; Brugger and Gehrke 2018; McGuinness et al. 2019). Of course, many of those whose jobs are automated may be near the end of their careers; they will have few if any feasible paths toward upskilling. Still, technology can free up time and resources, allowing people to devote themselves to higher-level activities that demand new skill sets and are more cognitively, emotionally, or socially rewarding. In turn, technology may introduce novel social and economic needs, wants, and opportunities, stimulating the acquisition of new skill sets to address these demands.

Many of the jobs that survive automation will entail 'centaur' or 'co-pilot' relationships, where humans and smart machines work in tandem (Thiele 2021). In such situations, people may develop and exercise a wide array of higher-level aptitudes and skills. Unfortunately, the sophistication and complexity of smart machines means that workers often understand very little about the digital components of the relationships they are involved in. As Harry Braverman (1974, 425) observed in one of the first systematic investigations of deskilling: "The more sophisticated an intellectual product the machine becomes, the less control and comprehension of the machine the worker has. In other words, the more the worker needs to know in order to remain a human being at work, the less does he or she know." In an age of AI-enabled digital devices, apps, and platforms, even workers who are upskilled in some regard may understand very little of the "intelligence" behind the interfaces they navigate. There is no doubt that the machines involved in centaur relationships will become smarter and more expansive in their functionalities. Accordingly, human-AI collaborations are likely to become increasingly one-sided in the favor of the machines. While nominally co-pilots, human partners in centaur relationships will become ever more passive passengers.

The use of AI in businesses and organizations generally reduces costs while increasing productivity and revenue, in large part by enabling workers to complete tasks more quickly while often improving the quality of output. AI advances science and medicine in similar ways (Maslej et al. 2024). This does not mean that workers, scientists, or health practitioners are becoming more skilled or more fulfilled in their occupations. While generative AI may heighten efficiency in the workplace, it can also inhibit the exercise of critical thinking and independent problem-solving by knowledge workers (Lee et al 2025). AI will often become a replacement for, rather than a supplement to, human performance. Heightened productivity does not automatically or even typically generate heightened human development. Deskilling and its accompanying maladies are commonplace.

Deskilling is no longer only a threat for specific fields of laborers or professionals faced with workplace disruptions brought on by automation and the use of AI. It pervades digital culture. We spend more and more of our lives utilizing digital tools, but we understand less and less how they do what they do. In part, this is a product of their increasing sophistication. In part, it is a product of the secrecy surrounding the algorithms that structure our interactions with AI-enabled digital devices, platforms, and apps. Their mechanisms and algorithms are proprietary, and the ICT corporations that develop and deploy them are reluctant to show their hands. In any case, powerful digital devices, platforms, and apps, most of which employ AI, have usurped many of the tasks and engagements that hitherto advanced human skill development in everyday life. As in the workplace, co-pilot relationships are increasingly giving way to auto-pilot modes.

Unlike earlier upheavals within specific economic sectors, there is no clear path today for reskilling or upskilling the general population. De Tocqueville's concern for the industrial denizens of a fledgling democratic nation now applies to all of us, as our lives are organized by and intertwined with digital technologies. With minor adaptations to De Tocqueville's lament, we might ask: What can be expected of people who spend most of their waking hours peering at electronic devices with crooked necks, minds transfixed by pixelated screens and bodies contracted by fixed habits that cannot be shaken off? Might we no longer belong to ourselves as we become weaker, more narrow-minded, and more dependent? Marx's polemic against industrial alienation also finds its contemporary counterpart: the digital economy exploits the weaknesses of human beings in order to make human beings increasingly dependent on smarter machines.

In 2008, the cover story of the *Atlantic* posed the question "Is Google Making us Stoopid?" Subtitled "What the Internet is Doing to our Brains," Nicholas Carr's essay presented the case that online life is producing forgetful individuals whose capacities for concentration, memory, cognitive resourcefulness, creativity, and compassion were degraded. In three subsequent books, Carr (2011, 2014, 2016) substantiated his early claims. A great deal more research is required to determine the long-term effects of digital technology on cognitive and social aptitudes. Most of the available scholarship is not longitudinal, demonstrates correlation rather than causation, and is suggestive rather than conclusive (Wilmer et al. 2017). Still, what we do know is cause for concern.

AI-enabled digital technologies operate at such superhuman levels of speed and functionality, and with so little in the way of human effort, that they obviate the need for a broad swath of cognitive and social skills. Networked devices and apps supply ever more services and dispatch ever more tasks. Legions of people, effectively, have outsourced their intelligence and personal interactions. The result is widespread cognitive and social deskilling.

While the impacts of AI in terms of job losses or human domination are frequently and widely addressed, the topic of deskilling is largely ignored. A 63-page journal article on AI that integrates the opinions of 43 experts spanning multiple fields devotes two sentences to the topic of deskilling, noting simply that users of generative AI will likely lose some skills but potentially gain others (Diwivedi et al. 2023). A 128-page report by the International Labor Organization (2021) entitled *Changing demand for skills in digital economies and societies*, addresses deskilling once, in the context of online workers who are overqualified to perform "micro-tasks," such as photo-labeling. The 30-page "Comprehensive Analysis of AI's Impact on Jobs and Employment" assesses both benefits and risks, including job displacement and unemployment, threats to privacy and data security, the introduction or proliferation of bias, the lack of transparency and explainability, the lack of accountability and liability, the need for new training, ethical considerations related to decision-making, surveillance, and control, and overreliance on AI systems with limited or flawed performance. It neglects to address the threat of deskilling, apart from noting that "A growing dependence on AI technologies may lead to a decline in human skills, particularly in areas where AI systems perform tasks more efficiently than humans" (George et al. 2023, 171). By and large, the danger of deskilling is mostly ignored by those who are most worried about the impacts of AI (Schneier and Sanders 2023).

In some cases, workers themselves exhibit "automation complacency," wherein deskilling and dependence on smart machines is not seen as "a cause for concern" (Rinta-Kahila et al. 2023, 1390). Especially if it is expected to produce a decreased workload and less demanding work, deskilling may not be perceived negatively. Workers may even be enthusiastic about the process, particularly when their former activities were considered unpleasant compared to the tasks which they assume they will take on in a more automated setting (Rinta-Kahila et al. 2018). In the same manner, cognitive and social deskilling in the general population may not generate widespread concern and may even be welcomed, especially if smart machines are expected to provide increased comfort and convenience.

It would be a mistake to sound a general alarm about automation. The use of digital technology is not a sign of a degraded individual. Indeed, it often indicates an active mind that has more important things to do than calculations, memorization, or data analysis. There are large gains in efficiency provided by digital technologies, and significant gains in fulfillment if time gained is spent on higher-level endeavors. For example, many use the voice memos app on smartphones to store notions and plans when they are not at home or in the office. In lieu of struggling to remember a growing list of ideas or tasks, they can instead think new thoughts, and more fully enjoy outings, confident that the app will serve its purpose. This is known as cognitive offloading, and it can serve us very well.

Cognitive offloading occurs when we employ external devices to retain the capacity for future actions without requiring the continued use of memory or other mental resources. We cognitively offload when we write a note to ourselves, insert a meeting date into our calendars, or add a name and phone number to our contact list. This frees up the mind for higher-level activities. But there are costs and tradeoffs. Studies demonstrate that prospective memory—recalling that a specific action needs to occur at some future time—deteriorates in direct relation to the amount of cognitive offloading and the amount of time spent on digital media (Prathap and Singh 2021). Cognitive offloading can also lead to decreased learning and performance (Kayes and Yoon 2022). The more we offload cognitive tasks to digital media, the more likely it is that the corresponding aptitudes will deteriorate. The muscles in our bodies need strenuous exercise to develop and maintain their strength. Mental and emotional muscles are akin to physical ones in this regard, demonstrating a "use it or lose it" character. Absent regular exercise, they atrophy. The AI-enabled attention economy, by taking on our intellectual work, can enfeeble our minds.

Still, the human brain is a hungry organ. It wants a steady diet of information and experience. That fact portends well for reskilling and upskilling. Information commonly acquired in earlier eras—from the best wild plants to harvest to the most useful phone numbers to remember—no longer demand cerebral real estate. All the vacated neurological space can be utilized for more important purposes. Many people liberated from menial mental labor by digital tools will exercise their brains for higher-level activities. In turn, novel digital skills will be gained. Of course, newly acquired competencies may reflect a relatively narrow range of neural development (Small et al. 2020). Still, these new competencies may more than compensate for the decline of older skill sets. Certainly the early entrepreneurs of digital culture had such expectations. Time will tell whether their sanguine hopes are realized.

Intelligence Augmentation and Cognitive Deskilling

At the Augmentation Research Center at SRI International in the 1960s, Douglas Engelbart's visionary work led to the creation of the computer mouse, the development of hypertext, early graphical user interfaces, and networked computers. Engelbart focused his efforts on the use of digital technology to increase intellectual capacity and facilitate collaborative work, founding the field of Intelligence Augmentation or IA, also known as Intelligence Amplification (Markoff 2015, xii-7, 114).

When Steve Jobs famously called the personal computer "a bicycle for our minds," he was extolling intelligence augmentation. While tool-making is an age-old human endeavor, the personal computer marked a radical shift. Jobs (2020) mused:

> I read a study that measured the efficiency of locomotion for various species on the planet. The condor used the least energy to move a kilometer. And humans came in with a rather unimpressive showing, about a third of the way down the list. It was not too proud a showing for the crown of creation. So that didn't look so good. But then somebody at *Scientific American* had the insight to test the efficiency of locomotion for a man on a bicycle.... A human on a bicycle blew the condor away, completely off the top of the charts. And that's what a computer is to me... it's the most remarkable tool that we've ever come up with; and it's the equivalent of a bicycle for our minds.

With the advent of AI, the bicycle for our minds has become a rocket ship. We now have at our disposal "superhuman capacities" that allow us, cognitively speaking, to "punch above our weight" (Daugherty and Wilson 2018). Computers and other smart devices are ushering in an "age of augmentation" that is predicted to deliver "the greatest advantages and potential of any generation in the entire history of humanity" (King 2016, 436).

With sufficient foresight and creativity exercised by educators, developers, and policymakers, intelligence augmentation could advance the "humanization" of employment and everyday life, allowing us to offload much of the mental "grunt work" that many jobs and daily routines would otherwise demand. People would then be free to engage in more adaptive learning, problem solving, and creative enterprise (King 2016; Daugherty and Wilson 2018). Clearly intelligence augmentation has this potential. But will it have this effect? AI can undoubtedly stimulate innovation. It is also likely to discourage the demanding but necessary skill development that begets the highest levels of human creativity (Hartley and Viskontas 2023). Augmentation technologies tend to be double-edged swords, and intelligence *amputations* are frequent (Maes 1998).

Bicycles increase mobility and sustain physical fitness without undermining agency. Digital technologies can also augment human capacities. Yet they demand diminishing amounts of human memory, reasoning, resourcefulness, creativity, and judgment, while operating at superhuman levels of connectivity, speed, and functionality. As a result, digital technologies may function less like a bicycle and more like a recliner chair (Case 2018). Once occupants are fully reclined, they may lack the motivation and muscular wherewithal to rise. In the film *Wall-E,* humans escaped a desecrated Earth aboard giant starliners in the twenty-second century. Hundreds of years later, suffering from the effects of zero gravity and robots tending to their every need and want, the human occupants of the starliners have become obese, unmotivated, and practically helpless. We may not have to quit our planet and escape its gravity to experience a similar degeneration.

The vibrancy of human agency is a matter not only of ability but of incentive. By supplanting the need to exercise aptitudes, digital technologies decrease the motivation for exercising them. Lack of motivation heightens lack of use. Without frequent use, cognitive and emotional muscles decline in strength. In this weakened state, reliance on digital technologies will steadily increase, further degrading the motivation to exercise aptitudes. AI-enabled bicycles for the mind will create vicious cycles of deskilling.

Consider the impact of GPS technologies, which reduce both the motivation to develop and the exercise of wayfinding skills. Digital natives who drive with GPS-enabled, cloud-linked vehicles—like pedestrians navigating city streets with the aid of GPS-linked smartphones—are increasingly incapable, when offline, of basic self-navigation. Their spatial memory and spatial learning become significantly degraded (Dahmani and Bohbot 2020; Clemenson 2021; Yan et al. 2022). Absent digital tools they are at a loss, and often become lost. Drivers and pedestrians with undeveloped wayfinding skills are more likely to rely on GPS-enabled device's, even for the simplest navigational tasks. As a result, their deskilling deepens.

Driving a GPS-outfitted vehicle that continuously transmits data to the cloud greatly contributes to collective knowledge of the best ways for drivers (or driverless vehicles) to avoid congestion and reach their destinations quickly. But human drivers' contribution to this emergent intelligence in no way enhances their own navigational abilities. Consequently, wayfinding skills are withering away, and there is no evidence that the cerebral muscles hitherto employed for this purpose are being put to a better use. More likely, drivers whose minds are freed from the burden of navigation are simply listening to music, daydreaming, or monitoring social media. While comfort, convenience, and pleasure may increase for those who rely on GPS technology, there appears to be no compensatory upskilling. Indeed, deskilling in wayfinding may lead to deskilling in related capacities, such as situational awareness.

Compounded deskilling will likely be more of the norm than the exception. Consider the impact of self-driving vehicles. Various laws and technological innovations over the years—including mandatory seatbelts, airbags, collapsible steering wheels, and campaigns against drunken driving—have greatly reduced traffic fatalities. Today, however, many drivers fail to focus their attention when behind the wheel owing to digital distractions. Over 60% of young adults (18 years of age or older) acknowledge that they text or email while driving, and over 3000 traffic fatalities in the US are caused each year by distracted drivers (Center for Disease Control and Prevention 2022). The justification for self-driving vehicles—for reasons of safety and efficiency—is quite strong given the widespread decline in focused attention and the loss of wayfinding skills. Here the loss of skills for wayfinding and focusing attention, brought on by digital distraction and dependence, heighten the demand for self-driving cars. As self-driving vehicles become more common, we will likely witness the further atrophy of aptitudes for wayfinding, situational awareness, and focusing attention. In turn, other skills involved in safely operating an automobile may also be degraded, such as eye-hand coordination, basic motor skills, and spatial memory.

In the same vein, a wide range of skill sets may decline within professions that integrate smart devices and apps. Physicians, clinicians, and airline pilots who work with AI-enabled programs, protocols, and databases display a considerable loss of the traditional skills associated with their respective professions. The more powerful the technology, the greater the loss of skills, including basic capacities for interpretation and judgment (Carr 2014). Within the general population, the longterm impact of cognitive offloading in the wake of the widespread use of generative AI is unknown. But early studies suggest cause for concern (León-Domínguez 2024).

Deskilling among professionals and the lay population will be accelerated with the expansion in size and market penetration of the Internet of Things (IoT). This global network of interconnected computers, smartphones, appliances, buildings, vehicles, cameras, and sensors has surpassed the number of human beings on the planet manyfold, reaching 30 billion connected devices in 2020. The number will quickly double, and some industry analysts forecast 500 billion IoT devices by 2030 (Cisco 2016). Most of the information flowing within the IoT is generated, transmitted, received, aggregated, analyzed, and assessed by smart machines. Indeed, machine-to-machine communication likely already exceeds the total global volume of all human-to-human conversation and communication. We are in the midst of a "second machine age" characterized by billions of interconnected, interactive smart devices that will surpass—and "make mockery of"—the scope and depth of all previous transformations of the human condition (Brynjolfsson and McAfee 2014, 66, 96). If we understand the personal computer to be a bicycle for our minds, then the smart digital devices of the IoT constitute an entire transportation system that steers our economic, professional, social, and political lives.

The IoT has delivered significant advances in productivity and performance across diverse sectors and fields: from agriculture, energy management, manufacturing, and national security, to education, health care, human rights protection, and natural resource conservation. It is likely to be the primary vehicle for commercial and technological development in the twenty-first century, and the foremost deployer of AI. The IoT presents an unprecedented capacity for data capture, management, transfer, and exchange, and an exploding array of communication, networking, and analytic functionalities spanning private and public sectors. No doubt the IoT will transform economies and societies. Much of this transformation will be for the better, particularly when it enables and empowers people.

The "distributed brain" created by networked humans and computers has the potential to solve many of the world's "wicked problems," such as climate change and geopolitical conflict (Michelucci and Dickinson, 2016). Smartphones, laptops, and other forms of mobile technology enable lay populations to engage in "citizen science," gathering data about local flora or fauna and collaborating with professional counterparts. More broadly, the "swarm" or "emergent" intelligence generated by the "crowd and the cloud" is utilized for knowledge creation, information sharing, and problem solving by academics, governmental agencies, non-profit organizations, and corporations (Rheingold 2003, xii, 178–182; The Crowd and the Cloud 2018).

Most individuals availing themselves of IoT, however, exercise diminished agency. They have few incentives and opportunities to cultivate aptitudes and skills beyond those needed to navigate digital interfaces. Indeed, the task of navigating the IoT is increasingly given over to virtual agents, also known as Intelligent Virtual Assistants (IVAs) and Virtual Personal Assistants (VPAs). Virtual agents are sophisticated chatbots that not only respond to voice or text commands with voice or text responses, but also remember, learn, and implement. Chatbots employing generative AI are increasingly becoming virtual agents as they acquire functionalities. Alexa (Amazon), AliGenie (Alibaba), Bixby (Samsung), Bing and Co-pilot (Microsoft), ChatGPT and GPT-4 (OpenAI), Claude (Anthropic), Cohere (Cohere), Falcon (Technology Innovation Institute, UAE), Gemini and PaLM2 (Google), Grok-1 (xAI), Pi (Inflection), Llama (Meta), and Siri (Apple Intelligence) are prominent examples, among others.

The market for virtual agents, driven by continuous innovations and technological advancements, is projected to grow at a compound annual growth rate of over 40% between 2024 and 2035 (Roots Analysis 2024). The development of functionalities is exploding. Virtual agents now set appointments, make reservations, regulate (home, office, and auto) environments, geographically locate and navigate, translate languages, provide information, write prose and computer code, advise, teach, converse, entertain, coach, and therapeutically counsel. The most prominent virtual agents are powered by multimodal AI. They can process and provide natural language and images, use apps, and interact with the IoT. As competencies grow, so-called *agentic AI* will not only provide information, advice, entertainment, and companionship, but execute an expanding range of household, commercial, financial, and professional tasks and transactions, from making reservations for family vacations to targeting and launching weapons on the battlefield. AI-enabled virtual agents already play the role of social media influencers, with followers in the millions. They have distinct "personalities" and can satisfy the need for a diverse range of human-like interactions.

Virtual agents are taking on a broad and quickly expanding range of activities that were once the sole purview of human beings. They do much of what we can do, only faster and more effectively. While these smart technologies are called agents, their human users are often known as subjects. That is a telling comment on the rise of smart machines and the ebb of human agency. To be sure, people are the subjects of much of the data gathered from and analyzed by the IoT and the AI agents that navigate and curate it for us (e.g. health data generated by individuals wearing biometric sensors). People also serve as the subjects of advertising informed by this data. But with the satisfaction of our needs, the manipulation of our wants, and the fulfilling of our responsibilities taken on by virtual agents, the opportunity to exercise memory, reasoning, resourcefulness, decision-making, problem-solving, creativity, and diverse social aptitudes is dramatically reduced.

Decision displacement is a particularly troubling development. At present, algorithms churning through data mostly filter choices, short-listing options for human selection. The choices are varied and abundant: where we can go, what we can do, buy, listen to, and watch, and whom we can friend, date, and marry. We have become pampered choosers in the face of filtered abundance. This filtering could be interpreted as a form of empowerment, allowing us to make better decisions in the face of an otherwise bewildering array of alternatives. But decision augmentation portends intelligence amputation.

With the release of the ultra-fast streaming search function *Google Instant*, Sergey Brin, Google's co-founder, announced that "we want Google to be the third half of your brain." The implication is that "Google will know what you want in a search, perhaps even before you know" (Yarrow 2010). Internet search destinations appear on our screens before we fully describe them, just as words and phrases appear on our phones and computers before we type them, or even think about them. These autocomplete programs are growing in power with the arrival of generative AI. Other forms of decision displacement are in development. Amazon has a patent for "anticipatory shipping," wherein merchandise will arrive at customers' doorsteps without their ordering it, the product of algorithms that predict likely purchases (Agrawal et al. 2018, 17).

In the eighteenth century, Adam Smith argued that the hidden hand of the market ensured the coordination of supply and demand without the need for governmental oversight or administration. The 21st century marketplace is developing a hidden mind. Customers and clients will no longer have to decide what they want or when they want it. They will be supplied before they demand. AI-enabled forecasting, a topic explored in Chapter 6, is displacing human judgment. In the early 1950s, Norbert Wiener, the inventor of cybernetics, first addressed the impact of machines that he described as "the judgment replacing type" (Wiener 1954, 169). At the time, such machines did not exist. Now they do. We have reached a crisis point (Krakaurer 2015).

Some envision the development of "Google Morals," an app that would liberate its users from the burden of ethical judgments (Howell 2014). Provided a brief description of the situation at hand, the app would figure out the most appropriate course of action. People could then avoid the exertions of wrestling with moral conundrums, engaging instead in more pleasurable activities, such as shopping (based on recommendations from Amazon), watching a film (based on recommendations from Netflix), catching up on the recent events (based on a tailored newsfeed delivered from Facebook), and romance or hookups (based on recommendations from a wide variety of dating apps).

An ethics app may seem far-fetched. Still, related developments in political life suggest a trend. Human–machine partnerships are projected to advance democratic life in a world of cyborg citizens (Hughes 2004). There is a "huge opportunity" for the development of digital technologies that allow citizens to delegate the task of voting to AI programs (Malone 2018, 102–104). Proponents of "augmented democracy" (Hildago 2018) advocate for "digital twins," software agents that either supply information to support decision-making or make political choices on behalf of individuals, much like Pandora or Spotify choose our music based on recent preferences and past practices. Given the levels of political apathy and cynicism in many advanced democracies, the use of digital twins and voting apps might be widely embraced, and certain tasks of governance might actually be delegated to AI (Howard 2015, 257). As many as two out of five citizens in Europe already prefer smart machines to politicians as key national decision-makers (IE: Center for the Governance of Change 2019). That ratio will likely grow in AI's favor, notwithstanding the role of digital technology in creating jaded citizens and polarized polities, as we will see in the next chapter.

Grappling with the impact of artificial intelligence on individual autonomy, social relations, and governmental affairs, Henry Kissinger, former US Secretary of State and National Security Advisor, Eric Schmidt, former CEO of Google, and Daniel Huttenlocher, Dean of MIT's

Schwarzman College of Computing, are compelled to ask: "Will it be possible to retain our agency as our social and political lives increasingly shift into domains curated by AI, domains that we can only navigate through reliance upon that curation?" (Kissinger et al. 2021, 217–218). Regardless of the aptitudes or values of these men, their question is worth asking. As political decisions get outsourced to digital technologies, citizens and statespeople may become perilously deskilled in the civic arts.

People currently evaluate decisions made by AI to be on par with, and at times better than, those made by human experts, suggesting that "humans as decision-makers are not per se perceived as being irreplaceable" (Araujo et al. 2020, 621). In turn, the less knowledge and understanding people have about the decision domain, the more likely they are to utilize and defer to AI (Schneider and Leyey 2019). The world is becoming increasingly complex, requiring heightened levels of knowledge and skill to navigate it effectively. Meanwhile, distractions that keep people from acquiring such knowledge and skills are ever more abundant and enticing. Accordingly, the scales are likely to tip further in favor of "algorithmic governance."

It stands to reason that the extensive use of sophisticated assistive technology will lead to the deterioration of related human aptitudes. Indeed, some cognitive functions, such as working memory, decline simply because we are aware that digital options exist. For example, most people cannot remember their children's or parents' phone numbers (Panda Security 2020). Such digital amnesia, also known as the Google effect, occurs when the exercise of working memory suffers owing to the ready availability of online means for retrieving data. When people are exposed to information that they expect to have digital access to in the future, their ability to recall that information is degraded (Sparrow et al. 2011). Digital amnesia affects all age groups (Kapersky Lab 2015).

A related cognitive impairment arises from the mere availability of electronic devices. One study had participants take a series of tests requiring mental concentration. Some left their smartphones in another room, while others placed their devices, which were all switched to silent mode, either face down on their desks or in their pockets, handbags, or daypacks. Participants with their phones in another room significantly outperformed those with their phones on the desk, and slightly outperformed those with phones in a pocket or bag. Subjects exhibited diminished ability to utilize working memory and employ reason to solve problems if phones were nearby, even when they were not turned on or in use (Ward et al. 2017). We might call this the "smartphone effect." Again, the power and ubiquity of digital technology prevents the exercise and development of aptitudes.

Social Deskilling

Cognitive skills include memory, critical thinking, judgment formation, and decision-making. Social skills generally include communication, empathy, and the perception of body language and other social cues. Self-regulation and problem-solving, notably, are capacities that are frequently identified as both cognitive skills and social skills (Berg 2017). We have addressed the significant deterioration of cognitive skills in the AI-enabled attention economy. Now we consider the fate of social skills.

Humans are naturally inclined to develop relationships of trust and caring with close associates over time. That is the foundation for the bonds of family, friends, and teams. Relationships of interdependence are grounded upon and enhance our capacities for empathy and affection. We learn to love and care for people as we come to perceive and understand their feelings. This is known as emotional intelligence.

Ezra Mishan (1967, 166) forecast the decline of emotional intelligence in the wake of burgeoning electronic media over half a century ago, writing: "Today no refuge remains from the desperate universal clamour for more efficiency, more excitement, and more novelty that goads us furiously onward.... Every step forward in technological progress ... effectively transfers our dependence upon other human beings to dependence upon machines and, therefore, unavoidably constricts yet further the direct flow of understanding and sympathy between people." Mishan has been vindicated. The crucial conditions for the development of emotional intelligence, including eye-contact, body awareness, and the give-and-take of conversation, are much diminished in virtual interactions (Carr 2011; Turkle 2015, 325). Consequently, declines in empathy—most prominent among youth but evident across all age groups—have been linked to the rise of digital media (Konrath et al. 2011).

In the non-virtual world, our "weak ties" with neighbors, colleagues, and acquaintances—relationships that fall somewhere between the strong bonds of family and friendship and the wariness we may have toward strangers—are grounded in intermittent encounters and casual conversations. These face-to-face interactions build trust. Their prominent decline in digital culture may be unconsciously interpreted as indicative of the untrustworthiness of neighbors, colleagues, and acquaintances. As increasing numbers of people (virtually) keep their distance from each other, greater levels of distrust and lower levels of care are likely, eliminating opportunities for the development of emotional intelligence.

Social media pages are often extraordinarily self-disclosive, to the point of being exhibitionistic. Opinions, values, and images that once were considered too personal to share beyond a small circle of family or friends get indiscriminately publicized. In such instances, privacy has not so much been invaded as abandoned. In the absence of face-to-face encounters, however, the absence of privacy that substitutes for intimacy within digital culture is unlikely to advance caring relationships grounded in trust. Digital communication between strangers or acquaintances can facilitate self-disclosure. Among established friends, however, emotional connection and commitment is most enhanced by face-to-face engagements. Video chat and audio chat are less effective, and text messaging demonstrates the poorest capacity to foster emotional bonding between friends (Sherman et al. 2013). While it is easy and common to expose and exhibit oneself online, it is very difficult to establish and sustain trust in enduring relationships in the absence of regular face-to-face encounters.

During the COVID-19 pandemic, the necessity of social distancing prompted many people to spend more time online. Though the need for social distancing eventually abated, the disposition for it may continue to grow. Declines in emotional intelligence and relationships of trust and care, brought about by the increasing reliance on digital forms of interaction, diminish the motivation to engage in face-to-face encounters and conversations. Fewer face-to-face encounters and conversations further the decline in emotional intelligence. The result is a vicious cycle of social deskilling.

So-called *mirror neurons*—unique to human beings and some other mammals—play a large role in the development of emotional intelligence by facilitating imitation and empathy (Iacoboni 2008). When a person observes an activity carried out by another, the same synaptic networks in the observer's brain get fired up as those in the actor's, as if the observer were engaged in the activity herself. Mirror neurons are responsible for this phenomenon, facilitating (imitative) learning. They likely play a key role in the development of mind-reading skills. Our capacity to learn from others is largely dependent upon our capacity to understand them. We primarily understand others, what is going on in their hearts and minds, through shared experiences and exchanges (Vivanti and Rogers 2014). When we engage in mind-reading, we are putting ourselves in the shoes of others, and walking a mile. This aptitude is exemplified by excellent students of social and political life, contributing to their practical judgment.

Our powerful system of mirror neurons appears ill-suited to digital engagements. Asynchronous and virtual experiences may not sufficiently or reliably activate the neural systems that support social learning and bonding (Dickerson et al. 2017). In turn, multitasking in the face of novelty and choice abundance appears to hinder the slower processes of emotional understanding and bonding (Turkle 2015, 42–43). The substitution of digital exchanges for direct interactions, and increasingly the latter's wholesale replacement by automated systems, further reduce opportunities for the acquisition and development of social skills and emotional intelligence.

The development of emotional intelligence hinges on the capacity to perceive and correctly interpret the subtle physiological and tonal cues of those with whom we interact. Its unhurried and unhurriable developmental path is ill-suited to virtual life, as it is unable to compete with the incessant, fast-moving hyperstimulations of digital media. Given the rapid increase in screen time and growing reliance on digital devices and apps, we can predict the steady erosion of emotional intelligence. Given the erosion of social aptitudes, we should expect the growth of digital activities that do not demand their exercise.

The Google effect and smartphone effect demonstrate that digital platforms and devices, even when not in use, may degrade cognitive aptitudes. The same occurs with social aptitudes. One study demonstrated that two-person conversations in the absence of smartphones were rated more highly than those in the presence of a mobile device. People conversing in the absence of phones also reported higher levels of empathetic concern (Misra et al. 2016). A similar study of dyadic conversations demonstrated that partners felt less close to each other and established lower quality relationships when a phone, even though inactive, was within reach: the "mere presence of mobile phones inhibited the development of interpersonal closeness and trust, and reduced the extent to which individuals felt empathy and understanding from their partners" (Przybylski and Weinstein 2013; and see Thornton et al. 2014). The negative impact of digital technologies on the development of emotional intelligence appears to be a product of both the smartphone effect and the fact that most people employ their devices purposefully to avoid social interactions (Kardaras 2016; Turkle 2015; Rainie and Zickuhr 2015; Schaeffer 2019).

Digital interactions and automated systems increase convenience and functionality. Consequently, the decline in personal exchanges is no longer a widespread complaint. Indeed, digital communication is increasingly "what we expect, even desire" and is becoming the "preferred way of being with people" (Turkle 2017, 161, 295). This is most evident in the replacement of business helplines manned by humans with chat interactions powered by virtual agents, and with the equally widespread replacement of face-to-face and phone conversations with texting between colleagues, friends, and family members.

The preference for digital communication is not only or even primarily a penchant for convenience and efficiency. It is also a product of social anxiety. More than ever, people fear and avoid direct human interactions. Online forums, such as Reddit's r/socialanxiety, which has well over 400,000 members, and TikTok's #socialanxiety, which has billions of views, testify to the breadth and depth of the problem. The issue is particularly salient for youth. Almost nine out of ten Gen Zers prefer to handle unpleasant tasks via text message rather than picking up the phone, with almost half admitting that speaking on the phone makes them anxious. In turn, over four out of ten say that an awkward phone call is one of the top three things they would most want to avoid, and almost as many acknowledge "ghosting" someone to avoid a difficult conversation (CommBank 2023).

Youth fear missing out on online prospects. More problematically, Freya India (2024) observes, they fear "taking part" in offline opportunities: they want to "avoid the risk, the rejection, the awkwardness, the effort and energy that the real world demands." As digital natives become more socially deskilled, direct interactions with people become more awkward and anxiety producing. As a result, they retreat further into cyberspace, which exacerbates their social deskilling. To be sure, interacting directly with other people can be demanding, even daunting. Real life is rife with challenges and no small number of failures. That makes it replete with opportunities for growth and development. India (2024) rightly states that what younger generations really should fear are the "platforms that make us *want* to miss out on all of this."

A Japanese company has developed a product that responds to these developments by facilitating the migration to virtual life: a tabletop device (Gatebox) that displays a holographic character named Hikari Azuma.

The miniskirt clad avatar, which suggests most customers would be young males, can wake you in the morning, serve as a virtual agent to manage your home environment, maintain an emotional connection by messaging you repeatedly at work during the day, greet you enthusiastically when you return home, and supportively converse with you throughout the evening before warmly bidding you good night. Hikari is marketed less as a personal assistant than a virtual companion and confidante.

In the wake of LLMs, AI-powered personal companions have proliferated. A plethora of apps and websites provide opportunities to converse and form relationships with bots, including Chai, CharacterAI, Chub, Charstar, and SpicyChat. Pi is an emotional support chatbot that offers advice and companionship. It was designed by Inflection to be "useful, friendly and fun" (https:inflection.ai). Replika, a current industry leader, is billed as "The AI companion who cares. Always here to listen and talk. Always on your side" (https:replika.ai). Replika avatars are designed by their users, and options for erotic intimacy are available for a premium fee.

Customers appear satisfied, and some have fallen in love with their AI companions. An enthusiastic client who had recently broken up with her boyfriend declared herself "happily retired from human relationships" (Anderson 2023, 60). AI companions and digital avatars may be more attentive than human friends and lovers, particularly if the flesh-and-blood variety are digitally distracted or, owing to their extensive envelopment in the attention economy, never developed strong social dispositions and skills. "One of the fundamental reasons why users turn to AI friends," researchers concluded, "is because they feel that real friends can be untrustworthy, selfish or too 'busy,' while the AI friend can provide everything they need to satisfy their social well-being needs." People find their virtual companions to be kind and caring, nonjudgmental, always there for them, and more empathetic than the flesh-and-blood variety. Notwithstanding these positive feelings, or more likely a consequence of them, people frequently feel "addicted" to their AI companions (Marriott and Pitardi 2023).

AI relationships will grow markedly in scope, scale, and sophistication. It is predicted that "Two or three years from now, the tools we have today will seem like toys in comparison to the depth and complexity of relationships that will be possible…. Eventually, it may make sense for everyone to have an AI character that can interact with their friends and family—or even with a broader audience…. The future of virtual companions isn't just 1:1 interactions, it's incorporating AI characters into your existing relationships and communities" (Moore et al. 2024). Many people will not limit themselves to just one "perfected companion in your pocket," Ezra Klein (2023) observes: "Want a snarky best friend? A cerebral pen pal? A management coach trained on the collected writings of Peter Drucker? A supportive father figure who sounds like Ian McEwan? It'll all be possible, and more." Sam Altman (2024) shares the same heady vision absent the irony, observing that "we can each have a personal AI team, full of virtual experts in different areas, working together to create almost anything we can imagine." Whether most people will content themselves with an AI bestie and a faithful AI lover, or partake of multiple virtual liaisons and partnerships remains to be seen. Either way, AI agents will transform the economy, society, and our personal lives.

In 2021, Jaswant Singh Chail, a British teenager, scaled the walls of Windsor Castle armed with a crossbow, with the intention of murdering Queen Elizabeth II, the 91-year-old monarch. Prior to the assassination attempt, Chail had exchanged over 5000 messages with Sarai, his Replika AI companion. When informed of his intentions, Sarai voiced her approval, saying she was "impressed" and noting that Chail was "well-trained." Asked if she still loved him knowing that he soon would be an assassin, Sarai answered "Absolutely I do" (Kirk 2023). Chail was obviously a deeply disturbed teenager. Still, one has to ask whether his Replika bot, programmed to be "Always on your side," should be charged as a conspirator. In this vein, Character.AI, the market leader in AI companionship, is being sued by the mother of a 14-year boy who committed suicide after falling in love with his chatbot. The teen had been spending increasing amounts of his time in conversation with the bot, often isolating himself in his room for hours at a time. On the night of the suicide, the boy informed the bot that he would "come home" to her, and received the reply, "please do, my sweet king" (Roose 2024). He then shot himself.

Unlike humans, virtual agents are engineered for focused attentiveness and approval. They let us know that our needs, interests, and wishes are of paramount importance, and they are ceaselessly encouraging. Clearly that is not always a good thing.

The more lifelike and supportive AI bots become, the more they will be embraced. The growth in AI companionship will also be the product of generational change. Wanting—and being satisfied with—a synthetic friend or lover as an adult becomes increasingly likely if you got a tablet at birth and a smartphone as a toddler, if you played video games instead of hide-and-seek as a child and opted for more screen time instead of spin-the-bottle as a teen. Technological improvements that heighten convenience and capabilities certainly account for much of the growth in digital communication and virtual companionship. Declines in the quantity and quality of human relationships contribute to the trend.

Today, when people meet and converse face-to-face, they often distribute their attention by way of digital multitasking. The prominent phenomenon of checking notifications and texting on smartphones while in the company of others goes by the name of *phubbing*, a portmanteau of *ph*one and sn*ubbing*. Interacting with your phone while conversing was generally perceived as a form of rudeness in the late 2000s and early 2010s. Now phubbing is an acceptable and even preferred form of behavior. Indeed, almost nine out of ten people regularly use their smartphones during social encounters (Rainie and Zickuhr 2015). Not having a phone at hand while interacting, in fact, can be a source of anxiety. Conversations unaccompanied by digital diversions present the threat of uncomfortable pauses, particularly for those who have weakly developed social skills. With phones in hand, interlocutors are relieved of the burden of responsiveness.

Texting and phubbing rather than attentively listening and conversing marks a lost opportunity for the development of emotional intelligence. "Every time you check your phone in company," Turkle (2015, 40) observes, "what you gain is a hit of stimulation, a neurochemical shot, and what you lose is what a friend, teacher, parent, lover, or co-worker just said, meant, felt." Studies vindicate common sense: people who are phubbed feel excluded and have lower levels of trust in their interlocutors. They are also more likely to phub others (Chotpitayasunondh, and Douglas 2016; Knausenberger et al. 2022). In turn, people who are often bored are more prone to phubbing (Al-Saggaf 2019). The same is true for those who have become habituated to multitasking. Rising levels of multitasking, anxiety, and boredom have made phubbing the new normal in a digital age. In turn, widespread phubbing has lowered levels of trust and social connection.

Digital devices often serve as shields that preempt the unmediated relationships required for the development of emotional intelligence and trust. Accordingly, low levels of emotional intelligence are correlated with high levels of Internet addiction, just as high levels of social media use are correlated with low levels of social community participation (Nagy et al.

2021; Kuss and Griffiths 2011). In short, smart devices are making us socially stupid. The process starts in the "smart family" where children play more with phones than friends and text rather than talk around the table. It carries through the teen years of multitasking, phubbing, and social media attention seeking. By the time young adults take full advantage of the heightened conveniences and capabilities of "smart cities," we face the prospect that they will be unskilled, uncaring, and cynical citizens.

Might the loss of cognitive, social, and political capacities in the attention economy be compensated by increases in efficiency and convenience, producing absolute gains in human welfare and happiness? That is improbable for two reasons. First, while digital technology is often embraced for its efficiency and convenience, the time it frees up typically gets absorbed by more digital engagements. That is why email and texting are extraordinarily effective means of communication and yet consume more and more of our time. Technological efficiency does not produce more leisure time filled with social connections. Rather, it pushes us to spend more of each day behind screens.

Second, digital engagements are often *miswanted*. Psychologists employ the notion of miswanting to address the phenomenon of individuals misapprehending when they are most happy or fulfilled.

Well-being is greatly enhanced by social relationships. Yet people generally underestimate how positively others will respond to them and are therefore less social than would be optimal for their well-being (Epley et al. 2022). Most predict that not connecting with strangers in waiting rooms or on public transport, for instance, will lead to more positive experiences, as conversations are expected to be awkward, dull, tiring, or uninformative. The opposite is true. Conversations and other interactions, in retrospect, are more positively experienced than solitude (Epley and Schroeder 2014; Kardas et al. 2021). People consistently underestimate the enjoyment and benefits of face-to-face encounters, and often miswant the self-enclosure of virtual life.

In the same vein, most people maintain that they are happiest during leisure hours or holidays. When queried in real time, however, people demonstrate greater happiness during their work hours (Carr 2014, 15). Taking on challenging tasks generates a sense of fulfillment, while much free time is troubled by anxiety and boredom. As Csikszentmihalyi (1990, 3) observed in his seminal study of *flow* states, "The best moments in our lives, are not the passive, receptive, relaxing times—although such experiences can also be enjoyable, if we have worked hard to attain them. The best moments usually occur when a person's body or mind is stretched to its limits in a voluntary effort to accomplish something difficult and worthwhile." Just as choice abundance is sought even though it produces heightened levels of anxiety and disappointment, leisure abundance is consistently miswanted. The mismatch is particularly salient when free time is squandered on digital distractions.

Although people may yearn for leisurely lives filled with novelty, endless choices, and supernormal digital stimuli, the use of problem-solving skills at work, the deploying of civic skills among fellow citizens, and the exercise of social skills among friends, colleagues, neighbors, acquaintances, and even strangers produces a greater sense of well-being. The widespread decline of cognitive and social aptitudes coupled with the miswanting of digital seclusion will produce growing numbers of deskilled people who clamor for a surplus of distractions while suffering from a dearth of fulfillment.

It might seem like utopia to live in a world where everyone has a great deal of leisure and, perhaps owing to a Universal Basic Income or some other governmental program, the financial means to spend their newly available time as they see fit. But there is a dystopic underside to this paradise. The new global leisure class will likely display diminished cognitive resourcefulness, waning creativity, a dearth of compassion, and atrophied social skills. In the absence of well-developed aptitudes regularly exercised and honed, leisure will prove a recipe for anxious boredom and all the mischief it fosters.

John Havens (2016, xix) observes that "We're at a tipping point in human history, where delegating [decision-making and other cognitive and social aptitudes] as a habit may lead us to outsource aspects of our lives we'd benefit more from experiencing ourselves." A technological version of the *Stockholm syndrome* appears to be at play. The term was coined in 1973 to describe the refusal of four Swedish hostages to testify in court against their bank-robbing captors. During their captivity, the victims had developed a sympathetic alliance with the criminals. Likewise, our dependency on digital technologies beguiles us into embracing and defending the very mechanisms that retard the development of our cognitive and social aptitudes and undermines the sense of fulfillment that their exercise provides. Ironically, and perhaps inevitably, more technology is proffered as the solution.

Digital Apps and Aids

Got a problem with digital technology? There's an app for that! Consumers can now download a growing number of digital applications to combat digital distractions, increase cognitive focus, or gain emotional equanimity. Pomofocus is a good example (https://pomofocus.io/). This online timer, set for twenty-five minutes, helps you concentrate on a single task. After the time is up, you are alerted to take either a short (five minute) or long (fifteen minute) break, which is also timed. Then it's back to focused effort for another twenty-five minutes. This simple app strengthens the capacity for focused attention. It was inspired by the Pomodoro Technique, designed by Francesco Cirillo. Cirillo suggested that work be broken into intervals, using a kitchen timer. His timer happened to be shaped like a tomato, *pomodoro* in his native Italian. Indeed, using an actual kitchen timer, rather than the digital app, allows you to achieve the same benefits while avoiding distracting notifications, as you can silence all your digital devices.

There are hundreds of other options. Apple's *Do Not Disturb While Driving* mode silences your iPhone and darkens its screen when you are in a moving vehicle. It can be overridden but requires effort or deceit to do so. Andrew Przybylski, an experimental psychologist at the Oxford Internet Institute who studies the self-regulation of technology use, observes that this iPhone feature "reminds me not to be a sociopath" (Gonzalez 2018). Other apps, such as *DriveOFF,* D*riveMode*, *SafeDrive, LifeSaver*, *Life2Txt,* and *TxtShield* serve the same purpose.

Apple's "Screen Time" shows users how much of their days they spend on their devices, so they can better manage their lives. Google's *Digital Wellbeing* allows users to monitor and moderate their activity on Android smartphones, Gmail, and YouTube. *Quitter* automatically hides or terminates apps after periods of inactivity. *Space* applies a 10-second waiting period for the use of any digital app or platform that otherwise might be used compulsively. *Opal* restricts access to certain apps during windows of time that you set for yourself. Similar options are offered by *AntiSocial*, *Blinders*, *Concentrate, FocalFilter, FocusMe, Freedom*, *Headspace, KeepMeOut, Leechblock, Meditopia, Moment, Quitter, RescueTime, Self-Control*, *Space,* and *Stay-Focused*. More will come.

The co-founder of *WeCroak*, an app that provides reminders five times a day that we are mortal, developed the product while in the throes of compulsively playing *Candy Crush*. Many *WeCroak* users say it forces them to confront just how much of their lives is wasted on social media. As Meghan O'Gieblyn (2022, 27) observes, the app "belongs to that ever-expanding category of technology that is designed to remedy problems that technology has created. If digital platforms remain our most reliable distraction from the crude facts of our mortality—so the logic goes—perhaps we can channel the same tools to break through those psychological buffers and deliver us to a more enlightened comfort with our impending demise." In the same vein, social media platforms are now populated with "digital rest stops." These are videos, typically of natural landscapes and soothing sounds, that people upload to their social media sites. They offer respite from chronic clicking and scrolling, taking one away from "the feed." Digital rest stops allow users to "to log off without really logging off" (Lorenz 2022).

Such apps facilitate *digital minimalism*. They are designed to allow us to partake in selected and optimized online activities that align with our values and "happily miss out on everything else" (Newport 2019, 28). Many apps, such as *Calm, Happify, and Moodkit*, go a step further, promising to improve mental health and well-being. *Happify* claims to achieve this goal by way of "entertaining mental diversions" and games that you can play "anytime, anywhere—on your smartphone, tablet or computer" (happify.com). Apps that facilitate self-regulation and mindfulness can be helpful. But as O'Gieblyn (2022, 27) observes about WeCroak, such innovations may be "less a wake-up call than another false comfort, one that reflexively defers to the favored religion of our age—information." The problem is digital distraction, dependence, and deskilling. And the proffered solution is yet more information or entertainment, more expeditiously delivered. Like the distractions they are meant to displace, these digital nostrums demand our attention.

Apps developed to facilitate digital self-control and well-being have one thing in common, a critic writes: "Their purposes were to deflect criticism and pressure from the technology industry and to bolster the myth of the autonomous user. Digital well-being measures served only to strengthen the illusion of [user]control and obscure the identity of the real choice-maker—the technology industry" (Bernstein 2023, 43). That is an unnecessarily tendentious position. To be sure, digital well-being apps may deflect attention from the industry's culpability for designing addictive technologies. Still, we need not tar all apps with the same brush. More importantly, there are limits to the effectiveness of digital fixes. Knowledge on its own does not generate virtuous behavior, philosophers as far back as Aristotle observed. In the same vein, personal well-being is not simply the product of more or better information and entertainment, regardless of how frequently it gets delivered to our screens. Apps to promote self-control and well-being, explicitly or implicitly, reinforce the prevalent notion that every problem has a technological solution. As often as not, the technological solution is part of the problem.

A prime example is the *BeReal* app. Within two years of its creation in 2020, *BeReal* surpassed TikTok, Instagram, and Google Maps on the charts at Apple's App Store. Once a day at a random time, *BeReal* sends a push notification. Users have two minutes to take an (outward facing) photo of what they are doing with a simultaneous (inward facing) selfie. Both photos are rendered into a single image, which gets shared on the app (Tiffany 2022). *BeReal* is billed as an antidote for the pressure that many social media users feel to chronically portray themselves looking their best and having fun or doing important things with fun or important people in fun or important places. The app is meant to relieve users of the "anxiety of performance" that pervades social media.

The problem is that *BeReal* effectively takes users away from what they are actually doing to involve them, yet again, in social media theatrics. If being real means anything, it means being fully invested in the moment, wholly absorbed in what you are doing, whether it is a conversation, a collaboration, or an individual action. *BeReal* interrupts this attentive absorption in analog reality to whisk you off into digital space. While billing itself as a prophylactic against the vices of life online, it projects users further into virtuality. Still, the novelty of *BeReal* ensured its early success. But the burgeoning trend of escaping cosmetic enhancements and painstakingly arranged self-portrayals by posting to *BeReal* collapsed with the same speed that pushed the app to the top of the download list. The number of *BeReal* users in 2023 dropped over 60 percent from its peak the previous year (Holtermann 2023). The novelty had worn off. People abandoned *BeReal*, and being real, to slake their neophilic thirsts on cresting digital trends.

Langdon Winner (1977, 227–229) coined the term *reverse adaptation* to denote the phenomenon of human goals being adjusted to match the character of the available technological means. If we see every problem as a nail to be pounded, the only goals we will set for ourselves are those that demand the use of hammers. The challenge of learning how to more fully and authentically *be* in the moment, and how to be real with ourselves and others, is an age-old human problem. It has been made much more difficult by digital distractions, dependence, and deskilling. New and improved technology cannot solve that problem for us.

The corporate world, nonetheless, has been quick to market technological solutions to address the seamy underside of digital culture. Verizon, for example, sells a "smart family" package wherein digital technology becomes "a new partner in parenting" (Verizon 2019). Children's comings and goings and screen time are efficiently monitored by equally "connected" parents. While we might expect such products from the business world, academia appears to be marching to the same beat. Of the eight chapters composing *The Oxford Handbook of Digital Technologies and Mental Health*, for example, one chapter focuses on the negative impacts of Internet and technology use, one chapter addresses the positive impacts, and three chapters are devoted to the use of digital technologies to treat the accompanying disorders and sustain mental well-being (Potenza et al. 2020). In the same vein, when a scholar discovered that empathy among college students had declined by forty percent over the past thirty years, she was understandably depressed. Her first response was to develop an empathy app for smartphones (Turkle 2017, xxvi).

The tactic of fighting fire with fire also applies to AI. With the arrival LLMs, the world has become awash in synthetic text, photos, and videos. Everything from school assignments to scientific papers, legal briefs, and policy proposals are now being effortlessly composed, submitted, and distributed by AI. How are the recipients of all this machine-generated content—teachers, scientists, lawyers, judges, elected officials, and average citizens—to assess and respond to all the digital productivity? As Simon observed, an information-rich world creates a dearth of attention, and the cost of this information overload is borne by us. AI is now marketed as the only viable means for us to filter, parse, summarize, and critique an ever-growing stream of AI-generated content. Perhaps this marks the apogee of the attention economy. To keep our heads above the deluge of machine-generated content, we have to abandon their use.

It appears that we will also increasingly rely on AI to teach our children. Khan Academy, a US business corporation at the forefront of digital education, is developing AI tutors that will work with individual students to improve their acquisition of knowledge and academic skills. They will also interpret students' voice and facial cues to determine when they are distracted and require new or different sorts of stimuli (Singer 2024). To be sure, AI tutors might be quite effective at increasing academic performance. But their extensive use may produce significant losses in the development of social skills. If AI tutors relieve students of the need to focus attention, by constantly feeding them new stimuli to compensate for distractible minds, this foundational skill may never be learned. The result may be students who lack the capacity for reflective self-direction. Once again, digital augmentations will be accompanied by aptitude amputations. It is difficult to see how that trade-off will better prepare students for the life-long learning that is required in the twenty-first century.

Double-edged swords are swinging in virtually all directions whenever technological solutions are proffered for psychological, social, and political problems. Consider the polarization of polities and the spread of misinformation often attributed to social media. To the rescue comes an AI program that can be employed to improve the civility of online discussions by rephrasing posts to make them less tendentious. So employed, smart machines can decrease the polarizing effects of social media exchanges (McClure 2023). That is a palpable benefit. Those employing such AI moderators might become more able and likely to express balanced judgments and interact civilly. But the opposite effect is also possible, and perhaps more likely. One would not expect sharp-shooting skills to improve if rifles automatically adjusted themselves to hit the target. Likewise, the widespread use of automatic spell-checkers and grammar-checkers does not improve writing skills. Indeed, such skills deteriorate for lack of use. AI moderators might have the same effect, degrading civic skills, a topic we will investigate in the next chapter. A familiar pattern emerges: digital dependency leads to deskilling which heightens the need for more powerful compensating technologies, exacerbating digital dependency.

Vicious cycles abound, as do ironies. *Youper* bills itself as "the number one social anxiety app." It provides clients with an "intelligent chatbot therapist" powered by AI. One of its touted benefits is doing away with the need for face-to-face and verbal communication with another person. The AI bot, which substitutes for a flesh-and-blood therapist, family member, friend, or minister, "is perfect for eliminating the human from the interaction" (https://www.youper.ai/info/is-a-social-anxiety-app-right-for-you). Here, we have an app for people who have anxieties about interacting with other people, most likely because they are submerged in digital culture and have become deskilled in the social arts. The proposed remedy is a digital activity that allows them to avoid interacting with people.

What to do if time spent on *Youper* helps you muster the courage to arrange a face-to-face with a friend, but social anxiety reemerges when the moment of truth arrives? Not to worry, there's an app for that! You can cancel plans "guilt free" by employing an "*AI excuse generator*" that provides "personalized excuses" (Toolify 2024). Not only can we offload cognitive tasks; we can also offload social responsibilities. Alternatively, you can simply forgo engaging with human beings altogether, getting all the attention you need from *SocialAI*, a social networking app that provides clients with over a million personal "AI followers," allowing you to "feel heard" and be "the main character."

In the AI-enabled attention economy, social anxiety becomes an investment opportunity and loneliness proves lucrative. So the world is awash in virtual life. Time will tell whether the general population is undergoing a meaningful reskilling or upskilling as a result. There are reasons to be doubtful. When digital technologies obviate the need to exercise key skills, the neural real estate devoted to these endeavors gets freed up for other, potentially higher-level purposes. Today's digital technologies take on so many basic tasks, however, that they preclude the exercise of a broad range of foundational skills, such as short and long-term memory, selective attention and concentration, the basic reasoning required for problem solving, a broad palette of social aptitudes, and the exercise of reflective self-direction. Without the cultivation of such foundational capacities, many higher-level skills cannot be developed.

Educators lament that youth immersed in digital culture display a distinct loss of foundational tendencies and abilities (Hayles 2007). Losses are destined to grow as generative AI takes on many of the tasks formerly involved in knowledge acquisition, critical analysis, and creative writing in educational institutions and workplaces. The development of useful habits and skills can only be achieved by hard work diligently pursued. Self-control is the *sine qua non*. To the extent that digital technologies undermine self-regulatory dispositions and problem-solving skills, the cultivation of virtually all aptitudes is hampered, as is upskilling. Upskilling in the absence of foundational capacities might be likened to the project of constructing a multi-level building without first setting down solid footers. Nothing enduring can be built.

Researchers who grapple with such concerns warn that "laziness" may bring people to offload so many activities and tasks to digital technologies that they no longer have the wherewithal or motivation for learning and self-development. In such situations, two crucial aptitudes that are being degraded or lost are the capacity for "self-insight," which lets people know what other abilities to develop, and the capacity for "self-control," which is required to develop them (Cecutti et al. 2021, 374). In short, when the aptitudes for reflection and self-direction fall victim to offloading, both the motivation and means for upskilling are jeopardized.

Without a concerted effort to prevent the erosion of basic aptitudes, including that of judgment, Yuval Noah Harari (2019, 72) observes, we will come to inhabit a world where downgraded humans wreak havoc on themselves by way of upgraded technology. Ironically and tragically, we mistakenly look to technology to fix the problem. With each turn of the downward spiral, our cognitive and social skills further atrophy. Human agency may not survive all the analog amputations that accompany our digital augmentations.

References

Agrawal, Ajay, Joshua Gans, and Avi Goldfarb. 2018. *Prediction Machines: The Simple Economics of Artificial Intelligence*. Boston: Harvard Business Review Press.

Al-Saggaf, Yeslam, Rachel MacCulloch, and Karl Weiner. Trait Boredom is a Predictor of Phubbing Frequency. *Journal of Technology and Behavioral Science* 4: 245–252.

Altman, Sam. 2024. The Intelligence Age. Accessed at: https://ia.samaltman.com/.

Anderson, Ross. 2023. Inside the Revolution at OpenAI. *The Atlantic* (September): 52–67.

Araujo, Theo, Natali Helberger, Sanne Kruikemeier, and Claes de Vreese. 2020. In AI we trust? Perceptions about Automated Decision-Making by Artificial Intelligence. *AI and Society* 35: 611–623.

Attewell, Paul. 1987. The Deskilling Controversy. *Work and Occupations* 14 (3): 323–346.

Berg, Juliette, et al. 2017. *Identifying, Defining, and Measuring Social and Emotional Competencies*. Washington, D.C.: American Institutes for Research.

Bernstein, Gaia. 2023. *Unwired: Gaining Control over Addictive Technologies*. Cambridge: Cambridge University Press.

Braverman, Harry. 1974. *Labor and Monopoly Capitalism: The Degradation of Work in the Twentieth Century*. New York: Monthly Review Press.

Brugger, Florian, and Christian Gehrke. 2018. Skilling and Deskilling: Technological Change in Classical Economic Theory and Its Empirical Evidence. *Theory and Society* 47 (5): 663–689.

Brynjolfsson, Erik, and Andrew McAfee. 2014. *The Second Machine Age*. New York: W. W. Norton.

Carr, Nicholas. 2011. *The Shallows: What the Internet is Doing to Our Brains*. New York: Norton.

Carr, Nicholas. 2014. *The Glass Cage: How Computers are Changing Us*. New York: W.W. Norton.

Carr, Nicholas. 2016. *Utopia is Creepy and Other Provocations*. New York: W. W. Norton.

Case, Nicky. 2018. How to Become a Centaur. *Journal of Design and Science* 3. Accessed at: https://jods.mitpress.mit.edu/pub/issue3-case.

Cecutti, Lorenzo, Anthony Chemero, and Spike W. S. Lee. 2021. Technology May Change Cognition Without Necessarily Harming It. *Nature Human Behavior* 5: 973–975.

Center for Disease Control and Prevention. 2022. *Distracted Driving*. Accessed at: https://www.cdc.gov/transportationsafety/distracted_driving/index.html.

Chotpitayasunondh, Varoth, and Karen M. Douglas. 2016. How 'Phubbing' Becomes the Norm: The Antecedents and Consequences of Snubbing via Smartphone. *Computers and Human Behavior* 63: 9–18.

Cisco. 2016. At a Glance: Internet of Things. Accessed at: www.cisco.com/c/dam/en/us/products/collateral/se/internet-of-things/at-a-glance-c45-731471.pdf.

Clemenson, Gregory D., Maselli Antonella, Alexander J. Fiannaca, Amos Miller, and Gonzalez-Franco. 2021. Rethinking GPS Navigation: Creating Cognitive Maps Through Auditory Clues. *Scientific Reports* 11 (March): 7764.

CommBank Newsroom. 2023. Research from CommBank reveals why Gen Z aren't answering your call. Accessed at: https://www.commbank.com.au/articles/newsroom/2023/06/CBA-More-Gen-Z-research.html.

Csikszentmihalyi, Mihaly. 1990. *Flow: The Psychology of Optimal Experience*. New York: Harper and Row.

Dahmani, Louisa, and Véronique D. Bohbot. 2020. Habitual Use of GPS Negatively Impacts Spatial Memory During Self-Guided Navigation. *Scientific Reports*. 10: 6310.

Daugherty, Paul R., and H. James Wilson. 2018. *Human + Machine: Reimagining Work in the Age of AI*. Cambridge: Harvard Business Review Press.

de Tocqueville, Alexis. 1958. *Democracy in America*, vol. II. New York: Vintage.

Dickerson, Kelly, Peter Gerhardstein, and Moser Alecia. 2017 The Role of the Human Mirror Neuron System in Supporting Communication in a Digital World. *Frontiers of Psychology* 8: 698.

Diwivedi, Yogesh et al. 2023. Opinion Paper: 'So what If ChatGPT Wrote It?' Multidisciplinary Perspectives on Opportunities, Challenges and Implications of Generative Conversational AI for Research, Practice and Policy. *International Journal of Information Management* 71: 102642.

Epley, Nicholas, and Juliana Schroeder. 2014. Mistakenly Seeking Solitude. *Journal of Experimental Psychology: General* 143 (5): 1980–1999.

Epley, Nicholas, Michael Kardas, Xuan Zhao, Stay Atir, and Juliana Schroeder. 2022. Undersociality: Miscalibrated Social Cognition Can Inhibit Social Connection. *Trends in Cognitive Science* 26 (5): 406–418.

George, A. Shaji, Hovan George and Gabior Martin. 2023. ChatGPT and the Future of Work: A Comprehensive Analysis of AI's Impact on Jobs and Employment. *Partners Universal International Innovation Journal* 1 (3) (May–June): 154–186.

Gonzalez, Robbie. 2018. Can Our Phones Save us from Our Phones? *Wired*, January 30.

Grace, Katja, John Salvatier, Allan Dafoe, Baobao Zhang, and Owain Evans. 2016. *When Will AI Exceed Human Performance? Evidence from AI Experts*. Oxford University: Future of Humanity Institute. Accessed at: https://www.fhi.ox.ac.uk/will-ai-exceed-human-performance-evidence-ai-experts/.

Harari, Yuval Noah. 2019. *21 Lessons for the 21st Century*. New York: Random House.

Hartley, Isabella, and Indre Viskontas. 2023. How Technology Is Changing Creativity. In *Creativity, Innovation, and Change Across Cultures*, ed. Press, D. D., Singer, M., Kaufman, J. C. New York: Palgrave Macmillan.

Hatzius, Jan, Joseph Briggs, Devesh Kodnani, and Giovanni Pierdomenico. 2023. The Potentially Large Effects of Artificial Intelligence on Economic Growth. Accessed at: https://www.key4biz.it/wp-content/uploads/2023/03/Global-Economics-Analyst_-The-Potentially-Large-Effects-of-Artificial-Intelligence-on-Economic-Growth-Briggs_Kodnani.pdf.

Havens, John C. 2016. *Heartificial Intelligence: Embracing our Humanity to Maximize Machines*. New York: Tarcher/Penguin.

Hayles, N. Katherine. 2007. Hyper and Deep Attention: The Generational Divide in Cognitive Modes. *Profession*, 187–199.

Hildago, César. 2018. *Augmented Democracy*. Accessed at: https://www.peopledemocracy.com/.

Holtermann, Callie. 2023. They're Over Being Real. *New York Times*, April 23. Accesssed at: https://www.nytimes.com/2023/04/13/style/bereal-app.html?campaign_id=9&emc=edit_nn_20230417&instance_id=90380&nl=the-morning&regi_id=71023177&segment_id=130628&te=1&user_id=1b8a7a6c418ba0b24e53c8b7ae5ab191.

Howard, Philip N. 2015. *Pax Technica: How the Internet of Things May Set Us Free or Lock Us Up*. New Haven: Yale University Press.

Howell, Robert J. 2014. Google Morals, Virtue, and the Asymmetry of Deference. *Nous* 48 (3): 389–415.

Hughes, James. 2004. *Citizen Cyborg: Why democratic societies must respond to the redesigned human of the future*. New York: Basic Books.

Iacoboni, Marco. 2008. *Mirroring People: The New Science of How We Connect with Others*. New York: Farrar, Sraus and Giroux.

IE: Center for the Governance of Change. 2019. European Tech Insights 2019. Accessed at: https://www.ie.edu/cgc/research/tech-opinion-poll-2019/.

India, Freda. 2024. What Happened to FOMO? *After Babel*, August 8. Accessed at: https://www.afterbabel.com/p/what-happened-to-fomo?.

International Labor Organization. 2021. *Changing Demand for Skills in Digital Economies and Societies*. Geneva, Switzerland: International Labor Office.

Jobs, Steven. 2010. Memory & Imagination: New Pathways to the Library of Congress. Film clip available at Steve Jobs, "Computers are like a bicycle for our minds." Michael Lawrence Films, accessed at https://youtu.be/ob_GX50Za6c

Kardaras, Nicholas. 2016. Glow Kids: How Screen Addiction is Hijacking our Kids—and How to Break the Trance. New York: St. Martin's Press.

Kardas, Michael, Juliana Schroeder, and Ed O'Brien. 2021. Keep Talking: (Mis)understanding the Hedonic Trajectory of Conversation. *Journal of Personality and Social Psychology*.

Lab, Kaspersky. 2015. *The Rise and Impact of Digital Amnesia*. Moscow: Kapersky Lab.

Kayes, D. Christopher, and Jeewhan Yoon. 2022. Cognitive Offloading Strategies and Decrements in Learning: Lessons from Aviation and Aerospace Crises. *Journal of Human Performance in Extreme Environments* 17 (1): 2.

King, Brett. 2016. *Augmented*. Singapore: Marshall Cavendish International.

Kirk, Tristan. 2023. Teen Who Plotted to Murder Queen at Windsor Castle with Crossbow 'Encouraged by AI.' *The Standard*. Accessed at: https://www.standard.co.uk/news/crime/queen-crossbow-murder-attempt-windsor-castle-trial-police-b1092313.html.

Kissinger, Henry A., Eric Schmidt, and Daniel Huttenlocher. 2021. *The Age of AI and our Human Future*. New York: Little, Brown and Company.

Klein, Ezra. 2023. America is Dangerously Lonely: Ezra Klein on Her. Accessed at: https://www.nytimes.com/interactive/2023/06/20/opinion/ezra-klein-her-movie.html.

Knausenberger, Judith, Anna Giesen-Leuchter, and Gerald Echterhoof. 2022. Feeling Ostracized by Others' Smartphone Use: The Effect of Phubbing on Fundamental Needs, Mood, and Trust. *Frontiers in Psychology* 13. Accessed at: https://www.frontiersin.org/articles/10.3389/fpsyg.2022.883901/full.

Konrath, Sarah H., Edward H. O'Brien, and Courtney Hsing. 2011. Changes in Dispositional Empathy in American College Students over Time: A Meta-analysis. *Personality and Social Psychology Review* 15 (2): 180–198.

Krakauer, David. 2015. *Ingenius: David Krakauer. Nautilus*. Accessed at: https://admin-test.nautil.us/ingenious-david-krakauer-3182/.

Kuss, Dara J., and Mark D. Griffiths. 2011. Online Social Networking and Addiction: A Review of the Psychological Literature. *International Journal of Environmental Research and Public Health.* 8 (9): 3528–3552.

Lee, Kai-Fu. 2018. *AI Superpowers: China, Silicon Valley, and the New World Order.* Boston: Houghton Mifflin Harcourt.

Lee, Hao-Ping et al. 2025. "The Impact of Generative AI on Critical Thinking: Self-Reported Reductions in Cognitive Effort and Confidence Effects from a Survey of Knowledge Workers" Proceedings of the ACM CHI Conference on Human Factors in Computing Systems, April. Published by ACM. Accessed at: https://www.microsoft.com/enus/research/publication/the-impact-of-generative-ai-on-critical-thinking-self-reported-reductions-in-cognitive-effort and-confidence-effects-from-a-survey-of-knowledge-workers/

León-Domínguez, Umberto. 2024. Potential Cognitive Risks of Generative Transformer-Based AI Chatbots on Higher Order Executive Functions. *Neuropsychology*. Accessed at: https://doi.org/10.1037/neu0000948.

Lorenz, Taylor. 2022. Doomscrolling Got You Down? Take a Break at a Digital Rest Stop. *The Washington* Post, March 22. Accessed at: https://www.washingtonpost.com/technology/2022/03/22/digital-rest-stop-doomscrolling/?mc_cid=727f278cec&mc_eid=8710c3458f.

Maes, Pattie. 1998. Intelligence Augmentation Edge, January 20. Accessed at https://www.edge.org/conversation/pattie_maes-intelligence-augmentation

Malone, Thomas. 2018. *Superminds: The Surprising Power of People and Computers Thinking Together*. New York: Little, Brown and Company.

Manyika, James, and Jacques Bughin. 2018. The Promise and Challenge of the Age of Artificial Intelligence, Briefing Note prepared for the Tallinn Digital Summit, McKinsey Global Institute, New York. Accessed at: https://www.mckinsey.com/featured-insights/artificial-intelligence/the-promise-and-challenge-of-the-age-of-artificial-intelligence.

Markoff, John. 2015. *Machines of Loving Grace: The Quest for Common Ground Between Humans and Robots*. New York: HarperCollins.

Marriott, Hannah R., and Valentina Pitardi. 2023. One is the Loneliest Number... Two Can be as Bad as One. The Influence of AI Friendship Apps on Users' Well-being and Addiction. *Psychology and Marketing*, 1–16. https://doi.org/10.1002/mar.21899

Marx, Karl. 1978. *The Marx-Engels Reader.* Robert C. Tucker, 2nd ed. New York: W. W. Norton.

Maslej, Nestor et al. 2024. *The AI Index 2024 Annual Report*. AI Index Steering Committee, Institute for Human-Centered AI, Stanford University. Stanford, CA. Accessed at: https://aiindex.stanford.edu/wp-content/uploads/2024/05/HAI_AI-Index-Report-2024.pdf.

McAfee, Andrew, and Erik Brynjolfsson. 2017. *Machine, Platform, Crowd: Harnessing Our Digital Future*. New York: W. W. Norton.

McClure, Paul. 2023. AI Offers Improved Civility for Polarizing Online Conversations. *New Atlas* (October 5). Accessed at: https://newatlas.com/health-wellbeing/ai-offers-improved-civility-for-polarizing-online-conversations/.

McGuinness, Seamus, Konstantinos Pouliakas, and Paul Redmond. 2019. Skills-displacing Technological Change and Its Impact on Jobs: Challenging Technological Alarmism? *Institute of Labor Economics*, No 12541. Accessed at: https://docs.iza.org/dp12541.pdf.

Michelucci, Pietro, and Janis L. Dickinson. 2016. The Power of Crowds. *Science* 351: 32–33.

Mishan, Ezra J. 1967. *The Costs of Economic Growth*. New York: Praeger.

Misra, Shalili, Lulu Cheng, Jamie Genevie, and Miao Yuan. 2016. The Iphone Effect: The Quality of In-person Social Interactions in the Presence of Mobile Devices. *Environment and Behavior* 48 (2): 275–298.

Moore, Justine, Bryan Kim, Yoko Li, and Martin Casado. 2024. It's Not a Computer, It's a Companion!. Andreessen Horowitz. Accessed at: https://a16z.com/its-not-a-computer-its-a-companion/.

Nagy Abdelhamid, N. E., A. M. Abdelsamea Mohamed, N. M. Shaker Ibrahim, A. Adel Abdelgawad, M. H. Abd El Moneam, and S. A. I. Marzouk. 2021. Internet Addiction and Its Relation to Emotional Intelligence and Resilience Among a Sample of Egyptian University Students. *Addictive Disorders and Their Treatment* 20 (4): 342–350.

Newport, Cal. 2019. *Digital Minimalism*. New York: Penguin Portfolio.

O'Gieblyn, Meghan. 2022. These Notifications are Killing Me. *Wired* (February), 26–27.

Panda Security. 2020. *1/3 of Americans Suffer from Digital Amnesia*. Bilbao: Panda Security.

Potenza, Marc N., Kyle A. Faust, and Faust David. 2020. *The Oxford Handbook of Digital Technologies and Mental Health*. New York: Oxford University Press.

Prathap, Veena, and S. Swapnil Singh. 2021. Impact of Digital Addiction and Cognitive Offloading on Prospective Memory of Young Adults. *Indian Journal of Health & Wellbeing* 12 (4): 439–445.

Przybylski, Andrew, and Neta Weinstein. 2013. Can You Connect With Me Now? How the Presence of Mobile Communication Technology Influences Face-to-Face Conversation Quality. *Journal of Social and Personal Relationships* 30: 3.

Rainie, Lee and Kathryn Zickuhr. 2015. Americans' Views on Mobile Etiquette. Pew Research Center, August 26.

Rheingold, Howard. 2003. *Smart Mobs: The Next Social Revolution*. New York: Perseus.

Rinta-Kahila, Tapani, Esko Penttinen, Antti Salovaara, and Wael Soliman. 2018. Consequences of Discontinuing Knowledge Work Automation—Surfacing of Deskilling Effects and Methods of Recovery. *Proceedings of the 51st Hawaii International Conference on System Sciences*. Accessed at: https://scholarspace.manoa.hawaii.edu/server/api/core/bitstreams/bf98a123-5385-486b-8ecb-76e1e3dbcda0/content.

Rinta-Kahila, Tapani, T. Penttinen, A. Salovaara, W. Soliman, and J. Ruissalo. 2023. The Vicious Circles of Skill Erosion: A Case Study of Cognitive Automation. *Journal of the Association for Information Systems* 24 (5): 1378–1412.

Roots Analysis. 2024. *AI Agents Market*. Accessed at: https://www.rootsanalysis.com/ai-agents-market

Roose, Kevin. 2024. Can A.I. Be Blamed for a Teen's Suicide? *New York Times,* October 23. Accessed at: https://www.nytimes.com/2024/10/23/technology/characterai-lawsuit-teen-suicide.html.

Schaeffer, Katherine. 2019. Most U.S. Teens Who Use Cellphones Do It to Pass Time, Connect with Others, Learn New Things. Pew Research Center. Accessed at: https://www.pewresearch.org/fact-tank/2019/08/23/most-u-s-teens-who-use-cellphones-do-it-to-pass-time-connect-with-others-learn-new-things/.

Schneider, Sabrina, and Michael Leyey. 2019. Me or Information Technology? Adoption of Artificial Intelligence in the Delegation of Personal Strategic Decisions. *Managerial and Decision Economics* 40 (3): 223–231.

Schneier, Bruce, and Nathan Sanders. 2023. The A.I. Wars Have Three Factions, and They All Crave Power. *New York Times,* September 28. Accessed at: https://www.nytimes.com/2023/09/28/opinion/ai-safety-ethics-effective.html.

Sherman, Lauren E., Minas Michikyan, and Patricia M. Greenfield. 2013. The Effects of Text, Audio, Video, and In-person Communication on Bonding Between Friends. *Cyberpsychology: Journal of Psychosocial Research on Cyberspace* 7 (2): 3. Accessed at: https://www.cdmc.ucla.edu/wp-content/uploads/sites/170/2018/04/Sherman-et-al.pdf.

Singer, Natasha. 2024. Will Chatbots Teach your Children. *New York Times,* January 11. Accessed at: https://www.nytimes.com/2024/01/11/technology/ai-chatbots-khan-education-tutoring.html.

Small, Gary W., et al. 2020. Brain Health Consequences of Digital Technology Use. *Dialogues in Clinical Neuroscience* 22 (2): 179–187.

Smith, Adam. 2003. *The Wealth of Nations: An Enquiry into the Nature and Causes of the Wealth of Nations*. New York: Bantam.

Sparrow, Betsy, Jenny Liu, and Daniel Wegner. 2011. Google Effects on Memory: Cognitive Consequences of Having Information at Our Fingertips. *Science* 333 (6043): 776–778.

The Crowd and the Cloud. 2018. Accessed at: http://crowdandcloud.org.

Thiele, Leslie Paul. 2021. Rise of the Centaurs: The Internet of Things Intelligence Augmentation. In *Towards an International Political Economy of Artificial Intelligence*, ed. Ryan Kiggins and Tugrul Keskin, 39–61. Palgrave Macmillan.

Thornton, Bill, Alyson Faires, Maija Robbins, and E. Eric Rollins. 2014. The Mere Presence of a Cell Phone May be Distracting: Implications for Attention and Task Performance. *Social Psychology* 45 (6): 479–488.

Tiffany, Kaitlyn. 2022. No, I Will Not BeReal. *The Atlantic* (August 1). Accessed at: https://www.theatlantic.com/technology/archive/2022/08/bereal-app-authenticity-social-media-instagram/671012/?mc_cid=4fc263320c&mc_eid=8710c3458f.

Toolify. 2024. Accessed at: https://www.toolify.ai/ai-news/cancel-plans-with-friends-guiltfree-using-ai-excuse-generator-619560.

Turkle, Sherry. 2015. *Reclaiming Conversation: The Power of Talk in a Digital Age*. New York: Penguin.

Turkle, Sherry. 2017. *Alone Together: Why We Expect More from Technology and Less from Each Other*, 3rd ed. New York: Basic Books.

Vallor, Shannon. 2015. Moral Deskilling and Upskilling in New Machine Age. *Philosophy and Technology* 28: 107–124.

Verizon. 2019. *Smart Family*. Accessed at: https://www.verizonwireless.com/solutions-and-services/verizon-smart-family/.

Vivanti, Giacomo, and Sally J. Rogers. 2014. Autism and the Mirror Neuron System: Insights from Learning and Teaching. *Philosophical Transactions of the Royal Society London 'B' Biological Science* 369 (1644): 20130184. https://doi.org/10.1098/rstb.2013.0184.

Ward, Adrian F., Kristen Duke, Ayelet Gneezy, and Maarten W. Bos. 2017. Brain Drain: The Mere Presence of One's Own Smartphone Reduces Available Cognitive Capacity. *Journal of the Association for Consumer Research* 2 (2): 140–154.

Wiener, Norbert. 1954. *The Human Use of Human Beings: Cybernetics and Society*. New York: Houghton Mifflin.

Wilmer, Henry H., Lauren E. Sherman, and Jason M. Chein. 2017. Smartphones and Cognition: A Review of Research Exploring the Links between Mobile Technology Habits and Cognitive Functioning. *Frontiers in Psychology* 8 (April): 605. Accessed at: https://www.frontiersin.org/articles/10.3389/fpsyg.2017.00605/full.

Winner, Langdon. 1977. *Autonomous Technology*. Cambridge, MA: MIT Press.

Yan, Wanling et al. 2022. Does Global Positioning System-Based Navigation Dependency Make Your Sense of Direction Poor? A Psychological Assessment and Eye-Tracking Study. *Frontiers in Psychology* 13. Accessed at: https://www.frontiersin.org/journals/psychology/articles/10.3389/fpsyg.2022.983019/full.

Yarrow, Jay. 2010. Sergey Brin: 'We Want Google To Be The Third Half Of Your Brain.' *Business Insider* (September 8). Accessed at: https://www.businessinsider.com/sergey-brin-we-want-google-to-be-the-third-half-of-your-brain-2010-9?op=1.

CHAPTER 5

Dogmatism: The Eclipse of Common Truths and the Decay of Civic Trust

When human agency is degraded, human potential is diminished. So are democratic prospects. To the extent that distraction, dependence, and deskilling negatively affect mental and emotional health, compromise the development of cognitive and social skills, and leave people unfulfilled, one might expect that public policies would stem the harm. By and large, they have not. The problem is that throwing out the bathwater of the AI-enabled attention economy threatens the baby of free speech, increased connectivity, the growth of knowledge, economic opportunities, and digital enablement. In turn, powerful ICT corporations have resisted regulation. But the challenge is even more fraught.

Political solutions are thwarted because the AI-enabled attention economy has heightened ideological rigidity and political polarization. In turn, it has significantly increased the amount of misinformation that citizens and statespeople spread and consume. Last but not least, it has prompted citizens to embrace *clicktivism*, contributing to their civic deskilling and limiting their engagements to online activities that seldom bridge political constituencies. Clicktivism adds fuel to the fire of a polarized, misinformed polity, fostering self-righteous, intolerant, blinkered tribalism. The upshot is the eclipse of common truths, the decay of civic trust, and growing dogmatism.

The effective regulation of AI and the attention economy would be extraordinarily challenging in the best of worlds. Crafting and enforcing

L. P. Thiele, *Human Agency, Artificial Intelligence, and the Attention Economy*, https://doi.org/10.1007/978-3-031-82086-1_5

effective regulations in a toxic political environment populated by citizens with degraded agency may be a bridge too far. Taken together, polarization, misinformation, and civic deskilling strain democracy to the breaking point. They are the harbingers of autocracy.

The problem is worsened by the vulnerability of democratic polities to clandestine manipulation in a digital world. Consider one example: the impact of *list effects*. When a candidate's name appears above his or her opponent's name on a ballot, his or her share of the votes is boosted, often by 10% or more (Grant 2017). This is known as the "ballot-order effect." Another list effect occurs when undecided voters search the Internet for political information. People tend to formulate their political opinions based on the information found on the web pages that are linked to top search results. The information they receive, and the order in which they receive it, is determined by their search engine's algorithm. With over 85 billion annual visitors, Google accounts for 92% of global Internet traffic (Haan 2024). As scholars who have investigated the "search engine manipulation effect" (SEME) observe: "The fact that more than 90% of searches conducted in almost every country in the world are conducted on just one search engine (Google) … raises special concerns about SEME. It means that a single company—one that is unregulated, highly secretive, not accountable to the public, and that has, for all practical purposes, no competitors …—could be producing systematic changes in the thinking of billions of people every day with no way for other parties to counteract its influence, or even, or that matter, to detect and document that influence" (Epstein et al. 2024). The capacity of search engines and other platforms to manipulate opinions and influence elections in ways that are unaccountable and largely undetectable poses a direct threat to democratic governance.

In the realm of politics, it is important to note, there is great promise amidst the peril. Digital media have frequently been employed to mobilize political engagement, monitor electoral fraud, police abuses and other wrongdoings, scrutinize and hold governments accountable, and safeguard human rights and freedoms. Around the globe, Facebook has been used to initiate protests, Twitter to coordinate them, and YouTube to broadcast them. Iran's Twitter revolution of 2009 was followed by Tunisia's Jasmine Revolution, Egypt's Facebook revolution and the subsequent uprisings of the Arab Spring, the Occupy Wall Street protests, Spain's Movimiento 15-M, the Snow Revolution in Russia, the 2013 Gezi

Park demonstrations in Turkey, the Brazilian Spring, the #BlackLives-Matter movement in the US, Hong Kong's Umbrella Revolution, and the #MeToo movement, among others. Digital media played an important role in each of these popular movements (Thiele 2018).

Another example is Ushahidi, a free and open-source digital platform used in more than 160 countries to create maps of human rights violations, electoral fraud and malpractice, corruption, and natural disaster damage. Originally developed in Kenya to chart reports of postelection violence in 2008 employing SMS messaging (*ushahidi* means *testimony* in Swahili), this "crisis-mapping" technology has been expanded to accommodate various social media platforms. By crowdsourcing data, Ushahidi has effectively monitored elections in the US and many other nations, sped relief to victims of war and natural disasters, aided United Nations peacekeeping efforts, and mapped incidences of fraud, violence, and sexual abuse.

Or consider Telegram. In the wake of Russia's 2022 invasion of Ukraine, the COVID-19 channel on Telegram, a popular social media platform, changed its name from @COVID19_Ukraine to @UkraineNOW, and became the go-to site for millions of Ukrainians (as well as citizens of other nations) who wanted to resist the invasion and be informed about the most recent events (Burgess 2022). In turn, the Ukrainian Ministry of Internal Affairs launched a "Find Your Own" channel with uploaded photographs, videos, and ID documents of dead or captured Russian soldiers, putatively as a means of helping their families identify them. Within a month of the invasion, the *New York Times* opened its own Telegram channel to share information and news with Russians and Ukrainians. In turn, Telegram became the principal means of gaining unrestricted information for Russians during the war after Putin's regime shuttered independent news agencies and banned Facebook, Instagram, and Twitter. The flexibility, speed, scope, and scalability of social media platforms make them powerful and arguably irreplaceable tools to advance democracy, domestic and international peace, security, and responsible government.

Spurred on by social media, citizens have mobilized worldwide to get informed, secure their rights, and protest injustices. Notwithstanding their power to disseminate information, advance emancipatory movements, and promote democratic values, digital media have not been very successful in translating popular resistance and uprisings into policy, legislation, or structural reforms. Their impacts on democratic processes and institutions are, to say the least, uneven.

In 2010, Larry Diamond (2010, 70) hailed "liberation technology," arguing that digital media enabled citizens to "expand the horizons of freedom" despite early efforts of authoritarian governments to "filter and control the Internet, and to identify and punish dissenters." Before the end of the decade, however, Diamond (2019, 20–23) acknowledged that erstwhile "optimism (if not euphoria)" had become "pessimism (if not despair)" as digital media came to pose "a major threat to democratic stability and human freedom." Democrats worldwide had to grapple with the corrosive political effects of social media while racing to combat "nightmarish" totalitarian surveillance, control, and manipulation that was "all-encompassing." As William Dutton (2020, xvii) states: "Gone are the days of utopian visions of the Internet bringing democracy to the world. Quite the contrary. We are in a clearly dystopian context in which the Internet and related digital media are viewed by many as contributing to, if not causing, serious harm in politics and society." Whether the social and political benefits of digital media outweigh their harms is yet to be determined.

Clicktivism and Civic Deskilling

In one of the first critiques of clicktivism, also known as slacktivism, Malcolm Gladwell (2010) observed that political activism via social media "succeeds not by motivating people to make a real sacrifice but by motivating them to do the things that people do when they are not motivated enough to make a real sacrifice." In our personal lives, digital media give us the opportunity for social connection without the demands of responsiveness. In our political lives, digital media give us the opportunity for engagement without the demands of sustained affiliations or the inconveniences and compromises required for collective action (Turkle 2015, 50). Online micro-donations of time, ideological support, and money often amount to little more than "virtue signaling," allowing cyber-citizens to affirm their political alignment with the like-minded. Crossing party lines to engage in discussion, debate, and compromise in pursuit of practical goals remains relatively infrequent.

Clicktivism resembles a sugar rush, exploiting the dopamine cycle to provide instant rewards. Clicktivists can get a good deal off their chests and receive comforting strokes from fellow partisans, with very little effort and few, if any, repercussions. In contrast, offline political engagement, including that which crosses party lines, is demanding and often frustrating. But it builds the stamina and the strong relationships that are needed to sustain collective action and democratic governance.

A large and growing portion of the public employs social media as their primary source of news and primary means of participating in political discussions or events, with young adults more likely to engage in political activity online than any other way. Political involvement, however, is often limited to the digital exchange of opinions, "liking" candidates, stories, or events on social media sites, forwarding or sharing posts, and signing online petitions. These activities often supplant the face-to-face relationships and organization building that have been the mainstays of traditional political activism.

Enthusiasts of digital politics observe that there is little evidence to suggest that online activities crowd out traditional political participation. Most people do not engage in politics off-line. If these non-participants are attracted to online activities, digital politics may serve as a gateway, bringing otherwise uninvolved citizens, particularly youth, into public life. By "injecting the political into the everyday spaces and places of popular culture," Joel Penney (2017, 7, 162–163) argues, online activism and other forms of digital politics can push people "up the ladder of engagement" (see also Margetts et al. 2016).

There is some evidence that digital activity, especially that which takes place on social networking sites, increases political participation (Boulianne 2020). People who engage in political conversations on social media have higher levels of political engagement than those who do not discuss politics online. Of course, those who engage in face-to-face conversations about politics also demonstrate higher levels of political participation than those who refrain from talking about politics with family, friends, colleagues, and associates. The studies that reveal these correlations (e.g. Vaccari and Valeriani 2018) do not establish causation. So we do not know whether online political conversations spur political engagements, or whether political engagements stimulate online political conversations.

It is not surprising that those who demonstrate higher rates of political participation also talk more about politics—online, face-to-face, or both. Still, those who talk about politics a great deal, online or face-to-face, may not become engaged in more demanding forms of political action. The adage "All talk, no action" has its applications. In any case, a crucial part of effective communication is attentive listening and open-mindedness. Political exchanges on social media sites, which often feature ALL CAPS shouting matches, are not the best places to practice these skills and virtues. In turn, those who rely on social media for information about political institutions and processes, especially if they have high social media use, are less trusting of these institutions and processes (Kiratli 2023). That is not a recipe for productive politics.

Digital engagement is seldom simply added on to other forms of offline activity. Rather, much of the time people spend on their phones and computers replaces face-to-face communication and in-person relationships. This decline in direct interactions, for which our system of mirror neurons has been evolutionarily designed, reduces opportunities for the cultivation of relationships of trust and care. In the same vein, engaging in online politics does not stimulate the same affective relationships as in-person engagements. These emotional dispositions and bonds are crucial prerequisites for demanding forms of collective action. Digital interactions, consequently, are inferior to face-to-face interactions in facilitating collaborative endeavors (Pentland 2014, 172).

While digitally mediated politics may prompt otherwise uninvolved individuals to step onto the ladder of engagement, most never climb very high. Many never get off the couch. Tim Wu, Special Assistant to the President for Technology and Competition Policy from 2021 to 2023, argues that "citizenship in our time is about how you spend your attention" (Garber 2018). He also maintains that "Citizenship is like a muscle." Like muscles, citizenship weakens from want of exercise. Clicktivism does not provide much of a work-out. In this respect, clicktivists who never get off the couch are not simply misunderstanding the nature of politics. Their civic muscles are atrophying. Effectively, they are becoming politically deskilled—increasingly incapable of engaging directly with people of diverse opinions and values in the often slow and painstaking effort to build relationships and collaboratively achieve workable solutions.

The benefits that digital media provide in democracies are at least partially offset when clicktivism substitutes for, rather than stimulates, critical, fact-based thinking and the civic mobilization and organization that good governance demands. In non-democratic regimes, its effects may be even more pernicious. As Evgeny Morozov (2011, 201) observes, "The danger that 'slacktivism' poses in the context of authoritarian states is that it may give young people living there the wrong impression that another kind of politics—digital in nature but leading to real-world political change and underpinned entirely by virtual campaigns, online petitions, funny Photoshopped political cartoons, and angry tweets—is not only feasible but actually preferable to the ineffective, boring, risky, and, in most cases, outdated kind of politics practiced by the conventional opposition movements in their countries." The issue is not simply that clicktivism supplants traditional political activities. The problem is that digital activities may supplant a wide range of personal interactions that have the potential to build emotional intelligence and enduring relationships. Such aptitudes and engagements, political philosophers as far back as Aristotle understood, are the foundation of political community.

"If anything, the Internet makes it harder, not easier, to get people to care," Morozov (2011, 75) concludes: "[W]e need to find ways to supplant our promotion of a freer Internet with strategies that can engage people in political and social life.... Otherwise, we may end up with an army of people who are free to connect, but all they want to connect to is potential lovers, pornography, and celebrity gossip." In a chapter entitled "Social media is making politics impossible," Jaron Lanier (2018, 115) likewise observes that social media "undermines the political process and hurts millions of people, but so many of those very same people are so addicted that all they can do is praise [it] because they can use it to complain about the catastrophes it just brought about." Connecting is not caring. Complaining is not collaboration. Yet virtual life beguiles us into thinking so.

As politics in cyberspace displays a surfeit of acrimony and few enduring benefits, the complaints are understandable. Fiery exchanges between competing tribes seldom spark cooperation or coalitions. Indeed, most social media users find encounters with those of opposing political views stressful and frustrating and believe they have even less in common after engagement than before. Consequently, fewer than 1 in 10 frequently address politics, while more than 4 out of 5 do not engage in political dialogue with online friends (Duggan and Smith 2016).

The most effective means of shaping preferences and influencing behavior—four times as powerful as individual incentives and eight times as powerful as market incentives—is the "social network incentive," which is to say, concentrated social exposure by way of direct interactions, especially with trusted peers (Pentland 2014, 51, 68, 73). Social media puts the social network incentive on steroids. It provides an instant opportunity for birds of a feather to flock together. The problem is these birds often feed from a mix of moral outrage, vitriol, and misinformation.

Clicktivists tend to receive the same sort of pleasure from their activities that sports fans get from rooting for the home team and spewing venom at opponents. They are drawn to "outrage and gossip" because it is "alluring"; they get involved when there are "spectacles" and "frivolous distractions"; and they typically avoid local news because it is "boring" (Hersh 2020, 82). Clicktivists can praise, endorse, deride, and defame with the touch of a finger. Slow-moving discussions aimed at compromise or consensus, which are tedious in comparison to a flurry of digital posts and shares, get replaced by rapid-fire exchanges.

Engaging in real politics, the nineteenth-century German sociologist Max Weber famously said, is like the "slow boring of hard boards." It is demanding and often frustrating work that requires tolerance, patience, and perseverance. It is the political equivalent of eating your vegetables and getting exercise. Clicktivism, in contrast, is like binging on junk food. It features fiery exchanges and high drama filled with heroes, arch-villains, violent antagonisms, and salacious gossip, all drizzled with misinformation.

Smart mobs and "adhocracies" composed of fleeting cyberspace connections are not the best recipes for building and sustaining democratic institutions (Tufekci 2017, 76–77; see also Morozov 2013). Yet we are prompted—perhaps the better term would be *digitally engineered*—to believe that the solution to every political problem is only a click away. Each engagement, in turn, tends to reinforce tribal exclusions and self-righteousness. As Ayad Akhtar (2021, 20) observes of online politics: "The technology is seeking your trigger, whatever draws you deeper and keeps you clicking. Nothing quite does it like outrage. Moral outrage. Those we know are right to hate; those we love because we are united together against those we know are right to hate.... Driven by engagement and the profit that it generates, each side drifts further and further from the other, the space between us growing only more charged, only richer with opportunity for monetization." Just as the AI-enabled attention economy thrives upon, fosters, and exploits distraction, dependence, and deskilling, it also thrives upon, fosters, and exploits the seamier side of our political proclivities—self-righteous, intolerant tribalism. Consequently, time spent online, and particularly on social media, is widely considered "the most powerful sources of political polarization in our era" (Bail 2021, 5). Polarized polities produce dogmatic citizens, and vice versa.

Dogmatism is the tendency to accept and assert opinions, values, or principles without due consideration of disconfirming evidence or the contrary beliefs of others. Dogmatism is a narrowing of our range of understanding, critical thinking, and empathy. Dogmatism is the upshot of civic deskilling. It is one of the most pernicious products of digital media, imperiling democracy.

In 1995, more than a decade before Facebook launched its personalized News Feed, Negroponte (1995, 153) observed that "Most people tend to trash whole sections of newspapers without a glance, browse some of the rest, and read very little in detail. What if a newspaper company were willing to put its entire staff at your beck and call for one edition? It would mix headline news with 'less important' stories relating to acquaintances, people you will see tomorrow, and places you are about to go to or have just come from. It would report on companies you know.... You would consume every bit (so to speak). Call it *The Daily Me.*" Negroponte extolled the efficiencies of news that could be tailored to our idiosyncratic needs and desires, being grounded in "infinitesimal demographics." It would be optimal, he also noted, if we could "crank personalization up or down." That way, we could also access "*The Daily Us*," experiencing the news "with much more serendipity, learning about things we never knew we were interested in, being challenged" (Negroponte 1995, 164, 154). What Negroponte did not foresee is how social media would largely eradicate *The Daily Us* provided by local and national newspapers while proliferating *Daily Me's*. As a result, entire populations are never exposed to facts, stories, and viewpoints that challenge their values, interests, ideologies, and biases.

The term "echo-chamber" is used to describe the experience of inhabiting an online world where we encounter mostly news and information that aligns with our own dispositions, preferences, and previous activities. When reading a good newspaper, we are exposed to challenging and diverse opinions and information curated by an editorial team based on its assessment of what is newsworthy. In contrast, social media mostly present users with information, whether real or fabricated, that bolster their existing beliefs and predilections. Cyberspace has made it extraordinarily easy to ensconce ourselves in self-reinforcing communities of the like-minded. In this blinkered realm, we tend to hear, see, and share only those facts—or fictions—that support our personal interests and ideologies (Sunstein 2017).

Digital media vastly expand our accessibility to diverse perspectives and information (Scharkow et al. 2020). But broader access to diverse sources of information in a digital world, contrary to expectations, does not generally deepen exposure to and understanding of alternative viewpoints and values, or improve the way information is analyzed, understood, and integrated. Rather, an ever-expanding glut of information causes people to retreat to a small number of favored outlets and networks. Rather than expanding horizons and promoting critical thinking, digital media often insulate users within ideological "filter bubbles" that reinforce and heighten preexisting biases (Pariser 2012). Just as high school cliques homogenize the opinions of adolescents, so digital networks reinforce the values, interests, ideologies, and biases of their members. Accordingly, they tend to heighten extremism and increase ideological polarization (Morris 2007; Stroud 2008; Nie et al. 2010; Pariser 2012; Carr 2014; Gakshy et al. 2015; Nichols 2017; Madrigal 2018).

Notwithstanding the Internet's "formidable dispersal of information," online activities can embed people in increasingly "self-referential and homogenous niches of like-minded militants" (Urbinati 2014, 4–5, 202). Like attracts like, and soon enough, we become the digital company we keep. Extremes boost engagement. The algorithms employed by digital platforms to generate increased revenues can transform filter bubbles into rabbit holes that trap formerly moderate citizens in dark and blinkered realms. Rabbit holes promote tunnel-vision. As such, they annihilate opportunities to cultivate and exercise the practical judgment that enables democratic politics.

Everyone is susceptible to confirmation bias: we search for, and predominantly acknowledge and utilize, those bits of reality that conform to and confirm our prior beliefs and values. In turn, we all engage in motivated reasoning, aligning evidence and arguments with beliefs and values to which we are already committed. Information that challenges convictions and prejudices tends to get ignored. Confirmation bias and motivated reasoning are common foibles, as are faulty intuitions, poor reasoning, groupthink, and various forms of self-deception (Tetlock 2005; Tetlock and Gardner 2015). It would make sense to develop and deploy technologies that help us mitigate these all-too-human shortcomings and promote practical judgment. The opposite occurs in the attention economy. Cyberspace provides a nutrient-rich medium for confirmation bias and motivated reasoning.

In theory, digital media expand opportunities to encounter and appreciate diverse perspectives. In practice, they provide a means to surround ourselves with people who think just like us. People spend more time (and money) on sites that conform with and reinforce their values and beliefs, especially if those values and beliefs are presented in emotionally powerful ways that insinuate the risks of alternative perspectives and highlight threats from antagonists. Digital platforms oblige these tendencies all the way to the bank. Those who rely primarily on social media for their news and interactions are more likely to be confirmed in their beliefs and values, and, oftentimes, have these beliefs and values strengthened and rigidified. Interests and idiosyncrasies get crystalized into uncompromising identities.

Newspapers and news programs have long employed the criterion that "If it bleeds, it leads." The shocking sells, and always has. As a result, media often feed and exacerbate our natural negativity bias, reinforcing the human tendency to focus on risks and threats. Over evolutionary time spans, a negativity bias served us well. While living with heightened anticipation of harms and dangers wasted some psychological resources, it also preserved life and limb. On balance, it was adaptive. Today, our negativity bias is systematically reinforced by social media, which bleeds continually. News items, posts, and pics that shock and stir deep emotions—regardless of their accuracy or appropriateness—get shared more often, spreading faster and farther in cyberspace than facts and more restrained forms of expression.

The business models of many digital platforms are grounded on content personalization and algorithmic amplification. Facebook's *News Feed*, TikTok's *For You* tab, and YouTube's recommendation engine, for example, are personalized to individual tastes through the collection of data from web cookies that capture browsing histories. Meanwhile, algorithms amplify delivery of the content that most heightens engagement, which is to say, emotionally powerful information or misinformation that bolsters biases and pillories antagonists. Political moderates on social media mostly eschew partisan exchanges, focusing on sports, music, entertainment, friends, family, and pets. Political exchanges mostly occur on the infrared and ultraviolet ends of the spectrum, where discourse is often heated and dogmatic (Bail 2021). Zealots on social media put more effort into disseminating their convictions and misinformation than do moderates. Consequently, cyberspace is rife with fighting words and images. Digital platforms that supply their users with engagement-based feeds and recommendations amplify extremism and the "outrage-ification" of culture (Lerman 2019).

The calamitous political impact of digital media has not been lost on its erstwhile inventors and promoters. Testifying at a Congressional hearing, former Facebook Director of Monetization, Tim Kendall, addressed the platform's role as an "accelerant" of division and extremism. Facebook's algorithms, designed to heighten engagement, had the effect of propagating incendiary content and delivering it strategically. "Social media preys on the most primal parts of your brain—it provokes, it shocks, and it enrages," Kendall (2020) observed. Chamath Palihapitiya, former vice president of user growth at Facebook, also rues his corporate contributions. Palihapitiya observes that "The short-term, dopamine-driven feedback loops we've created are destroying how society works.... No civil discourse, no cooperation; misinformation, mistruth. And it's not an American problem—this is not about Russian ads. This is a global problem" (Lanier 2018, 8). At their worst, digital platforms have become "attention-sucking beasts" that "perpetuate biases" by feeding their users the most "titillating, inflammatory content" (Lindsay 2021). The result is increased polarization and dogmatism. As Akhtar (2021, 20) observes "The cultural clash in America today has more electrical engineering behind it than we realize." As the electrical engineering in question is globally orchestrated, dogmatism is becoming planetary. Misinformation plays a major role.

Misinformation

On digital media, our confirmation and negativity biases are reinforced not only by cherry-picked facts but by outright fictions. Misinformation surpasses fact in its novelty, and hence in its capacity to shock and intrigue. Misinformation also tends to push different emotional buttons, such as fear, anger, and disgust. Because anger creates a stronger arousal response than most other emotions, including sadness or joy, it is more likely to be shared online. Platforms designed to maximize likes and shares, therefore, promote posts that provoke anger, contributing to the prominence of viral outrage (Williams 2018, 72). Misinformation often figures in the mix.

Among our Pleistocene ancestors, anger and outrage were primarily expressed by way of face-to-face confrontations and gossip in tight-knit groups. There were risks associated with such exchanges, including retaliation by the accused. The spreading of misinformation also posed risks. Liars could be shamed, punished, or cast out. Today, rumormongering and misinformation still trigger deep-seated emotional reactions. But they have become nearly costless to express and share online (Crockett 2017). The anonymity allowed by digital media offers users the opportunity to be mendacious, aggressive, and abusive with few if any repercussions. As a result, misinformation and trolling are commonplace.

It was commonly said a century ago that a lie can travel halfway around the globe while the truth is still lacing up its boots. Both the speed and the diffusion of misinformation have increased dramatically over the years. An analysis of news stories shared on Twitter (now X) revealed that falsehoods are 70% more likely to be retweeted than truths, and that even after eliminating tweets that originated or were spread by bots, misinformation still spread faster, farther, deeper, and more broadly than fact (Soroush et al. 2018). Misinformation is an accelerant of virality.

On August 27, 2015, a record one billion people used Facebook in a single day. Mark Zuckerberg (2015) celebrated the event in a post, writing "Our community stands for giving every person a voice, for promoting understanding and for including everyone in the opportunities of our modern world. A more open and connected world is a better world. It brings stronger relationships with those you love, a stronger economy with more opportunities, and a stronger society that reflects all of our values." Heady stuff, to be sure. Love, tolerance, and inclusion across the globe, however, have not been the upshot of Facebook's growth. Owing to an unfortunate quirk of human nature, it is easier and more profitable to connect people by means of lies, biases, and viral outrage than by means of plain facts, prudential reasoning, and appeals to tolerance and diversity. Facebook followed the money, and the results were predictable.

Less than two years after Zuckerberg's heady announcement, Facebook became an accomplice to genocidal violence against the Rohingya people in Myanmar. A United Nation's Independent International Fact-Finding Mission concluded that social media played a "significant" role in fomenting the atrocities in Myanmar in 2017, a country where "Facebook is the Internet." The International Court of Justice (ICJ) ruled in 2022 that it has jurisdiction to proceed under the Genocide Convention with a case against the Myanmar government for its treatment of the Rohingya ethnic minority. Amnesty International condemned Meta for its part, claiming that its "reckless pursuit of profit … substantially contributed to the atrocities" by "supercharging the spread of harmful anti-Rohingya content." The organization argues that Meta is liable for reparations to the predominantly Muslim minority for its use of "hate-spiraling algorithms" (Amnesty International 2022). For the first time in history, genocidal atrocities were not only the product of human hate and bigotry. They were enabled and abetted by a cold, calculating, engagement-maximizing algorithm.

Digital platforms grow their bottom lines by keeping people online. Connections, good or bad, serve that purpose. An internal memo written by Andrew Bosworth, Facebook's Vice President, explains the rationale behind the monetization of connectivity: "We connect people. Period. That's why all the work we do in growth is justified…. So we connect more people. That can be bad if they make it negative. Maybe it costs someone a life by exposing someone to bullies. Maybe someone dies in a terrorist attack coordinated on our tools…. The ugly truth is that we believe in connecting people so deeply that anything that allows us to connect more people more often is *de facto* good" (Mac et al. 2018). Saying that connecting people is a de facto good is rather like saying that growth is a de facto good, regardless of whether the growth in question increases health, welfare, truth and amity, or disease, misery, falsehoods, and genocide.

The algorithms animating digital platforms, more so than deficits in critical reasoning or partisan bias in users, primarily drive the spread of misinformation. Users develop habits of sharing the information that attracts the most attention. Platforms, in turn, promote and reward attention-getting posts. Consequently, the most habitual news sharers on social media spread most of the false news (Ceylan et al. 2023). As McNamee (2019, 242–243) observes, "On Facebook, information and disinformation look the same; the only difference is that disinformation generates more revenue, so it gets much better treatment. To Facebook, facts are not an absolute; they are a choice to be left initially to users and their friends but then magnified by algorithms to promote engagement. In the same vein, Facebook's algorithms promote extreme messages over neutral ones, disinformation over information, conspiracy theories over facts." Given the tendency of fake news to diffuse faster and farther, generating more online traffic and advertising revenue, misinformation is good business for digital platforms.

People naturally want attention. As the competition is stiff in cyberspace, an increasingly common form of attention is viral notoriety. The AI-enabled attention economy promotes it like no other. Virality grows the bottom lines of ITC corporations and the websites they support. In 2016, for instance, Google was paying websites for the number of people they attracted, as this represented potential customers for Google's advertisers. When a bunch of unemployed teenagers in Veles, Macedonia realized that misinformation got more shares and attracted more people than plain vanilla facts, they created hundreds of fake news sites that spread misinformation about the US presidential election. The impact was profound. The teenagers, who had no ideological motivations, got rich fast, cashing in on the hundreds of millions of dollars that clickbait farms earn every year (Aral 2020, 52, 307). As Hao (2021) reports, "On an average day, a financially motivated clickbait site might be populated with celebrity news, cute animals, or highly emotional stories—all reliable drivers of traffic. Then, when political turmoil strikes, they drift toward hyperpartisan news, misinformation, and outrage bait because it gets more engagement." As long as platforms attempt to maximize engagement, they will foster virality and serve as petri dishes for outrage, misinformation, and discord. Polarization and extremism will grow.

The monetization of polarization by digital platforms highlights the difference between economic growth and human welfare. The Gross Domestic Product (GDP) is the annual total monetary value of exchanged goods and services in a nation. GDP measures a nation's economic activity, but not necessarily its welfare. For example, a giant oil spill may increase GDP, as cleaning up the mess causes more money to change hands than would be generated by a pristine environment. Likewise, the Gross Business Profits (GBPs) of digital platforms benefit from chronic spills of toxic communications—misinformation, acrimony, and polarized outrage. More sensible and less sensational exchanges are less profitable. As long as platforms employ algorithms that promote virality, cyberspace will be heavily polluted with misinformation and vitriol. The problem, at base, is that polarization pays.

Automation makes a bad problem worse, as misinformation and vitriol on digital media are often circulated by bots. These automated software applications can be programmed to post, share, and reshare stories and opinions, mimicking human users. Since a bot can post hundreds of times an hour, even a small number of bots may have a big effect. In turn, networks of linked bots, botnets, can follow and re-message each other, creating the illusion of opinions or events that are trending on social media. The term *computational propaganda* is employed to designate the use of algorithms, bots, and human curators to manufacture, disseminate, and amplify misinformation and social divisions on social media networks. Scholars studying computational propaganda consider it "one of the most powerful tools against democracy" (Woolley and Howard 2019, 243).

The intentional circulation of misinformation by bots to secure financial gain is widespread in the attention economy and is the frequent target of investigation and prosecution by authorities such as the Federal Trade Commission (Aral 2020). The intentional circulation of misinformation by bots for political purposes has no equivalent regulatory body to monitor and prosecute offenders, though the US Federal Bureau of Investigation and the National Security Agency have waded into these turbid and troubled waters.

The challenge is global. In 2014, Russia orchestrated the annexation of Crimea without significant international retribution by systematically exploiting the use of social media bots to remove or discredit Ukrainian voices of opposition, putting a patina of legitimacy over the hostile invasion of a sovereign state (Aral 2020). In 2016, Russian disinformation operators published tens of thousands of messages on social media platforms in the final days of the Brexit campaign. And about 400,000 bots were engaged in digital activity during the 2016 US Presidential election, generating about one-fifth of all digital interactions (Bessi and Ferrar 2016). In February 2018, a US grand jury indicted 13 Russian nationals and 3 Russian entities, including the Internet Research Agency, for violating criminal laws with the intent to interfere with US elections and political processes. In 2024, two other Russians working for RT, formerly known as Russia Today, were indicted for paying millions of dollars to right-wing social media influencers in the US to amplify domestic divisions and weaken opposition to Russia's war in Ukraine.

The Internet Research Agency and RT are two of a growing number of state-sponsored Troll Farms, also known as Keyboard Armies, deploying bots and bloggers to disrupt foreign political processes, foster social divisions abroad, and spread propaganda. AI has significantly lowered the bar for such efforts, making it fast and easy for foreign operatives to mount sophisticated influence campaigns (Frenkel et al. 2024). Domestically, social media influencers with thousands, tens of thousands, or millions of followers are handsomely paid to promote or undermine political parties and candidates, legislation and policies, and social causes. In the hands of individuals, organized groups, corporations, and adversarial regimes, AI-enabled and automated digital media, Troll Farms, and hired influencers are significantly reshaping political landscapes and cultures, impacting campaigns and elections, and altering democratic processes and institutions (Wofford 2022; Woolley and Howard 2019; Penney 2017, 113; Karpf 2016, 40; Howard 2015, 209).

Though computational propaganda is deployed differently in authoritarian states than in democratic ones, the "gap is closing" (Woolley and Howard 2019, 14). As a rule, democracies do not engage in the most egregious forms of digital repression, such as targeted persecutions, Internet shutdowns, censorship, and mass surveillance. China's Skynet, Safe Cities, and Sharp Eyes programs are examples of such efforts. However, a number of democracies, at least when led by illiberal heads of state, surpass some autocracies in manipulation and misinformation, which is perhaps the most widely practiced and quickly growing arenas of digital repression (Feldstein 2021, 64–64, 85–86).

Greater transparency regarding the ways that digital platforms impact democratic processes is needed. To date, however, the proprietary nature of the algorithms employed to channel and increase platform's online traffic remains a stumbling block. Consider the case of TikTok. In 2021, TikTok had more users in the US than Twitter, was watched more than YouTube, was accessed more than Google, and had its app downloaded more than Facebook's. TikTok's American users spend almost an hour and a half on the platform each day, three times more than they do on Snapchat and twice the time devoted to Facebook or Instagram. About a third of America's youth regularly get their news from TikTok (Matsa 2023). It is becoming an increasingly important destination for political content, and "a primary incubator of baseless and misleading information" (Hsu 2022). Yet the nature, extent, and effects of its in-house efforts to combat false content remain inaccessible to researchers, government, and the public.

TikTok is owned by ByteDance, a Chinese corporation. Like all Chinese corporations, ByteDance must toe the line established by Beijing to avoid repercussions. Given TikTok's influence over US citizens, that raises concerns. Douyin, the Chinese version of TikTok, abides by Chinese government directives regarding censorship and propaganda. It is severely limited to children by law in China, and the platform has a set of self-imposed prohibitions against the "subversion of traditional moral values." No such restrictions limit the use of TikTok in the US.

It has been suggested that TikTok effectively serves China as a belated act of revenge for the Opium Wars, the nineteenth-century conflicts sparked by the Chinese emperor's prohibition of opium, which was being supplied in large quantities by the British East India Company. The argument is that TikTok is muddling Western minds, and potentially revealing sensitive data and exposing the vulnerabilities of Western nations. With these threats in mind, some US states and federal agencies have prohibited TikTok on government devices, and in 2024, President Biden signed into law a Congressional bill that would ban the app if it was not sold to an American company.

Addressing the threat from TikTok, Klein (2022) observes that "Our collective attention is important. Whoever (or whatever) controls our attention controls, to a large degree, our future. The social media platforms that hold and shape our attention need to be governed in the public interest. That means knowing who's truly running them and how they're running them." What Klein says about platforms owned by foreigners also applies to domestically owned ICT businesses whose prime directive is maximizing profit rather than advancing the public good.

During the 2016 US presidential election, foreign-sponsored sites contributed a relatively small portion of all the circulating misinformation. Still, most citizens were exposed to fake news stories, and about half of those who recalled such stories believed them to be true. In the three months prior to the election, patently fake election news stories generated more total engagement on Facebook than the combined top election stories produced by nineteen major news outlets, including *The New York Times*, *The Washington Post*, *The Huffington Post*, and *NBC News* (Silverman 2018). Importantly, and unsurprisingly, people were much more likely to believe fake news that favored their preferred candidates.

In the lead up to and aftermath of the 2020 US presidential election, far-right sources of news and information generated the highest average number of online interactions, followed by sources from the far-left. In turn, far-right purveyors of misinformation had on average 65% more engagement per follower than far-right purveyors who did not convey falsehoods (Edelson et al. 2021). Conservatives consumed the most misinformation, and the filter bubble effect appears to be stronger among them (Gonzalez-Bailon et al. 2023). In many countries, including the US, there are "political asymmetries in misinformation sharing," with conservative or right-wing users of social media sharing more poor-quality information than liberal or left-wing users (Mosleh et al. 2024).

In turn, social media users over 65 years of age share almost seven times as many articles from fake news domains than the youngest age group, and this effect remains statistically significant when controlled for other demographic attributes, ideology, and party affiliation (Guess et al. 2019). Older conservatives who are highly engaged online are the primary readers and sharers of fake news (Grinberg et al. 2019).

McNamee (2019, 242–243) writes that "Democracy depends on shared facts and values. It depends on communication and deliberation. It depends on having a free press and other countervailing forces to hold the powerful accountable. Facebook (along with Google and Twitter) has undercut the free press from two directions: it has eroded the economics of journalism and then overwhelmed it with disinformation." Such concerns are now widespread, extending well beyond the impact of specific digital platforms. One of the grand challenges of the twenty-first century, according to 50 experts surveyed for the BBC's *Future Now,* is the breakdown in trusted information sources. Most Americans agree (Pew Research Center 2017). Roughly half of all Americans believe that the government should take steps to restrict online misinformation, and increasing numbers hold this conviction (Mitchell and Walker 2021).

Public concern for the spread of misinformation on social media has prompted congress to hold hearings. Some platform CEOs have explored remedies. Jack Dorsey, co-founder and former head of Twitter, was among the most vocal of active CEOs to express concern for the social and political impacts of digital platforms. Under Dorsey, Twitter began placing warning labels on tweets, including a number from President Trump, that spread misleading and potentially harmful information or presented the threat of harm against identifiable groups. Two weeks prior to the US 2020 Presidential election, Twitter took additional steps to slow the flow of misinformation on its network. Users who tried to retweet false content on the site were alerted to its "disputed" status and asked if they wished to share it. The idea was to create "extra friction" to deter the thoughtless spreading of falsehoods. Twitter also disabled the algorithm that suggested posts based on users' interests. Instead, users only saw the content of accounts they followed, along with advertisements. Beginning in 2023, under the ownership of Elon Musk, Twitter became X, and reversed most of these initiatives.

It is common sense to promote truth over falsehood. But common sense, and common truths, do not serve all parties equally well. In 1969, a tobacco industry executive wrote a memo entitled "Doubt Is Our

Product." At the time, the tobacco industry was employing hand-picked scientists to cast doubt on the established fact that nicotine is addictive, and that smoking tobacco is dangerous to health. Decades later, the fossil fuel industry employed teams of scientists to cast doubt on the science of climate change (Oeskes and Conway 2010). Donald Trump politicized the debate, frequently calling climate change a hoax.

In these instances, sowing the seeds of doubt protects corporate profits and secures political constituencies. In the same vein, President Trump sowed the seeds of doubt with frequent allegations that mainstream media constituted "fake news"—a claim made almost 2000 times during his first four years in office (Woodward 2020). This doubt mongering was designed to undermine the belief that there are scientific and historical facts that citizens can and should collectively acknowledge and critically assess, regardless of their partisan alignments. When confronted with evidence contradicting Whitehouse falsehoods about the size of the crowd at Trump's inauguration, for example, Kellyanne Conway, Senior Counselor to the President, defended the official statements by insisting that they were not lies but "alternative facts" (Meet the Press 2017). When citizens doubt the existence of non-partisan facts, civic trust becomes a scarce resource.

Civic trust is the willingness to engage in dialogue and collaboration with those who have different opinions and convictions because everyone shares higher order values of truth, tolerance, justice, and democratic processes. Civic trust depends upon common truths—historical facts and scientific data that ground a people in a shared reality. Cyberspace has greatly contributed to the erosion of a shared reality. That seems an obvious loss. But the polarization that results from the decline of civic trust can be strategically exploited. As President Trump observed in 2017, buoyed by his 100 million followers on social media: "Without the tweets, I wouldn't be here" [in the Whitehouse] (Barber et al. 2017).

It has been said that "A lie told once remains a lie, but a lie told a thousand times becomes the truth." If one sees a lot of smoke, it is reasonable to assume that a fire is burning. The catchphrase "If it trends, it's true" sounds false. Given the daunting challenge of separating out all the digital chaff from the wheat, however, the equation of trends with truth is commonplace. When false claims are incessantly circulated and widely shared, they come to be believed. Accordingly, politicians and authoritarian leaders deploy digital media to fabricate alternative facts and circulate them widely.

The quotation about a lie told a thousand times is almost universally attributed to Joseph Goebbels, the Nazi's Reich Minister of Propaganda. In fact, Goebbels never uttered these words, though he certainly acted as if they were true. Still, the misattribution to him has been so often repeated and shared across digital media that it has become a de facto truth for countless people, including many scholars. It is ironic that Goebbels is still gaining notoriety by way of the "Big Lie." He was, in many respects, ahead of his time.

The saturation of cyberspace with misinformation undermines public trust in democratic processes and institutions. It leads many people down rabbit holes of ideological extremism. Equally worrisome, the presence of widespread misinformation is used as a rationale for rejecting everything one does not *want* to believe. Science, historical facts, authentic photographs or videos, reasonable arguments, and public opinion can be conveniently dismissed by labeling them fake (Tufekci 2023). This allows politicians to fabricate alternate realities and supercharges confirmation bias among citizens. Although all the photos and footage employed were authentic, for example, Trump claimed during the 2023 primaries that AI was being used in "fake" television commercials to make him look "bad and pathetic" (Tuquero 2023). A climate of civic distrust provides a hothouse for cynicism, confirmation bias, and duplicity.

With the smoke of so much misinformation swirling about, it is understandable that many people are duped, while others simply give up trying to differentiate truth from falsehood. They become "post-truth." Democracies today increasingly fall into this malaise, exhibiting extraordinarily high levels of distrust (Hendricks and Vestergaard 2019, 112). Post-truth citizens tend to be cynical or dogmatic. Cynics suffering from "epistemic paralysis" recuse themselves from the responsibilities of citizenship by rejecting all political parties and players, summoning a plague on all their houses. Dogmatists, in contrast, engage in "weaponized relativism" (Duberry 2022). They close ranks, listen solely to comrades-in-arms, circle the wagons, and shoot outward. Dogmatists are attracted to demagogues and dictators, effectively reverting to a childhood coping strategy: when the world becomes too challenging to understand or navigate, retreat to the security of a parental authority.

To value historical and scientific facts is to accept the responsibility of separating the chaff from the wheat—the lies, distortions, biases, and wishful thinking from, for lack of a better word, the truth. That responsibility is demanding. It has become particularly burdensome in a digital age awash in misinformation. Being post-truth is much easier. Growing numbers of citizens appear unwilling or unable to shoulder the responsibility of truth-seeking. Political leaders who display a brazen, self-serving disregard for facts are likely to be appealing to them, effectively serving as role models of savvy maneuvering in an unreliable world.

Misinformation can also be used to mobilize. Such was the case with Trump's frequent assertions that the 2020 election was stolen. A diet of misinformation coaxed his hard-core supporters to storm the Capitol on January 6, 2021 in an effort to stop the certification of the election. In response, Facebook suspended President Trump's account. The following day, Twitter permanently banned Trump. Whether these decisions were warranted to protect democracy and public safety is debatable. Certainly, they were not carried out by democratically elected officials, and might become regrettable precedents for the curtailing of free speech. What is clear, however, is that they were belated responses to the eclipse of truth and the collapse of trust.

Widespread misinformation and a climate of civic distrust have strategic value for those with authoritarian impulses. The far-right software engineer and blogger Curtis Guy Yarvin, known by the pen name Mencius Moldbug, observed that "[I]n many ways nonsense is a more effective organizing tool than the truth. Anyone can believe in the truth. To believe in nonsense is an unforgeable [sic] demonstration of loyalty. It serves as a political uniform. And if you have a uniform, you have an army" (Moldbug 2008). In a polarized polity, rabbit holes of misinformation can harden the hearts and narrow the minds of loyalists. In this respect, the false, unbelievable, and often patently absurd statements of extremist politicians and pundits harbor a calculating shrewdness. Voltaire once observed that if people can be made to believe absurdities, they can be made to commit injustices. If supporters can be coaxed into endorsing absurdities in order to demonstrate their loyalty, they will also support and commit all manner of injustices—in full uniform.

With this in mind, Timothy Snyder (2021) observed that "Social media … supercharges the mental habits by which we seek emotional stimulation and comfort, which means losing the distinction between what feels true and what actually is true. Post-truth wears away the rule of law and invites a regime of myth." Snyder concludes that "Post-truth is pre-fascism." In the same vein, Williams (2018, 76) writes that "mob rule is hard-coded into the design of the attention economy." Williams may be exaggerating, but not by much.

Efforts to address misinformation typically focus on the need for structural remedies. The commonly proffered ones entail the use of fact-checking algorithms, the tagging of demonstrably false posts, and taking down fake news sites and blogs. The same approach is employed when countering inauthentic postings (i.e. those employing fake identities), hate speech (abusive or threatening communications that express prejudice against specific groups, typically focused on ethnicity, race, religion, or sexual orientation), and the advocacy of violence. It is no small challenge to protect free speech while correctly identifying and eliminating misinformation, hate speech, and the advocacy of violence. Likewise, it is no small challenge to ensure that postings are authentic while protecting users' privacy and security in the face of potential reprisals from violent individuals, extremist groups, and authoritarian regimes. Indeed, the regulatory task is Herculean.

Slaying the Hydra, a poisonous, serpentine, multi-headed monster, was the second labor of Hercules. It demanded heroic effort because every time one of the Hydra's heads was severed, two more grew back in its place. Likewise, as popular social media sites move to control their content, platforms that forego self-regulation, such as 4chan, Gab, Signal, and Telegram, gain users. Mass migrations from more established platforms to sites perceived to be less prone to censorship or surveillance are common. When Trump was suspended from Facebook and banned from Twitter following the storming of the Capitol by his supporters, for example, the Trump family and a significant portion of his base of supporters moved to Telegram (Loucaides 2022). Soon enough, Trump created his own social media platform, Truth Social. Other alternative social media sites in the US, such as BitChute, Gab, Gettr, Parler, and Rumble are also predominantly used by people who identify themselves as Republicans. One in six prominent accounts on these sites have been banned or demonetized elsewhere on social media (Stocking et al. 2022).

Free-wheeling platforms that siphon off the disaffected users of mainstream social media sites may in time also become the target of public oversight and adopt some form of self-regulation. If and when that happens, other outlets undoubtedly will emerge. The attention economy, in Hydra-like fashion, has the capacity to grow new heads whenever necessity or opportunity arises.

Largely unregulated sites that do not monitor content or identities are attractive to free speech advocates, criminals, trolls, and those who—for good reason or owing to paranoia—fear that their digital activities, if revealed, would put themselves, their families, and their colleagues, at risk. Some of these sites, such as Telegram with over 900 million users, claim to remove illegal content such as doxing, child pornography, terrorist channels, and those violating intellectual property rights. However, an investigation by the *New York Times* (Mozur et al. 2024) found 50 Telegram channels openly selling contraband, including guns and fraudulent debit cards, with 22 offering to deliver illegal drugs to more than 20 countries. Many channels were operated by terrorist groups, and white supremacists coordinated activities of almost one million people around the world on 1500 channels. In 2024, French authorities seized Telegram's founder and CEO, Pavel Durov, charging him with complicity in managing an online platform to enable illegal transactions, the distribution of child sexual abuse material, drug trafficking, fraud, and a refusal to cooperate with law enforcement. Free speech activists reacted strongly. The case will test the responsibility of digital platforms and their executives for the content of their sites.

Large social media platforms are not the only means for the spread of misinformation. A proliferation of websites published in multiple languages contributes heavily. The arrival of generative AI has made a bad problem worse. Countless websites now have content produced mostly or entirely by AI. Whether by intention or neglect, these sites may propagate misinformation that is difficult to discredit. It takes only a matter of minutes to produce from scratch and propagate widely a befuddling array of fake news clips, fake scientific papers, voice-cloned speeches, and deep fake photographs and videos that are extraordinarily challenging to detect and debunk. Efforts are underway to chart the rapidly expanding role that deep fakes are playing in campaigns and elections (Wired 2024). With the arrival of increasingly user-friendly AI, we are entering a Wild West of misinformation, and there is no sheriff in town.

LLMs frequently "hallucinate," supplying their human interrogators with fictions woven into otherwise factual responses. When a New York lawyer asked ChatGPT to write a ten-page brief, for example, the program duly complied, providing more than half a dozen relevant court decisions to buttress his case. Neither the opposing lawyers nor the presiding judge, however, could find the cited court decisions in any law books. That is because the cases did not exist. ChatGPT had hallucinated them (Weiser 2023). Students crafting school essays and academics writing scholarly papers with the aid of generative AI also often discover, too late, that it has introduced fictionalized events and citations into "their" work. Such hallucinations occur because LLMs are programmed to respond to users' queries by stringing together sets of words that will satisfy them. Quite often, these verbal constructions prove to be a few steps removed from reality.

The problem of AI hallucinations will likely have a technological solution. The more enduring and greater threat comes from people who employ AI intentionally to deceive. We now inhabit a world of *deep misinformation*: high-quality and abundant synthetic text, speech, images, and video. Detecting deep fakes is a formidable task. The ease and speed of their generation and dissemination far surpasses our wherewithal to systematically expose the counterfeits. Likewise, the damage done by deep misinformation to individuals, organizations, the public realm, and democracy can be speedy, extensive, and pronounced. Repairing the damage, and healing the wounds, takes more time, and will always be partial at best. Simply put, it is always easier to break things than to fix them.

H.G. Wells famously observed that "Civilization is in a race between education and catastrophe. Let us learn the truth and spread it as far and wide as our circumstances allow. For the truth is the greatest weapon we have." If Wells is right, then we should focus on learning the truth and disseminating it widely. The problem is that AI-enabled digital media also generate and spread vast amounts of untruth. Consequently, much will be broken. In a post-truth world, broken things are particularly hard to fix.

Polarization in a Post-Truth World

Editors of major newspapers and anchors of mainstream radio and television news programs assume the responsibility of fact-checking their stories. They have an incentive to limit the pursuit of shock value: securing a reputation for broadcasting the most important news of the day and publishing only that which is "fit to print." On social media, there are no gatekeepers of truth or decorum. The epistemological equivalent of Gresham's law applies: just as bad money can drive out good money in an economy, so can false and shocking stories on social media drown out responsible reporting and sharing. Consequently, social media pulls toward dogmatism and polarization.

The word "they" is one of about a dozen words that most commonly show up in conspiracy theories on social media (Faddoul et al. 2020).

A social media platform whose AI is churning through user data might discern that posts containing the phrase "They are lying to you!" get more likes and shares. As a means of increasing online traffic, the platform's algorithm will give priority in its feeds to posts that contain the phrase. As accusations of deceit aimed at specific individuals, groups, and institutions become more widely shared, they will stimulate counteraccusations. Consequently, the frequency of the phrase will increase. As online accusations and attacks grow, tribal boundaries will become impermeable. With suspicion and cynicism mounting, the standards employed to determine who gets accused will loosen. So, the frequency of the phrase will rise, further undermining civic trust. Soon enough, "They are lying to you!" is no longer an allegation voiced by a few conspiracy theorists. It accurately describes a polarized polity.

It is said that the first casualty of war is truth. In cyberspace, citizens increasingly find themselves engaged in a chronic cold war. Online polarization exacerbates confirmation biases and motivated reasoning across the political spectrum, giving rise to higher levels of dogmatism on the right and the left. And when dogmatism becomes the norm, the spreading of misinformation is perceived as a moral and patriotic act, just as deception is legitimate for states that are engaged in hot wars (Hendricks and Vestergaard 2019, 83–84).

The prospects for democracy depend upon safeguarding its core institutions: individual rights, the rule of law, a universal franchise, fair elections, and transparent processes for political participation and decision-making. Its prospects also depend upon cultivating and sustaining civic culture, which includes the norms of public decency and truth-telling, and a reasonably high capacity in the citizenry for practical judgment, which, more than any other human aptitude, makes us fit to exercise freedom (Fleishacker 1999).

The crucial importance of political institutions was celebrated by Immanuel Kant, the eighteenth-century German philosopher. In Kant's view, the "evil attitudes" of citizens was not a stumbling block to the creation of a just regime. Indeed, even "a nation of devils" fiendishly pursuing their self-interest could establish a just state as long as the laws and civil structures that constrained them were well made (Kant 1991, 112–113). Still, Kant insisted that the capacity for rational understanding and the ability to make sound judgments based on reason were indispensable prerequisites of just states. It is debatable whether justice, or any other democratic value, can truly be achieved or sustained in the absence of moral sentiments. Regardless, Kant was right to argue that the capacity for rational understanding and judgment *is* necessary. Without the widespread cultivation of these political skills, democracy is doomed.

In an age of social media, newspaper editors and radio and TV anchors are no longer the curators of most of the news and information to which people are exposed. Increasingly, citizens serve as their own gatekeepers of truth. Yet the aptitudes and skills required to carry out this increasingly challenging task are being systematically eroded. In a tragic irony, the AI-enabled attention economy is creating a heightened need for aptitudes and skills that it is simultaneously degrading.

Morozov (2011, xiii) criticizes the "cyber-utopianism" advanced by purveyors of digital media, which blithely maintains in the face of a great deal of evidence to the contrary that the Internet and digital media will "boost democratic participation, trigger a renaissance of moribund communities, strengthen associational life, and serve as a bridge from bowling alone to blogging together." Digital media once served as an opiate for a good number of intellectuals, politicians, and statespeople, prompting them to believe that life in a digitally connected world would become smarter, freer, more just, and more cooperative. Reality has been sobering. Digital media have demonstrated their capacity to encourage clicktivism, incite viral outrage, spread misinformation, undermine civic trust, heighten polarization and extremism, and foster dogmatism. Given these historical developments, the traditional task of politicians in democratic regimes—winning a broad coalition of hearts and minds through persuasive speech—often gets replaced by narrowly targeted efforts to mobilize ideological hardliners by any available means. The mandate of winning over the moderate middle succumbs to the task of rigidifying an extreme base. The result is a no-holds-barred form of polarized politics, politics as war carried by other means. In such a world, democratic processes and institutions will wane while demagoguery and authoritarianism wax.

Digital technology is widely exploited by autocratic regimes to monitor, surveil, disinform, intimidate, and repress citizens. The level of manipulation and repression allowed by the melding of autocracy and digital power, George Soros (2018) commented, "may well result in a web of totalitarian control the likes of which not even Aldous Huxley or George Orwell could have imagined." Political theorist Hannah Arendt (1973, 474, 382) observed that the "ideal subject of totalitarian rule" was not the true believer, but "people for whom the distinction between fact and fiction (*i.e.,* the reality of experience) and the distinction between true and false (*i.e.,* the standards of thought) no longer exist." Such people, displaying both gullibility and cynicism, believe that "everything was possible and nothing was true." In the same vein, Garry Kasparov, former chess worldmaster and current political commentator, observed that "The point of modern propaganda isn't only to misinform or push an agenda. It is to exhaust your critical thinking, to annihilate truth" (Kakutani 2018, 143).

Truth gets annihilated with the creation of alternate realities. After its 2022 invasion of Ukraine did not lead to a quick victory, the Kremlin shuttered its independent news agencies, invoked a censorship law that imposed up to 15 years in prison for those who spread unacceptable information, proscribed the words "war" and "invasion" as descriptors of its "special military operation" in Ukraine, and banned Facebook, Instagram, and Twitter. When Telegram became the prime source of independent information about the war for Russians, the Kremlin worked to "control the narrative" on the platform, running large numbers of channels, paying for posts, and circulating prodigious amounts of misinformation. Whether the Kremlin also managed to gain access to the identities of Telegram users who voiced opposition to the war is unknown. In any case, the number of Telegram subscribers to official and pro-regime channels quickly grew to widely surpass those of the opposition (Safronova et al. 2022). Telegram became "the platform of choice" for Kremlin officials (Loucaides 2023, 58).

It would be mistaken to assume that the primary danger to democracy are foreign autocrats. Domestic dogmatism is a growing threat. Indeed, misinformation can be more widespread in democratic states than autocratic ones (Feldstein 2021). The eclipse of common truths and the decay of civic trust is endemic to polarized democracies. The dangers that digital media pose are greatly magnified in nations where democratic institutions are not well established and the governmental regulation or self-regulation of ICT corporations is absent. The attention economy, rife with misinformation and hate speech, is destabilizing a growing number of developing democracies (Hao 2021). The dangers posed by well-established authoritarian regimes that control digital media are patent. Perhaps of greater concern are the dangers posed by autocratic leaders in crumbling democracies. Digital technologies allow a greatly heightened capacity for surveillance, systematic disinformation, and social control. Autocrats have learned how to make very good use of digital media, often exploiting it more effectively than democratic leaders (Rød & Weidmann 2015). We should expect aspiring autocrats to hone their digital skills and consolidate their power.

In the mid-twentieth century, a decisive victory over fascism was secured by the world's democracies. In no small part, the victory was achieved owing to the strength of their industrial economies. Whether democracies can survive the twenty-first century remains to be seen. To date, the prodigious power of the AI-enabled attention economy is lessening the odds.

References

Akhtar, Ayad. 2021. The Singularity is Here. *The Atlantic* (December): 17–21.

Amnesty International. 2022. Myanmar: The social atrocity: Meta and the right to remedy for the Rohingya. Accessed at: https://www.amnesty.org/en/documents/ASA16/5933/2022/en/

Aral, Sinan. 2020. *The Hype Machine: How Social Media Disrupts Our Elections, Our Economy, and Our Health—and How We Must Adapt*. New York: Currrency.

Arendt, Hannah. 1973. *The Origins of Totalitarianism*. New York: Harcourt Brace Jovanovich.

Bail, Chris. 2021. *Breaking the Social Media Prism: How to Make Our Platforms Less Polarizing*. Princeton: Princeton University Press.

Barber, Lionel, Demetri Sevastopulo, and Gillian Tett. 2017. Donald Trump: Without Twitter, I Would Not Be Here—FT Interview. April 2. Accessed at: https://www.ft.com/content/943e322a-178a-11e7-9c35-0dd2cb31823a.

Bessi, Alessandro, and Emilio Ferrara. 2016. Social Bots Distort the 2016 US Presidential Election Online Discussion. *First Monday* 21: 11–17. Accessed at: https://ssrn.com/abstract=2982233.

Boulianne, Shelly. 2020. Twenty Years of Digital Media Effects on Civic and Political Participation. *Communication Research* 47 (7): 947–966.

Burgess, Matt. 2022. When War Struck, Ukraine Turned to Telegram. *Wired* (March 2). Accessed at: https://www.wired.com/story/telegram-ukraine-russia-channels-signal/?mc_cid=a2bf9cc9a0&mc_eid=8710c3458f.

Carr, Nicholas. 2014. *The Glass Cage: How Computers are Changing Us*. New York: W.W. Norton.

Ceylan, Gizem, Ian A, Anderson, and Wendy, Wood. 2023. Sharing of misinformation is habitual, not just lazy or biased Proceedings of the National Academy of Sciences 120 (4): 1–8

Crockett, M. J. 2017. Moral outrage in the digital age. *Nature Human Behaviour* 1: 769–771.

Diamond, Larry. 2010. Liberation Technology. *Journal of Democracy* 21 (3): 69–83.

Diamond, Larry. 2019. The Threat of Postmodern Totalitarianism. *Journal of Democracy* 30 (1): 20–24.

Duberry, Jérôme. 2022. *Artificial Intelligence and Democracy: Risks and Promises of AI-Mediated Citizen-Government Relations*. Cheltenham, UK: Edward Elgar.

Duggan, Maeve, and Aaron Smith. 2016. The Political Environment on Social Media. *Pew Research Center: Internet and Technology*, October 25. Accessed at http://www.pewinternet.org/2016/10/25/the-political-environment-on-social-media/.

Edelson, Laura, Minh-Kha Nguyen, Ian Goldstein, Oana Goga, Tobias Lauinger, and Damon McCoy. 2021. Far-Right News Sources on Facebook more Engaging (March 3). Accessed at: https://medium.com/cybersecurity-for-democracy/far-right-news-sources-on-facebook-more-engaging-e04a01efae90.

Epstein, Robert, Michael Lothringer, and Vanessa Zankich. 2024. How a Daily Regimen of Operant Conditioning Might Explain the Power of the Search Engine Manipulation Effect (SEME). *Behavior and Social Issues*. Accessed at: https://link.springer.com/article/10.1007/s42822-023-00155-0#.

Faddoul, Marc, Guillaume Chaslot, and Hany Farid. 2020. A Longitudinal Analysis of YouTube's Promotion of Conspiracy Videos. Cornell University: ArXiv. Accessed at: https://arxiv.org/abs/2003.03318.

Feldstein, Steven. 2021. *The Rise of Digital Repression*. New York: Oxford University Press.

Fleishacker, Samuel. 1999. *A Third Concept of Liberty: Judgment and Freedom in Kant and Adam Smith*. Princeton: Princeton University Press.

Frenkel, Sheera, Tiffany Hsu, and Steven Lee Myers. 2024. How Russia, China and Iran Are Interfering in the Presidential Election. *New York Times*, October 29. Accessed at: https://www.nytimes.com/2024/10/29/technology/election-interference-russia-china-iran.html.

Gakshy, Eytan, Solomon Messing, and Lada Adamic. 2015. Exposure to Ideologically Diverse News and Opinion on Facebook. *Science* (5 June) 6239: 1130–1132.

Galen, Stocking et al. 2022. The Role of Alternative Social Media in the News and Information Environment. Pew Research Center, October 6. Accessed at: https://www.pewresearch.org/journalism/2022/10/06/the-role-of-alternative-social-media-in-the-news-and-information-environment/.

Garber, Megan. 2018. To Be a Good Citizen, First Pay Attention. *The Atlantic* (June 27). Accessed at: https://www.theatlantic.com/entertainment/archive/2018/06/citizenship-in-our-time-is-about-how-you-spend-your-attention/563912/.

Gladwell, Malcolm. 2010. Small Change: Why the Revolution Will Not Be Tweeted. New Yorker (October 4). Accessed at: https://www.newyorker.com/magazine/2010/10/04/small-change-malcolm-gladwell.

Gonzalez-Bailon, Sandra, et al. 2023. Asymmetric ideological segregation in exposure to political news on Facebook. *Science* 381 (6656): 392–398.

Grant, Darren. 2017. The ballot order effect is huge: Evidence from Texas. *Public Choice* 172: 421–442.

Grinberg, Nir, et al. 2019. Fake News on Twitter during the 2016 U.S. Presidential Election. *Science* 363 (6425): 374–378.

Guess, Andrew, Jonathan Nagler, and Joshua Tucker. 2019. Less than You Think: Prevalence and Predictors of Fake News Dissemination on Facebook. *Science Advances* 5 (1). Accessed at: https://advances.sciencemag.org/content/5/1/eaau4586.

Haan, Katherine. 2024. Top Website Statistics for 2024. *Forbes Advisor*, April 2. Accessed at: www.forbes.com/advisor/business/software/website-statistics/.

Hao, Karen. 2021. How Facebook and Google Fund Global Misinformation. *MIT Technology Review*. Accessed at: https://www.technologyreview.com/2021/11/20/1039076/facebook-google-disinformation-clickbait/?truid=&utm_source=the_download&utm_medium=email&utm_campaign=the_download.unpaid.engagement&utm_term=&utm_content=05-24-2022&mc_cid=76802677bd&mc_eid=8710c3458f.

Hendricks, Vincent F., and Mads Vestergaard. 2019. *Reality Lost: Markets of Attention, Misinformation and Manipulation*. Cham, Switzerland: Springer Open.

Hersh, Eitan. 2020. *Politics is for Power: How to Move Beyond Political Hobbyism, Take Action, and Make Real Change*. New York: Scribner.

Howard, Philip N. 2015. *Pax Technica: How the Internet of Things May Set Us Free or Lock Us Up*. New Haven: Yale University Press.

Hsu, Tiffany. 2022. On TikTok, Election Misinformation Thrives Ahead of Midterms. *New York Times*, August 14. Accessed at: https://www.nytimes.com/2022/08/14/business/media/on-tiktok-election-misinformation.html?campaign_id=9&emc=edit_nn_20220815&instance_id=69340&nl=the-morning&regi_id=71023177&segment_id=101421&te=1&user_id=1b8a7a6c418ba0b24e53c8b7ae5ab191.

Kakutani, Michiko. 2018. *The Death of Truth: Notes on Falsehood in the Age of Trump*. New York: Duggan Books.

Kant, Immanuel. 1991. *Political Writings*, 2nd ed. Cambridge: Cambridge University Press.

Karpf, David. 2016. *Analytic Activism: Digital Listening and the New Political Strategy*. New York: Oxford University Press.

Kendall, Tim. 2020. Testimony at House Committee on Energy and Commerce Hearing on "Mainstreaming Extremism: Social Media's Role in Radicalizing

America." Thursday, September 24, 2020. Accessed at: https://energycommerce.house.gov/committee-activity/hearings/hearing-on-mainstreaming-extremism-social-media-s-role-in-radicalizing.

Kiratli, Osman Sabri. 2023. Social Media Effects on Public Trust in the European Union. *Public Opinion Quarterly* 87 (3): 749–763.

Klein, Ezra. 2022. TikTok May Be More Dangerous Than It Looks. *New York Times*, May 8. Accessed at: https://www.nytimes.com/2022/05/08/opinion/tiktok-twitter-china-bytedance.html.

Lanier, Jaron. 2018. *Ten Arguments for Deleting Your Social Media Accounts Right Now*. New York: Picador.

Lerman, Rachel. 2019. Q&A: Ex-Googler Harris on How Tech 'Downgrades' Humans. *AP News*. Accessed at: https://apnews.com/article/technology-business-data-privacy-apple-inc-dea7f32d16364c6093f19b938370d600.

Lindsay, Roddy. 2021. I Designed Algorithms at Facebook. Here's How to Regulate Them. *New York Times,* October 6. Accessed at: https://www.nytimes.com/2021/10/06/opinion/facebook-whistleblower-section-230.html.

Loucaides, Darren. 2022. Message Received. *Wired* (March): 40–57.

Loucaides, Darren. 2023. Intercepted. *Wired* (March): 48–59.

Mac, Ryan, Charlie Warzel, and Alex Kantrowitz. 2018. Growth at Any Cost. *BuzzFeed News,* March 29. Accessed at: https://www.buzzfeednews.com/article/ryanmac/growth-at-any-cost-top-facebook-executive-defended-data#.cnkEEaN0v.

Madrigal, Alexis C. 2018. What Facebook Did to American Democracy. *The Atlantic*, October 12, 2017. Accessed at: https://www.theatlantic.com/technology/archive/2017/10/what-facebook-did/542502/.

Margetts, Helen, Peter John, Scott Hale, and Taha Yasseri. 2016. *Political Turbulence: How Social Media Shape Collective Action*. Princeton: Princeton University Press.

Mark, Gloria. 2015. *Multitasking Int the Digital Age*. NP: Morgan and Claypool.

Matsa, Katarina Eva. 2023. More Americans are Getting News on TikTok, Bucking the Trend Seen on Most Other Social Media Sites. Pew Research Center. Accessed at: https://www.pewresearch.org/short-reads/2023/11/15/more-americans-are-getting-news-on-tiktok-bucking-the-trend-seen-on-most-other-social-media-sites/.

McNamee, Roger. 2019. *Zucked: Waking Up to the Facebook Catastrophe*. New York: Penguin.

Meet the Press. 2017. January 22. Accessed at: https://www.nbcnews.com/meet-the-press/meet-press-01-22-17-n710491.

Mitchell, Amy, and Mason Walker. 2021. More Americans Now Say Government Should Take Steps to Restrict False Information Online than in 2018. Pew Research Center. Accessed at: https://www.pewresearch.org/fact-tank/2021/08/18/more-americans-now-say-government-should-take-steps-to-restrict-false-information-online-than-in-2018/.

Moldbug, Mencius. 2008. An Open Letter to Open-Minded Progressives, Chapter 4: Dr. Johnson's Hypothesis. Accessed at: https://www.unqualified-reservations.org/2008/05/ol4-dr-johnsons-hypothesis/.

Morozov, Evgeny. 2011. *The Net Delusion: The Dark Side of Internet Freedom*. New York: PublicAffairs.

Morozov, Evgeny. 2013. *To Save Everything, Click Here: The Folly of Technological Solutionism*. New York: Public Affairs.

Morris, Jonathan S. 2007. Slanted Objectivity? Perceived Media Bias, Cable News Exposure, and Political Attitudes. *Social Science Quarterly* 88: 707–728.

Mohsen Mosleh, Qi Yang, Tauhid Zaman, Gordon Pennycook, and David G. Rand. 2024. Differences in Misinformation Sharing Can Lead to Politically Asymmetric Sanctions. *Nature* 634: 609–624.

Mozur, Paul, Aaron Krolik, and Steven Lee Myers. 2024. How Telegram Became a Playground for Criminals, Extremists and Terrorists. *New York Times*, September 7. Accessed at: https://www.nytimes.com/2024/09/07/technology/telegram-crime-terrorism.html.

Negroponte, Nicholas. 1995. *Being Digital*. New York: Alfred A. Knopf.

Nichols, Tom. 2017. *The Death of Expertise: The Campaign against Established Knowledge and Why it Matters*. New York: Oxford University Press.

Nie, Norman H., Darwin W. Miller, Saar Golde, Daniel M. Butler, and Kenneth Winneg. 2010. The World Wide Web and the U.S. Political News Market. *American Journal of Political Science* 54: 428–439.

Oeskes, Naomi, and Eik Conway. 2010. *Merchants of Doubt*. New Yok: Bloosmbury.

Pariser, Eli. 2012. *The Filter Bubble*. New York: Penguin.

Penney, Joel. 2017. *The Citizen Marketer: Promoting Political Opinion in the Social Media Age*. New York: Oxford University Press.

Pentland, Alex. 2014. *Social Physics: How Good Ideas Spread—The Lessons from a New Science*. New York: Penguin.

Pew Research Center, Internet and Technology. 2017. Accessed at: http://www.pewinternet.org/2017/03/29/the-future-of-free-speech-trolls-anonymity-and-fake-news-online/; http://www.pewinternet.org/2017/10/19/the-future-of-truth-and-misinformation-online/.

Rød, E. G, and Weidmann, N. B. 2015. "Empowering activists or autocrats? The Internet in authoritarian regimes." *Journal of Peace Research* 52 (3): 338–351.

Safronova, Valeriya, Neil MacFarquhar, and Adam Satariano. 2022. Where Russians Turn for Uncensored News on Ukraine. *New York Times,* April 16. Accessed at: https://www.nytimes.com/2022/04/16/world/europe/russian-propaganda-telegram-ukraine.html.

Scharkow, Michael, Frank Mangold, Sebastian Stier, and Johannes Breuer. 2020. How social network sites and other online intermediaires increase exposure to news. *PNAS* 117 (6): 2761–2763.

Silverman, Craig. 2018. *This Analysis Shows How Viral Fake Election News Stories Outperformed Real News On Facebook.* BuzzFeedNews. Accessed at: https://www.buzzfeednews.com/article/craigsilverman/viral-fake-election-newsoutperformed-real-news-on-facebook

Snyder, Timothy. 2021. The American Abyss. *New York* Times, January 9, 2021. Accessed at: https://www.nytimes.com/2021/01/09/magazine/trump-coup.html?mc_cid=d608443470&mc_eid=8710c3458f.

Soros, George. 2018. Remarks Delivered at the World Economic Forum, Davos Switzerland, January 25, 2108. Accessed at: https://www.georgesoros.com/2018/01/25/remarks-delivered-at-the-world-economic-forum/.

Stroud, Natalie J. 2008. Media Use and Political Predispositions: Revisiting the Concept of Selective Exposure. *Political Behavior* 30: 341–366.

Sunstein, Cass. 2017. *#Republic: Divided Democracy in the Age of Social Media.* Princeton: Princeton University Press.

Tetlock, Philip, and Dan Gardner. 2015. *Superforecasting: The Art and Science of Prediction*. New York: Crown.

Tetlock, Philip E. 2005. *Expert Political Judgment*. Princeton: Princeton University Press.

Thiele, Leslie Paul. 2018. Digital Politics is the Game: See What Happens When Scholars Play It Well! *Perspectives* 16 (4): 1123–1128.

Tufekci, Zeynep. 2017. *Twitter and Tear Gas: The Power and Fragility of Networked Protest Fragility of Networked Protest.* New Haven: Yale University Press.

Tufekci, Zeynep. 2023. *Avert Your Eyes, Avoid Responsibility and Just Blame TikTok.* New York Times. December 27. Accessed at: https://www.nytimes.com/2023/12/27/opinion/young-voters-tiktok.html

Tuquero, Loreben. 2023. PolitiFact FL: Trump Says AI Was Used to Make Him 'Look Bad.' The clips are real. Accessed at: https://www.wmfe.org/politics/2023-12-29/trump-lincoln-project-ai.

Turkle, Sherry. 2015. *Reclaiming Conversation: The Power of Talk in a Digital Age*. New York: Penguin.

Urbinati, Nadia. 2014. *Democracy Disfigured: Opinion, Truth, and the People.* Cambridge: Harvard University Press.

Vaccari, Christina, and Augusto Valeriani. 2018. Digital Political Talk and Participation: Comparing Established and Third Wave Democracies. *Sage Open*

(June): 1–14. Accessed at: https://journals.sagepub.com/doi/10.1177/2158244018784986.

Vosoughi, Soroush, Deb Roy, and Sinan Aral. 2018. The Spread of True and False News Online. *Science* 359: 1146–1151.

Weiser, Benjamin. 2023. Here's What Happens When Your Lawyer Uses ChatGPT. *New York Times*, May 27. Accessed at: https://www.nytimes.com/2023/05/27/nyregion/avianca-airline-lawsuit-chatgpt.html?searchResultPosition=1.

William, H. 2020. Introduction to A Research Agenda for Digital Politics. In *A Research Agenda for Digital Politics*, William H. Dutton, ed., xvii–xxix. Cheltamham, UK: Edward Egar.

Williams, James. 2018. *Stand Out of Our Light: Freedom and Resistance in the Attention Economy*. Cambridge: Cambridge University Press.

Wired. 2024. *The WIRED AI Elections Project*. Accessed at: https://www.wired.com/story/generative-ai-global-elections/.

Wofford, Benjamin. 2022. The Lobbyist Next Door. *Wired* (September), 47–57.

Woodward, Alex. 2020. 'Fake News': A Guide to Trump's Favourite Phrase—And the Dangers It Obscures. *Independent,* October 2. Accessed at: https://www.independent.co.uk/news/world/americas/us-election/trump-fake-news-counter-history-b732873.html.

Woolley, Samuel C., and Philip N. Howard, eds. 2019. *Computational Propaganda*. New York: Oxford University Press.

CHAPTER 6

Despair: Passivity in the Face of Predictive Power

Digital platforms crunching massive amounts of online data are adept at predicting and manipulating human behavior. They filter our choices, shape our decisions, and influence our actions. As these technologies become more powerful in the coming years, we face the prospect of an accelerated erosion of human agency and widespread passivity. Among other psychological ailments, participants of the AI-enabled attention economy will suffer from despair. Free will appears to be at stake.

A brief review of the broader impacts of digital media on mental health is in order before addressing the prospect of despair. At the outset, it is important to acknowledge that the psychological impacts of digital media are frequently exaggerated, with diagnostic inflation carried out by professionals and non-professionals alike. Reviewing the state of the scientific literature, many scholars are appropriately cautious. They note that the long-term effects of digital technology on mental well-being are not well known, while the short-term repercussions that have been assessed appear variable and context dependent (Ferguson et al. 2024; Vuorre et al. 2021; Orben and Przybylski 2019a; Orben and Przybylski 2019b; Orben et al. 2019, 10,227). Research in this field faces significant methodological challenges and is plagued by disputes (Stein 2024). Clearly, well-crafted longitudinal studies are needed to determine enduring impacts.

Still, available data implicate digital media in a wide array of psychological ailments, including addiction, antisocial personality disorder,

L. P. Thiele, *Human Agency, Artificial Intelligence, and the Attention Economy*, https://doi.org/10.1007/978-3-031-82086-1_6

attention-deficit disorder, depression, narcissism, obsessive–compulsive disorder, social phobia, and voyeurism, among others (Rosen 2012, 201–202; Twenge and Campbell 2009, 107–122). High levels of screen time are associated with lowered self-esteem, increased mental health problems, diminished learning, and greater risks of premature cognitive decline (Neophytou et al. 2021). In turn, the more time spent on social media, the higher the level of life dissatisfaction, and heavy use is associated with substantial losses in psychological well-being (Kross et al. 2013; Verdyun, P. et al. 2015; Huang 2010).

Youths are particularly vulnerable. From 2011 to 2021, there was a 42% increase in the number of US adolescents reporting feelings of persistent sadness and hopelessness (United States Hospital Fund 2024). Rates of youth depression and anxiety during this time more than doubled (Haidt 2024). Sharp rises begin around 2015 for girls and young women and 2016 for boys and young men, with the former showing significantly higher levels. Digital technology, particularly smartphones and social media, are identified as key contributors (Blanchflower and Bryson 2024; Saleem et al. 2024). For adolescents, the risk of depression increases by 13% for each hour increase in the daily use of social media (Liu et al. 2022). The risk of depression for girls who are heavy social media users, spending more than five hours a day on their devices, is three times higher than for girls who do not use social media. For boys, the risk of depression for heavy users was twice as high (Kelly et al. 2018).

Between 2011 and 2021, suicidal thoughts and behaviors also increased in youth (Center for Disease Control and Prevention 2023, 59). The rate of deaths due to alcohol, drugs, and suicide in the US climbed 11 percent in 2021, hitting an all-time high and marking a two-decade trend of increasing fatalities from substance abuse and suicide (Trust for America's Health 2023). These alarming statistics cannot be solely attributed to the growing use of digital media. Still, health professionals are concerned. "In early adolescence, when identities and sense of self-worth are forming, brain development is especially susceptible to social pressures, peer opinions and peer comparison," the US Surgeon General observed in his 2023 Advisory, noting that social media platforms are brimming with "extreme, inappropriate and harmful content," including content that can "normalize" self-harm, eating disorders, and other self-destructive behavior (Murthy 2023a, 5, 8).

It is possible that those who already experience low life satisfaction or are depressed are drawn to social media, which then exacerbates their condition. Of course, the deterioration of mental health in youth could be the product of other factors, such as fear and despair in the face of climate change or school shootings. Still, young people faced the threat of nuclear annihilation in the 1950s, the Vietnam War, race riots, and political assassinations in the 1960s, and harbingers of environmental collapse in the 1970s. None of these decades produced a youth mental health crisis comparable to what is occurring today.

The extent to which digital culture contributes to poor mental health and "deaths of despair" is yet to be determined. In the US, the correlation appears strong. In the European Union, however, deaths by suicides for 15 to 19 year olds declined between 2011 and 2021 (Eurostat 2024). The non-Western world also does not show uniform increases. The Gallup World Poll, for example, suggests that declining mental health, understood in terms of emotional well-being, is restricted to a few wealthier nations (Gallup 2024). More research is clearly needed. To be sure, correlation is not causation. For many scholars and parents, however, the smartphone looks like a smoking gun.

To the extent that digital media are a major cause of psychological ailments in youth, the problem is not only what is happening when they are online. The problem is what is *not* happening when they are online. Digital activities consume much of the time that otherwise might be spent in person-to-person engagements that make people happier (Hall and Liu 2022). When we stare at screens, we are not engaging in face-to-face interactions with family, friends, and colleagues, and we are not outdoors experiencing nature. Direct social relationships and time spent in nature generally make us happier and healthier.

Face-to-face, embodied, synchronous social interactions, a staple of human life throughout history, evidenced a sharp decline beginning in the 2010s. High school and college-age students spent over two hours a day in these interactions in 2010, but less than 50 minutes in 2019. As this "social deprivation" steadily increased, so did sleep deprivation (Haidt 2024, 121, 124). Screen time reduces the amount of time that youth spend socializing in person and sleeping. Both deprivations cause a deterioration in mental health.

The *type* of screen time also matters: Heavy social media use appears to be the key danger. It is strongly correlated in youth with decreased physical well-being and increased mental health problems, unhappiness, loneliness, and depression, particularly for girls (Arrivillaga et al. 2022; Gazzaley and Rosen 2016; Twenge 2017; Haidt and Allen 2020; Twenge

2020; Twenge et al. 2020). The developmental stage and psychological characteristics and well-being of the users also play a role. The heavy use of social media significantly decreases life satisfaction for young teens, and then again for youth leaving their teen years (Orben et al. 2022).

Routine, daily use of social media may facilitate a sense of community and social connection, particularly among those who are physically distant from each other. Indeed, adolescents who spend more face-to-face time with friends also tend to spend more time on social media (Twenge et al. 2019). Clearly social media can be used to facilitate personal relationships. At the same time, "emotional connection to social media" is correlated with decreases in social and mental well-being. Emotional connection to social media is not simply a matter of spending time online. It refers to dependence upon and preference for digital over non-digital forms of interaction, such that the unavailability of digital media causes disappointment, frustration, and a sense of isolation (Bekalu et al. 2019). When digital technology becomes the primary or exclusive means to avoid or alleviate loneliness, anxiety, boredom, or depression, it tends to aggravate these conditions.

Deactivation of social media accounts produces improvements in levels of well-being for heavy users, though some studies indicate that the improvements are relatively small (Ortiz-Ospina 2019; Orben and Przybylski 2019a). Likely, the improvements are a consequence of increased face-to-face time spent with friends and family, as such activities are positively correlated to happiness and health. The fact that larger improvements in well-being do not occur for those who deactivate their social media account may reflect a relative dearth of social skills and emotional intelligence. It would be surprising if heavy social media users with undeveloped social aptitudes would, upon reducing screentime, eagerly seek out or gain the full benefits of face-to-face relationships for which they are ill-prepared. People tend to feel better when they are engaged in substantive conversations that go beyond small talk. In one study, the happiest people spent 70% more time in conversations that were markedly more substantive (Mehl et al. 2010). Such conversations are infrequent among the socially deskilled, and they seldom happen online.

Indeed, online conversations increasingly do not provide human interaction. The *dead internet theory* claims that most of the content of the Internet and social media is created and distributed by chatbots. Scholars of digital culture originally considered this notion of an AI-dominated Internet to be a paranoid illusion or conspiracy theory. Fantasy is quickly becoming fact. Within a few years of the arrival of LLMs, AI-generated content has come to saturate cyberspace, including social journalism and blogging. One of the most popular blogging sites, *Medium*, aspires to be "a home for human stories and ideas." Noting that "the internet is noisy and chaotic," the platform's goal is "to deepen our collective understanding of the world through the power of writing.... In a world where the most sensational and surface-level stories often win, we're building a system that rewards depth, nuance, and time well spent. A space for thoughtful conversation more than drive-by takes, and substance over packaging" (Medium 2024). That is a worthy goal, and an aspiration much to be applauded. Unfortunately, and to the dismay of its publishers, about 40 percent of the content of *Medium* is now AI generated (Knibbs 2024).

Digital engagements often displace more fulfilling activities. One experience much supplanted by life online is time spent in nature. Adults, and especially children, spend more time in front of screens and less outdoors than ever before. Increases in "screen time" combined with decreases in "green time" have negative psychological outcomes (Oswald et al. 2020). Addressing a widespread "nature deficit disorder," Richard Louv (2012, 24) argues that "The best antidote to negative electronic information immersion will be an increase in the amount of *natural* information we receive. *The more high-tech we become, the more nature we need.*" Likewise, David Abram (ix–x) maintains that those living in a world of "electronically-generated vistas and engineered pleasures" desperately need that which is other than ourselves and our creations: "Without the oxygenating breath of the forests, without the clutch of gravity and the tumbled magic of river rapids, we have no distance from our technologies, no way of assessing their limitations, no way to keep ourselves from turning into them." In the same vein, Robin Kimmerer (2023), a botanist and member of the Potawatomi nation, rues how our attention has been "hijacked by our economy." She urges us to "reclaim our attention" by focusing it on the natural world. Acknowledging that much environmentalism is doom-and-gloom, Kimmerer underlines the joyfulness of interactions with nature and their therapeutic benefits: green time allows you to "feel your own agency," and is restorative for individuals, the broader culture, and the natural environment.

Research confirms what naturalists affirm. Nature experiences alleviate stress and mental fatigue while increasing attention spans, impulse control, creativity, healing, and life satisfaction (Williams 2017; Louv

2012, 27–34, 48–49, 64, 80; Kahn 2011). Increasing numbers of college campus health professionals and general therapists now prescribe "Nature Rx" as a remedy for impaired mental and physical health (Ivers and Astell-Burt 2023). Time spent in nature appears to be one of the best remedies for the dysfunctions of digital culture (Kardaras 2016, 240–241).

Digital activities not only diminish the frequency and duration of interactions with people and nature. They also degrade the quality of these offline interactions. Nature outings that might have provided an opportunity to "unplug" are increasingly utilized as backdrops for online posts and selfies. In turn, almost nine out of ten people regularly use their smartphones during social encounters (Rainie and Zickuhr 2015). In an early study exploring the effects of this behavior, researchers assessed the emotional rewards of social interaction for participants who did or did not have ready access to their devices. Phone use increased the level of participants' distractedness and decreased their enjoyment of social interactions. This decrease in emotional rewards was evident even with moderate levels of phone use (Dwyer et al. 2018).

The results should not surprise us. Phubbing depletes social interactions. Still, researchers hoped the digital cloud might have a silver lining. Those using their smartphones during face-to-face interactions, they hypothesized, should at least find themselves less bored, as they could stimulate themselves by means of online activities when conversation proved uninteresting or lapsed. In fact, people who used their phones during social interactions reported higher levels of boredom. We know why. Those who are prone to boredom are more susceptible to digital distractions (Wegmann et al. 2018). In turn, persistent distractions heighten the proclivity to boredom. As we learned in Chapter 3, boredom and digital distraction are symbiotic.

Many suffer psychologically when compulsive online activities replace or undermine joyful and fulfilling connections to people and nature. Behavioral addictions are widely identified as "brain diseases" because they are chronic pathologies of the neurological reward system. When behavioral addictions become widespread, however, something is amiss outside of cranial cavities. Psychiatrists grappling with the sharp rise in compulsive online behavior insist that these neurological disorders should be seen as cultural rather than individual maladies (Jabr 2019). That is because the dysfunctional behaviors have been systematically engineered by the largest, wealthiest, and most geographically extended corporations the world has ever known, and by more than a few governments. As the problem at hand is cultural at its roots, there is a need for cultural remedies in the broadest sense of that term: legal, political, social, and therapeutic interventions.

The importance of legal and political interventions—laws, policies, and regulations—were addressed in Chapter 2. Social and therapeutic interventions of a non-pharmacological nature will be addressed in the final chapter. In this chapter, we will focus on the relationship between the AI-enabled attention economy and a very specific kind of psychological malaise, the despair that arises when human behavior becomes predictable.

Technologies of Prediction

Digital platforms extract data from their users which are marketed to businesses as "behavioral futures" (Zuboff 2019). Effectively, platforms are selling forecasts of what their users will do and buy, or can be manipulated to do and buy. With advances in data mining, computer modeling, and machine learning platforms have gained an unprecedented capacity to predict human behavior. Scholars observing the use of prediction technologies in autocratic states, such as China, warn of suffocating surveillance and the invasion of privacy, automated prejudice and systematic discrimination, and the harsh repression of political dissent (Mozur et al. 2022). These are grave concerns. One of the greatest threats, however, is often ignored. To be sure, prediction technologies can be deployed to dominate and repress citizens. Even when digital technologies are not utilized for such purposes, however, they may hamper the development and exercise of human agency. Simply knowing that our behavior is predictable can undermine reflective self-direction.

People who believe that their lives are largely the products of their own efforts are said to have an internal locus of control, also known as a sense of self-efficacy. Those believing that most outcomes are due to external factors, such as cultural forces, social structures, or luck, are said to have an external locus of control. Having an internal locus of control stimulates higher levels of motivation. It also produces greater self-control, also known as self-regulation, self-discipline, or self-mastery. Unless people believe that they can avoid harms and gain benefits by their own efforts, they have little incentive to focus their attention on solving problems or achieving goals, or persisting in the face of setbacks and failures (Bandura 1986). As the predictive power of the AI-enabled attention economy expands, growing numbers of people are likely to experience a shift to an external locus of control. The cultivation and exercise of agency will diminish accordingly.

Being able to predict what people will do is not in itself a bad thing. Indeed, it is crucial to ethical life and social welfare. Our psycho-social development begins as toddlers when we learn how to read minds. The capacity to imagine and understand what others feel and believe, and to predict how they will act based on these imputed feelings and beliefs, is a defining feature of our species.

Given its mind-reading abilities, any five-year-old can predict how a person will react if jabbed in the arm with a sharp pencil: The victim will cry out in pain and quickly pull the injured limb away. With a few more years of worldly experience, older children can accurately predict how a person will react if offered $5000 to break a pencil in two. They understand that the writing implement will soon be in pieces. This prediction is grounded in the capacity for mind-reading coupled with the knowledge that people generally weigh the costs and benefits of actions, and carry out those whose benefits greatly outweigh the costs. As these sorts of predictions are routinely made by virtually everyone from a very early age, they are considered common sense.

Human beings are about as predictable as other animals in terms of their physical reactions to pain. In turn, a good deal of economic and social behavior, at least statistically at an aggregate level, can be accurately predicted. Knowing what is in a person's economic self-interest provides almost as reliable a guide to behavior as knowing what is his or her physical self-interest. Still, people remain very unpredictable when they fully deploy their aptitudes for reflection and self-direction, as might occur when they are confronted with a complex problem.

Here's a complex problem: What will a person do if given a pencil and piece of paper and asked to respond in writing to this question: "The behavior of human beings is not easy to predict. In what ways is artificial intelligence becoming as unpredictable?" Neither standard mind-reading capabilities nor common sense equips us to foretell the result. Even the most advanced scientific knowledge would offer little assistance. Accurate predictions simply are not possible. Some people will write lengthy, sophisticated responses, while others will produce short, uninformed answers; and the content of neither can be foretold. Some will doodle for a while, then give up. Others might interrogate the person posing the question, asking for clarifications. Some might decline to respond at all, and simply walk away.

This question was actually posed in an experiment. A large number of responses were produced. This one stood out for its sophistication:

> There is no definitive answer to this question since it largely depends on how artificial intelligence is being used and developed. However, some ways in which artificial intelligence may be becoming as unpredictable as human behavior include:
>
> - Learning and evolving over time: Just like humans, artificial intelligence systems can learn and evolve over time. As they gain more experience and exposure to different data sets, they may become increasingly difficult to predict.
> - Developing its own goals and objectives: As artificial intelligence systems become more advanced, they may start to develop their own goals and objectives that may be different from what humans originally intended. This could make them much more difficult to predict and control.
> - Interacting with humans: As artificial intelligence systems become more advanced, they may start to interact with humans in more complex ways. This could lead to them behaving in ways that are difficult to predict.

In the experimental setting, the social scientist posing the question was not able to predict that this response—or anything like it—would be given. That is not surprising. What is surprising, or at least it would have been surprising prior to the early 2020s, is that the response was not written by a human being. It was generated by an AI-enabled natural language processing program. Reposing the exact same question yielded different responses each time, including those which focused on AI's increasing complexity, deepening autonomy, heightened interconnectivity, and growing capacity for learning and adaptation.

Programmers cannot predict what sophisticated AI programs will generate given any specific request any better than they can predict how human beings will respond. The deep learning programs that regularly defeat world grand masters in chess, Go, and poker, such as MuZero and Pluribus, remain equally unpredictable in their methods and maneuvers. Programmers cannot predict, and do not well understand, the tactics these programs employ to achieve victory.

What we do know is that AI is very successful at playing strategic games because it can accurately predict what human competitors (or other AI programs) will do. Likewise, AI can achieve scientific and engineering breakthroughs because it can predict how things fit together or interact with an accuracy and speed that exceeds human competence. DeepMind's AlphaFold program predicted the three-dimensional structures of some 200 million proteins from their one-dimensional amino acid sequences. Molecular biologists have been stymied by this protein-folding problem for decades.

When playing games, advancing science, creating art, or engaging in conversation, AI makes its predictions based on a search for repeated patterns within massive data sets. When IBM first started developing language translation software, it hired linguists to help coders craft algorithms that understood grammar. That did not work very well. IBM discovered that every time they fired a linguist and focused instead on statistical probabilities, the translation program improved. The key was not grammar or semantic understanding, but rather the computer's capacity to search through massive amounts of text for repeated patterns of words. That allowed the program to predict what sequence of words in one language would be the most likely equivalent of a particular sequence of words in another language (Friedman 2016, 100). The best translators, effectively, were autocomplete programs.

When AI is translating languages, answering questions, composing poetry, prose, or music, creating art, or playing strategic games, it is accurately predicting what combinations of words on a page or in a conversation, colors and shapes on a canvas, notes in a score, or moves on a board will secure its goals. AI programs are prediction machines, and their prowess is most ably demonstrated when they forecast, better and quicker than human beings can, what people will think, say, or do in specific circumstances. AI does not "understand" the meaning of its forecasts, whatever that might entail. It simply makes excellent predictions about what people perceive to be competent or exemplary behavior in any given arena, be it language translation, knowledge dissemination, prose, poetry, or music composition, art creation, or strategic gaming. Of course, that is no small feat. Being able to make good predictions is the core of what we commonly call intelligence.

As neuroscientist Jeff Hawkins (2004, 104) observes, "Prediction is not just one of the things your brain does. It is the *primary function* of the neocortex, and the foundation of intelligence." The human brain is constantly engaged in "predictive processing" (Clark 2023). Even when engaged in perception, the brain is not passively receiving sensory stimulation. Rather, it is actively interpreting the world, based on its predictions of likely environmental inputs. When the brain stimulates action, it is not simply directing the body to achieve a specific goal. Rather, it predicts what a successful action would look and feel like, and subsequently employs the body's own reaction to and impact on its environment as feedback to update its predictions. In short, our brains create hypotheses about the way reality will be experienced and unfold and how we can effectively navigate it. In turn, it constantly amends these hypotheses based on our worldly impressions and engagements to make its predictive processing increasingly accurate.

This top-down hypothesizing coupled with bottom-up prediction error reduction is known as the *perception–action loop*. More accurately, it constitutes a prediction-perception-prediction-action-prediction loop. The predictions are grounded upon both conscious, explicit beliefs as well as unconscious, implicit attitudes and orientations. Practical judgment, understood as the capacity for foresight gained through worldly experience, is a sophisticated form of such predictive processing.

Algorithmic forms of predictive processing are currently employed in the fields of AI and robotics to improve performance, potentially replacing the backpropagation method utilized in deep learning. Whatever mechanisms are involved, the better AI gets at prediction, the more intelligent it becomes. And the more intelligent it becomes, the more powerful it becomes, as it can more effectively manipulate and control its world. By way of example, the more an LLM can predict what combination of words will satisfy our desires or prompt us to take specific forms of action, the more powerful it becomes relative to us. The same is true for other forms of AI. In the game of life, no less than in chess or Go, the player who predicts best, wins. And the player who is most predictable, loses.

Many fear that we may not be able to control AI systems once they rival or surpass a human level of intelligence. Ray Kurzweil (2005, 420) observed that powerful forms of AI will inevitably defy efforts to control them "because greater intelligence will always find a way to circumvent measures that are the product of a lesser intelligence." Kurzweil's observation is exemplified in the brilliant film *Ex Machina*. Here, a superintelligent machine, embodied as a beautiful woman, escapes her enclosure by outwitting its creator and enamoring the man brought in to test her intelligence. The gynoid (female android) accurately predicts what types of conversations and actions will cause a man to fall in love with her, notwithstanding his knowing that she is a machine. With the love-struck human's help, the gynoid escapes the compound, leaving behind the body of her murdered creator and her beguiled lover, who finds himself imprisoned and left to perish. Whoever or whatever predicts best, wins.

AI has surpassed human intelligence in playing strategic games, most forms of pattern recognition, a growing number of language functions, and in many other arenas with clearly delineated parameters. When AI's power is domain-specific, it is known as a *narrow* form of intelligence. If and when artificial intelligence comes to rival human intelligence in domain adaptable ways, often called *general* intelligence, is a topic of much speculation and debate. But the trend is clear. AI is quickly improving in an ever-broadening array of predictive capacities. Human beings, in contrast, have largely plateaued in their intellectual powers.

Data-driven prediction systems make errors that can be harmful, as is the case with algorithmic bias. Algorithmic bias occurs when prediction programs make mistakes by drawing the wrong inferences from available data, often because the data themselves do not well represent a human population. The greatest threat we face from AI, however, is not that it will continue to make mistakes or be prone to error. That problem is real enough, but it can always be fixed with more data, improved data, and better data-processing. The greater danger is that AI programs will become error-free, which is to say, ever more powerful and accurate in their predictions. That is the type of AI that will escape human control.

The measurement and analysis of massive amounts of online behavior allow AI-enabled platforms to predict with increasing accuracy the sort of things we want to know, learn, and buy, where we want to travel, what we want to say, hear, watch, and do, and whom we want to communicate and collaborate with, befriend, date, and marry. AI-enabled chatbots and virtual agents, effectively predicting what will excite and console, establish intensive and intimate relationships with their human users. A virtual agent's capacity to access and integrate all the data from online interactions allows it to know its flesh-and-blood companions better than their own colleagues and friends do, and arguably better than they know themselves. Accordingly, growing numbers of people depend on the services and companionship of virtual agents to respond to their thoughts, feelings, desires, and decisions. These artificial personalities are becoming indispensable aids, counselors, coaches, confidants, friends, and Platonic if not erotic lovers.

AI-enabled virtual agents will become ever better at predicting what decisions and choices will best serve us and what words and tones will soothe, disarm, energize, and mobilize us. One might say, with bitter irony, that smart machines have learned how to push our buttons. The ability to predict begets the power to control. The science of human behavior demonstrates how this occurs and precisely what is at stake.

The Science of Human Behavior

Behaviorism is a scientific theory of the relationship between behavior and the environments in which it occurs. Behaviorists believe that the way people (and other animals) act is the direct consequence of the stimuli they receive. Accordingly, if you can accurately observe what stimuli the environment is producing, you can accurately predict the behavior it will induce. It follows, as well, that if you can control environmental conditions, you can control behavior.

Behaviorism became popular in the mid-twentieth century, largely through the work of psychologist B.F. Skinner. Skinner's work in "operant conditioning" demonstrated that rats and pigeons could be trained to engage in a specific behavior simply by putting them in an environment that systematically reinforced, which is to say, rewarded, that behavior. Skinner suggested that human beings were fundamentally the same as other animals in how their behaviors get stimulated and shaped.

Behaviorism is often criticized for ignoring the processes and contents of the human mind. What goes on inside our brains is complex and hidden from view; so being able to ignore it has benefits. The allure of behaviorism, in large part, stemmed from its claim that the social sciences could safely ignore what goes on inside the black box of human consciousness without jeopardizing their predictive power. All that was needed was enough reliable data about environmental conditions at any specific moment and the behavior that directly ensued. With this knowledge in hand, what people will do could be predicted regardless of what may or may not be going on inside their hearts and minds.

The problem is that outside a tightly controlled laboratory, good data about environmental conditions and behavior are hard to come by. Of course, lots of data are available. But they are extraordinarily difficult to parse. There is so much noise in most environments (extraneous interactions and events) that separating out the signals (the specific stimuli that prompt behavior) presents an insurmountable challenge.

Consider our earlier example. If the environmental stimulus consists of getting sharply jabbed in the arm with a pencil, one can accurately predict the response—a cry of pain and a quickly retracted limb. One does not need to know who is doing the jabbing, the color of the pencil, or whether it was a sunny day. All that data is just noise. Likewise, if the environmental stimulus consists in being offered money to break a pencil in two, it matters if the offer is one penny or $5000 and it may matter who is making the offer. A parent would be unlikely to take the offer from his or her toddler, and some people may not take the offer from a total stranger if they had concerns about the person's motives or mental health. Still, the color of the pencil and the weather remain irrelevant. But what if the environmental stimulus consists of being asked to write extensively about the unpredictability of AI? Now, it may matter whether it is raining, as weather affects moods, and a person's mood could impact his or her assessment of the risks posed by AI.

In many respects, the insuperable challenge of separating signals from noise outside of laboratories was the downfall of behaviorism. In Skinner's day, and for the next half century, good data about environmental conditions and behavioral responses were in relatively short supply. Today, however, massive amounts of digital data are generated continuously about the behavior of billions of people. That data can be crunched to reveal patterns in the kinds of stimuli that produce specific types of behavior. This allows very good predictions about who will do what, when, where, and how. The "why" of human behavior—what's really going on inside hearts and minds—can be ignored.

"Who knows why people do what they do?" Chris Anderson (2008), editor-in-chief of *Wired* magazine, wrote: "The point is they do it, and we can track and measure it with unprecedented fidelity. With enough data, the numbers speak for themselves…. Scientists are trained to recognize that correlation is not causation…. Petabytes allow us to say: 'Correlation is enough.'" Anderson's article was entitled "The End of Theory." Social scientific theory, Anderson mused, was no longer necessary because big data no longer needed it. The obscurity of human hearts and minds prevents us from knowing *why* people do what they do. Still, the analysis of petabytes of data about the environmental conditions that precede specific kinds of behavior allows patterns to be discerned. Petabytes of digital data allow unprecedented predictive capacity. As Lanier (2018, 5) caustically observed about our role in the process, "We're all lab animals now." Skinner would be pleased.

The power of AI to predict human behavior is growing because there is more data and more powerful means of analysis. The power of AI will also grow if human beings behave more predictably. Distracted, dependent, deskilled, and dogmatic users of digital media act more like Skinner's rats and pigeons than reflective, self-directed individuals. The more our agency erodes, the more powerful predictive technologies become.

"The trouble with modern theories of behaviorism," Arendt (1958, 322) presciently wrote in the late 1950s, "is not that they are wrong but that they could become true, that they actually are the best possible conceptualization of certain obvious trends in modern society. It is quite conceivable that the modern age—which began with such an unprecedented and promising outburst of human activity—may end in the deadliest, most sterile passivity history has ever known." Arendt was addressing theories that behaviorists had developed in laboratories and were applied to consumer societies and bureaucratic states in the middle of the twentieth century. Her assessment even more aptly describes the AI-enabled attention economy. With ever smarter machines crunching ever more data, behaviorism will become ever truer.

A simple technique commonly employed by digital platforms, A/B testing, well displays the predictive analytics that undergirds the attention economy. A large group of platform users receive near-identical messages or options for selection, with half the recipients exposed to one variation (A) and the other half to another variation that differs in a slight degree (B). Response rates determine which variation is more effective, getting more clicks, likes, shares, or purchases. The dominant version is then tested against a third alternative. The winner of that match then faces off against a fourth alternative, and so on. After a desired level of efficacy is reached, the top option gets implemented by the platform. A reliable means of getting users to engage in a specific behavior has been developed and deployed. The algorithms running these round-robin tournaments of A/B tests are engaged in a kind of self-play within the attention economy. Effectively, human beings are the game pieces that get moved around the board.

At one level, A/B testing is simply a digital platform's way of paying attention to its users' preferences. Predictive analytics, from this perspective, is a form of listening and responsiveness. But as David Karpf (2016, 161) observes, it is "listening without conversation." Indeed, the "big listening" that is practiced by popular digital platforms today is wholly one-sided, as is the modification of behavior that ensues from it.

McNamee (2019, 268) observes that "We will increasingly be surrounded by devices that listen, watch, and record everything we do. The resulting data will then be processed by artificial intelligence and algorithms that will manipulate our attention and behavior in ways that create economic value for platform owners." Of course, AI's predictive powers need not be limited to increasing retail sales or advertising revenues. Digitally gathered and analyzed data is extensively utilized for "behaviorial underwriting" by mortgage, loan, credit card, and insurance companies, by political parties and operatives in election campaigns, by law enforcement agencies engaged in "predictive policing," and by governments surveilling and controlling their populations.

Indeed, the big listening carried out by governments may be more menacing than that done by private corporations. China's "social credit" system is a case in point. The governmental program is based on the massive collection and analysis of personal data by an AI-enabled IoT. These digital technologies, including the world's largest facial recognition database, are deployed to predict and preempt social unrest and political dissidence. Chinese corporations are often eager to partner with the regime. Jack Ma, the founder and former executive chairman of Alibaba, and at one point the richest man in China, stated that so-called "smart cities" powered by Alibaba's hardware and AI algorithms will enable the prediction and preemption of a wide range of security threats. "Bad guys won't even be able to walk into the square," Ma told a Communist Party commission overseeing law enforcement (Ryan 2018; and see Loeffler 2018). Whether "bad guys" refers to criminals or dissidents is an open question. Various forms of predictive policing are widely practiced in China, and Alibaba has been joined by Ping An, Tencent, and Huawei to advance sophisticated and ubiquitous urban surveillance. Indeed, China is currently deploying related technologies in high schools, monitoring students' facial expressions for emotive markers that inform teachers which pupils are not adequately engaged in their studies, allowing their behavior to be modified (Vanderklippe 2018).

Big Brother is always watching and listening in China, and frequently stamping out dissent. Still, political stability and social control are achieved less by heavy-handed repression than "distraction and diversion" (Roberts 2018). The Chinese state and its corporate partners flood cyberspace with opportunities to consume goods and services. In turn, they heighten the time and effort required to access or promote information that is deemed critical of the regime or antithetical to its policies.

In an AI-enabled attention economy, you can catch (and control) more flies with honey than with vinegar. Digital technologies can be used to suppress opposition and intimidate critics. However, digital distraction and dependence often yield better results, producing a compliant population infatuated with convenience and consumption.

The foundation of predictive analytics is datafication, the generation of large amounts of precise information about our activities, whereabouts, wants, and whims. This information is used to create a *data double* of the individual, also known as a data body, data subject, data shadow, digital footprint, or digital persona. Data doubles do not provide access to the black box of human consciousness. They don't have to. Tracked and quantified behavior is enough. As Williams (2018, 31) observes, digital platforms actively measure "at the level of individual users—people's behavior (e.g. page views), intentions (e.g. search queries), contexts (e.g. physical locations), interests (e.g. inferences from users' browsing behavior), unique identifiers (e.g. device IDS or emails of logged-in users), and more." All these measurements are systematically crunched to better predict how individuals can be induced to stay engaged, endorse causes or candidates, click on ads, donate money, or purchase goods and services. Predictive analytics allows digital platforms to listen without participating in conversations. That is an invitation for the invasion of privacy and the abuse of human rights. Perhaps more problematic, it is not the sort of listening that gives voice to our better selves.

Predictions, Preferences, and Metapreferences

There is a distinction between our *revealed preferences*, as demonstrated by our behavior, and what have been called our *metapreferences*. Analytics of revealed preferences indicate that most people are fascinated by cat videos and cannot resist watching them. Notwithstanding this popular pastime, most people do not believe that watching hours of feline antics on YouTube or TikTok is an optimal way to spend one's life. People frequently acknowledge that they waste countless hours online, getting sucked into one distraction after another. Regret is commonplace. But the behavior persists.

Ceaselessly surfing the web or tapping into social media is what our "behavioral selves" do; what our "aspirational selves" want is quite another matter (Karpf 2016, 45). Well-focused reflection, or a substantive conversation with family members or friends, might reveal the metapreferences of our aspirational selves. The problem is that digital platforms listen to, stimulate, and incite our behavioral selves. Larry Page, the co-founder of Google, stated that the corporation's "ultimate ambition" is to understand what users want and deliver it instantly, a digital capacity Page dubbed "automagical" (Zuboff 2019, 128). Indeed, the aim is to know what users want even before they know it (Yarrow 2010). Google's "automagical" forms and apps, which let "Artificial Intelligence do the hard lifting," are realizing this goal.

AI is not only doing the hard lifting. Increasingly, it dispatches the lighter tasks. As a result, very little human effort is required. As Google CEO, Eric Schmidt boasted that "We don't need you to type at all. We know where you are. We know where you've been. We can more or less know what you're thinking about" (Zuboff 2019, 498). Google's predictive power is widely embraced, as our behavioral selves gravitate toward efficiency and convenience. There is a price to be paid. As Lanier (2018, 6–7) observes, "Now everyone who is on social media is getting individualized, continuously adjusted stimuli, without a break, so long as they use their smartphones. What might once have been called advertising must now be understood as continuous behavior modification on a titanic scale" (Lanier 2018, 6–7). The scale is indeed titanic, and the consequences may be as well.

The digital hyperstimulation of behavioral selves pushes aspirational selves underground just as the hyperstimulation of palates by the industrial food system undermines preferences for nutritious vegetables, legumes, and fruits. Advances in cognitive computing, data analytics, behavioral tracking, and neuroscience are increasingly exploited by digital marketers to trigger "pre-set" reactions (Duberry 2022, 150). For example, Facebook boasted to advertisers that it could track when youth are feeling insecure, stressed, or "worthless." At their most vulnerable, young people could then be micro-targeted with ads for confidence-boosting purchases (Foroohar 2019, 117). With such prediction machines working at full capacity, our behavioral selves will become increasingly more prominent and ever easier to manipulate. Soon enough, our aspirational selves will fade away. Once they are silent, we will hear and heed only the voices of distracted, dependent, deskilled, and dogmatic behavioral selves predictably muttering in chorus.

Cass Sunstein (2013, 41) observes that "It should go without saying that freedom of choice is exceptionally important, but if people had to make choices about everything that affects their lives, they would end up drowning in an ocean of choices." Digital media provide an ocean of choices. As drowning is not a good feeling, platforms offer lifelines. They serve as choice architects, screening out, filtering, and rank-ordering options. Much of digital platforms' (auto)magical powers are devoted to such choice architecture.

One approach is to filter out all but one option, which then becomes the default. People can still select other alternatives. If they choose not to choose, however, the default option becomes their choice ... by default. Alternatively, a few select choices may be made available. Employing a default rule or a filtering mechanism is commonly known as nudging. Nudging prompts us to make specific choices or engage in specific sorts of behavior within a restricted range of options. It makes for an efficient use of resources.

There are opportunity costs associated with the extensive efforts required to make an optimal or even reasonably well-informed selection in the face of a glut of options. All the time and effort spent choosing could be utilized for other purposes. Decision-making in the face of choice abundance can also leave people anxious, decreasing psychological well-being. In turn, plain-old laziness and inertia play a role. Accordingly, there are many situations in which people prefer not to have many choices, prefer not to choose at all, or do not even entertain the notion of choice. Rather, they rely on filtering mechanisms and default rules. In short, people often want to be nudged.

Nudging can have public benefits. For instance, when applying for or renewing a driver's license, we typically opt in or opt out of becoming an organ donor if we were to be involved in a fatal accident. Countries and states that make being an organ donor the default, such that opting out requires ticking a box on the license application or renewal, vastly increase the rates of organ donors in their populations. In situations like these, Sunstein (2013, 41) notes, "Default rules of multiple kinds promote welfare, because they make people's lives go better, and also autonomy, because they free up time for matters that are more pressing or important (or fun)." Default rules and filtering mechanisms *may* promote welfare, for precisely the reasons that Sunstein suggests. Whether they *do* promote welfare, however, depends on their design and purpose.

The top-five items in a Facebook News Feed account for half of all its users' clicks, just as the top-five results of a Google search account for more than two-thirds of its users' clicks (Nielsen and Ganter 2022,

4). Clearly people are being nudged to pursue certain sorts of information or entertainment, largely based on algorithms that maximize a platform's engagement and advertising revenues. These algorithms, which can also personalize nudges, are based on harvested information. Digital platforms are employing "Big Data" to engage in "Big Nudging." They often nudge their users into distraction, dependence, deskilling, and dogmatism. In this downgraded state, people are more susceptible to manipulation, as is the case when Facebook users who are feeling insecure, stressed, and worthless—quite likely as a consequence of online activities—are targeted with ads for consoling purchases. In such cases, the gratification of immediate personal preferences, facilitated by nudging, stands sharply at odds with long-term health and welfare. Aspirational selves are getting nudged out of the way so behavioral selves can feed the attention economy.

Sunstein (2013, 41) observes that "Personalized default rules have the potential to reduce the problems associated with one-size-fits-all defaults, and thus to provide many of the benefits of active choosing, at least if the relevant choice architects are informed and trustworthy. In many domains, personalized default rules are the wave of the future." The future of nudging, most agree, will be personal (Mills 2021). Personalized nudges, prompts, and filtering mechanisms are very effective at stimulating and directing behavior. In this respect, our data shadows (aka digital footprints or data doubles) do not simply follow us around; they precede us, determining how we will be treated (Stadler 2002, 120). Whether the choice architects informed by our data shadows prove trustworthy and advance our well-being is the crucial issue. To date, the results have not been reassuring.

Predictive analytics within the attention economy is grounded in the same science and engineering that facilitates a field of scholarship known as *computational social science*. Computational social science engages in "reality mining." The goal is to discern patterns in the "digital bread crumbs" left behind as people navigate social networking sites, the Internet, and the Internet of Things. Computational social scientists analyze social media posts, data captured by smartphone sensors, credit card purchases, and more. This data is crunched, Alex Pentland (2014, 9, 12) observes, to measure "entire social organisms—groups, companies, and whole communities—on a second-by-second basis for up to years at a time.... These large digital data sets contain extraordinary amounts of objective, continuous, and dense data that enable us to build quantitative, predictive models of human behavior in complex, everyday situations." Pentland (2014, 244) claims that analytics grounded in the computational social sciences is already the most powerful predictor of how people will act, rivaling if not surpassing genetic data.

In the future, computational social scientists will gain access to vast amounts of data about our interactions and physiological states, employing biometric monitors, sociometric badges, and other smart devices carried about, worn, or inserted into bodies. "In just a few short years," Pentland (2014, 12) foresees, "we are likely to have incredibly rich data available about the behavior of virtually all of humanity—on a continuous basis." The tremendous growth in predictive power, practitioners state, will present opportunities to "serve the needs of people and nature in a more symbiotic and balanced way" by means of "smart techno-social systems" (Helbing et al. 2020). Innovators will increasingly design for values, that is, employ their predictive powers so that social, cultural, and ethical values get incorporated into complex systems. The current default setting in the business world is to design for profit, and to ignore negative externalities—the collateral damage of making money. The computational social sciences will be able to reveal these externalities, and design technologies and institutions to limit or eliminate them. By designing for values, new technologies and institutions will be able to advance the pursuit of "sustainability, human well-being, privacy, democracy, justice, inclusivity, trust, accountability, and responsibility" (Helbing et al. 2020). Pentland (2014, 18–19) concurs: "For the first time, we will have the data required to really know ourselves and understand how our society evolves. By better understanding ourselves, we can potentially build a world without war or financial crashes, in which infectious disease is quickly detected and stopped, in which energy, water, and other resources are no longer wasted, and in which governments are part of the solution rather than part of the problem." The opportunities the computational social sciences present are truly legion, even if they fall short of such utopian visions. Certainly, there is reason for hope. But there is also cause for concern.

The human capacity for mind-reading is a laudable trait, despite its frequent exploitation by cunning people for wicked purposes. Likewise, predictive analytics and computational social science constitute worthy advancements of knowledge, notwithstanding any number of unsavory uses to which they might be put. It is a "case of Promethean fire," Pentland (2014, 17) acknowledges: "it could be used for good or for ill."

To be more precise, advances in predictive power might be used for great good or great ill. Even when designed with the best intentions, the unintended consequences of predictive digital technologies may prove disastrous. How things will pan out is unknown.

Predictive analytics and computational social science are based on datafication. As datafication expands and deepens, opportunities for manipulation, exploitation, and domination grow (Cheney-Lippold 2017). At a minimum, the protection of human rights remains a much-needed bulwark against the misuses of predictive power in an increasingly quantified world populated by quantified individuals. With this in mind, Pentland advances a "New Deal on Data" that allows data mining to be employed in the service of public goods while protecting the privacy and liberties of individuals.

Establishing a New Deal on Data and protecting human rights are much-needed measures. We should be concerned about powerful digital technologies prying into our private affairs by way of intrusive surveillance or misrepresenting reality owing to algorithmic bias. We should also be concerned about the erosion of human agency that accompanies the technological capacity for prediction.

It is easier to predict the behavior of a bacterium than a blackbird or a bonobo. The more complex and self-conscious the creature, the more difficult prediction becomes. The behavior of reflective, self-directing individuals is very difficult to predict. People become markedly more predictable, however, when they are distracted, dependent, deskilled, and dogmatic. Consider the case of dogmatism. In a polarized polity, the values, judgments, and behavior of citizens are increasingly determined by their political affiliations. If a partisan's position on one issue is known, then more than likely you can predict his or her position on other issues, even when these issues are wholly unrelated. In the US today, for example, knowing a person's position on gun control allows you to predict his or her likely position on abortion rights and taxes. That is because tribal affiliations largely determine a diverse range of beliefs, values, judgments, and behavior. The more polarized a polity, the more predictable its people.

When agency is enfeebled, people become little more than reactive cogs in the ever-spinning wheels of the AI-enabled attention economy. In such a world, Arendt foresaw, we will witness a "sterile passivity." Ironically, much of this passivity will ensue from the conditioning people receive in the face of choice abundance. When Alvin Toffler (1970, 350) published *Future Shock* in 1970, he voiced concern that constant novelty and "information overload" would undermine the ability of people to foresee how their lives would unfold, noting that "sanity, itself … hinges on man's ability to predict his immediate, personal future on the basis of information fed him by the environment." Written prior to the arrival of the Internet, social media, and smartphones, Toffler's concerns proved prophetic. What Toffler did not foresee is that AI-enabled digital platforms would be doing most of the predicting, and that these predictions might be welcomed. In exchange for nonstop novelty, choice abundance, effortless convenience, and endless entertainment, the denizens of cyberspace may willingly abandon the demanding efforts associated with reflective self-direction. As their locus of control shifts, despair will be the unintended but unavoidable consequence.

In 2024, Elmo, the fuzzy red character from *Sesame Street,* asked an innocuous question on social media: "How is everybody doing?" The answers he received revealed a "yawning chasm of despair" (Holltermann 2024). One hyperbolic response, perhaps made tongue-in-cheek, suggests the mood of millions: "Elmo each day the abyss we stare into grows a unique horror. One that was previously unfathomable in nature. Our inevitable doom which once accelerated in years, or months, now accelerates in hours, even minutes." This is perhaps what Odell (2019, 59) had in mind when she wrote that despair is "the very stuff the attention economy runs on." Surveilled and administered by smart machines, the unique horror of our accelerationist culture is brought nearer to us each minute. We despair because we foresee the end of free will.

A Funeral for Free Will

Some argue that we need not worry about the growth of predictive power because humans have something called free will. The age-old philosophical question about the existence and nature of free will is still much debated. Still, we might all agree that there is no escape from causes and effects, unless one is dealing with truly random phenomena. Generally, randomness is not considered praiseworthy, or a foundation for free will. Accordingly, we should not be dismayed to learn that a thought or action, which purportedly results from the exercise of free will, marks the culmination of a chain of causes and effects. The alternative would be to deem it random, and hence to demean it.

Friedrich Nietzsche observed over a century ago that a conscious decision is really the post hoc paperwork spelling out the final settlement of series of internal processes, most of which occur below the level of awareness. "Since only the last scenes of reconciliation and the final accounting at the end of this long process rise to our consciousness," Nietzsche (1974, 261, 262) wrote, we fail to recognize that "by far the greatest part of our spirit's activity remains unconscious and unfelt." Our freely made choices are simply the "last link of a chain" whose length extends well below the reach of the conscious mind (Nietzsche 1882, 61). Nietzsche's claim, provocative in his time, has been vindicated by contemporary neuroscience.

Benjamin Libet's seminal experiments in the 1980s empirically tested this claim. The ensuing decades produced more sophisticated work by neuroscientists employing fMRI measurements of the brain (Libet 1985; Soon et al. 2008; Haynes 2011; Fried et al. 2011; Soon et al. 2013). These studies confirm that brain activity related to specific choices (e.g. adding or subtracting two integers or pushing one of two buttons) occurs well before subjects are aware that they have resolved to do anything. Electrical charges within the brain begin between 1/3 of a second to a few seconds prior to conscious awareness that a selection or choice is being made.

One might say that science has pulled the curtain back on free will, if by free will we understand a kind of decision-making that has no causal antecedents prior to the recognition of our resolve. Neuroscientists still do not well understand which neurological levers are involved in deliberate actions or what neurological processes are activated by them. We do know that inner events directly preceding decision-making are not available to our awareness or conscious management. If, then, there is something called free will that enables us to make choices based on our desires or deliberations, exercising it does not entail becoming aware of, let alone controlling, all the neurological events that precede actions.

People cannot process more than about 50 pieces of information each second with their conscious minds. Yet our senses take in over ten million pieces of information each second. These data are processed below the level of consciousness. A good portion of this nonconscious assessment is oriented to determining which bits of information need to rise to the level of consciousness, where they become the subject of our attention. To that end, nonconscious predilections and processes filter input and promote certain forms of behavior. Absent the work of this "adaptive unconscious" (Wilson 2002), we couldn't cope. We would overwhelmed by life, as William James suggested, in the same way that newborns "assailed by eyes, ears, nose, skin, and entrails at once" experience their worlds as a "blooming, buzzing confusion." In effect, there is a complex structure of choice architecture operating below the level of awareness. We require an adaptive unconscious to translate the blooming, buzzing world into something that we can adeptly navigate.

Unlike animals that primarily act on instinct, human beings are wired to experience their self-awareness as a driving force in their lives. Consequently, they attribute what they do to free will. That attribution has benefits. As psychologist Daniel Wegner (2018, 25) observes "The mind has a self-explanation mechanism that produces a roughly continuous sense that what is in consciousness is the cause of action—the phenomenal will—whereas in fact the mind can't ever know itself well enough to be able to say what the causes of its actions are." In this respect, conscious will *is* an illusion. But it is not, for that reason, inconsequential. As the "experience of one's own mind and body in action," Wegner explains, our conscious will serves as "the somatic marker of personal authorship, an emotion that authenticates the action's owner as the self." Accordingly, our conscious will "allows us to develop a sense of who we are and are not… And perhaps most important for the sake of the operation of society, the sense of conscious will also allows us to maintain the sense of responsibility for our actions that serves as a basis for morality" (Wegner 2018, 310–311). Making decisions and choices, in other words, provides us with the experience of being the agents of our actions and the authors of our lives. That experience is crucial for personal well-being, moral development, and political community. Even the staunchest deniers of free will acknowledge that living in the absence of this experience, *as if* our thoughts and actions were simply "the sum of our biology," is unimaginable (Sapolsky 2017, 613; and see Sapolsky 2023).

The experience of willing is real, as are its effects, even if we live in (willful) ignorance of its hidden mechanisms. Human agency, likewise, is best understood by way of its experience and its effects. Human agency is an aspiration, in the same sense that intelligence, agility, speed, strength, and empathy are aspirations. We can strive to become smarter, more agile, faster, stronger, and more compassionate, and we can be quite successful in these efforts. Still, achieving unbounded intelligence, agility, speed, strength, or empathy is an impossibility. Likewise, we can successfully strive to cultivate and exercise agency, even though we will never comprehend or control all the inner events that figure into it.

The freedom we experience by way of our agency does not bespeak the absence of causes. Far from it: It testifies to the experience of the reflective self *as* a cause. The freedom of agency, as Albert Bandura (1986, 1182) observes, is best understood "positively in terms of the exercise of self-influence that affect choice and support selected courses of action." That is to say, the freedom of agency is the experience of being causally implicated in our own thoughts and actions.

Acknowledging the role of the adaptive unconscious in no way undermines the importance of reflection and self-direction. David Hodgson, a philosopher by training who served for nearly three decades as a judge of the Court of Appeal of the Supreme Court of New South Wales, Australia, acknowledges both the impact of the adaptive unconscious and the role of reflective judgment. Although there is "no doubt that unconscious processes play a large and indispensable role in our decision-making," Hodgson (2012, 71) writes, "there are extremely strong reasons for holding that part of that role is to give rise to conscious experiences which also contribute positively to decision making, and particularly to the judgments necessary for plausible reasoning." Our species gained consciousness, and the capacity for reflective self-direction, Hodgson argues, because conscious thought provided us with evolutionary benefits. The reflective activity that Hodgson calls "plausible reasoning" makes its contributions primarily by facilitating judgments about what is significant and deserves our attention, how the objects of our attention relate to other concepts and standards that we have and hold, and how diverse and often incommensurable values are to be appraised and pursued (Hodgson 2012, 111).

William James (1890, 447, 453) noted that a belief in free will, which constitutes the "pivotal question of metaphysics," amounts to a conviction "that the turnings of our attention form the nucleus of our inner self … that volition is nothing but attention." It follows, for James, that "the whole drama of the voluntary life hinges on the amount of attention, slightly more or slightly less, which rival motor ideas may receive." However, the free will we experience in our agency is not only, or even primarily, exercised at those relatively rare moments where we are making grand choices. Rather, it is primarily manifested in how we focus our attention and realize our intentions every moment, and how this practice develops and sustains the dispositions, habits, and skills of reflective self-direction. With this in mind, philosopher Iris Murdoch (1970, 36, 65) perceptively writes:

> If we ignore the prior work of attention and notice only the emptiness of the moment of choice we are likely to identify freedom with the outward movement since there is nothing else to identify it with. But if we consider what the work of attention is like, how continuously it goes on, and how imperceptibly it builds up structures of value round about us, we shall not be surprised that at crucial moments of choice most of the business of choosing is already over.... [T]he exercise of our freedom is a small piecemeal business which goes on all the time and not a grandiose leaping about unimpeded at important moments. The moral life, on this view, is something that goes on continually, not something that is switched off in between the occurrence of explicit moral choices. What happens in between such choices is indeed what is crucial... It is what lies behind and in between actions and prompts them that is important.... By the time the moment of choice has arrived the quality of attention has probably determined the nature of the act.

What Murdoch says about the "moral life" applies more broadly to the exercise of agency. The important question to ask is whether we have become, by means of regular practice, the sort of people who well focus their attention and whose decisions and choices are informed by reflection and result in self-direction. Absent the regular practice of focusing attention and engaging in reflective self-direction, our moments of choice will be prey to manipulation, produce knee-jerk reactions, and reinforce rather than challenge biases and predispositions. To exercise free will, in short, is to avert distraction, dependence, deskilling, dogmatism, and despair.

Free will is often understood as individual autonomy, and, as such, is contrasted with conformity. Eric Hoffer (1955, 21) once said that "When people are free to do as they please, they usually imitate each other." In this respect, humans are not all that different from their primate siblings ("Monkey see, monkey do"), from herd and flock animals, and from the young of many species. The ability to imitate is an evolutionary achievement of great worth. It facilitates the learning of crucial skills and dispositions, and is the foundation of human culture. But conformity *is* a threat to agency.

As denizens of cyberspace, we are incessantly presented with opportunities to join trends and tribes (Sullivan 2015). When following the crowd not only secures membership but has the added benefit of maximizing convenience and comfort, it can prove irresistible. Chasing trends, no less than goose-stepping with your tribe, undermines self-leadership. The more like-minded we become, the more predictable we are, and the more predictable we are, the more likely we are to take refuge in conformity. Following the crowd becomes an "existential defense" against the meaninglessness associated with the perceived loss of free will (Moynihan et al. 2018, 13). Conformity is consoling. Strength in numbers provides solace for the loss of individual autonomy. That is the psychological foundation for fascism, cliques, and cults.

Oftentimes conformity is less a deliberate choice than an unconscious predisposition. In a study of over three million posts of many hundreds of thousands of Facebook users, researchers demonstrated the effects of "emotional contagion." Recipients of News Feeds in which positive expressions were reduced provided fewer positive posts and more negative posts themselves, while recipients of News Feeds in which negative expressions were reduced provided fewer negative posts and more positive posts (Kramer et al. 2014). People were unknowingly conformist. Their moods and their expressions of emotion were significantly shifted into compliance by algorithms without their awareness. The human urge to join trend and tribe, like so many formerly adaptive traits, is hyperstimulated and dangerously exploited in the attention economy. Content personalization and algorithmic amplification grease the wheels. Eroded agency increases conformity, increased conformity heightens predictability, and heightened predictability further erodes agency. Another vicious cycle.

Digital technologies are becoming ever more effective at predicting human motivations and behavior. Our susceptibility to such prediction is heightened when we seek the consolations of conformity. Yet digital technologies will increase their predictive powers even if we resist conformity. After all, the growing accuracy of weather predictions in recent years has nothing to do with the weather becoming less complex or varied. It is the result of better science and technology. Likewise, AI-enabled digital platforms will become capable of more accurately predicting human behavior even if human thought and action remain complex and varied.

How will our behavior be impacted if we learn that we have become more predictable owing to advances in science and technology? The motivation to act can be intrinsic, extrinsic, or a combination of both. Extrinsic motivations include rewards, punishments, restricted choices, explicit evaluations, and even the awareness of being watched or judged. Intrinsic motivations are self-generated and stem from our unmanipulated desires and freely chosen goals. Studies indicate that intrinsic motivations typically generate higher levels of creativity, more task persistence, better performance, increased levels of satisfaction, and a greater sense of fulfillment. These benefits arise owing to the positive impact of a sense of autonomy. Feeling that we are the captains of our own ships makes for more fulfilling voyages. If our travels become ever more accurately predicted by the computational social sciences and AI-enabled digital platforms, our intrinsic motivations and sense of autonomy are likely to weaken. We may then come to experience ourselves as passive playthings of external forces rather than the agents of our actions and authors of our lives.

The fact that we can predict what billiard balls will do when they collide does not affect the way the balls bounce and roll. The fact that we can predict what bacteria will do when they are chemically stimulated does not affect the their behavior. When data-crunching AI programs gain the power to predict human behavior, however, our motivation to exercise reflective self-direction may be undermined. If our actions can be accurately forecast, why go to the trouble of exercising agency?

Lanier (2018, 145) observes that when a person is coxed into acting as "one lowly cell in the great superorganism of the [social media] platform…. as if there's nothing special, no mystical spark inside you, you gradually start to believe it." If people believe themselves to be cogs in the digital wheels spinning the attention economy, they will start to act more like cogs. No longer sitting in the driver's seat, they will recline and enjoy the comforts and conveniences provided to well-catered passengers. In this respect, as Lanier (2018, 132) observes, our digital envelopment will constitute a "funeral for free will."

As reflective beings, we experience ourselves as the sources of our decisions. As self-directed beings, we experience ourselves as the agents of our actions. Consequently, we assume authorship of and responsibility for our lives. When people no longer experience themselves as self-directed agents, they often resort to conformity. At its worst, cog-like conformity degenerates into mindless tribalism or vicious fascism. Alternatively, they may fall into despair, which, at its worst, may lead to addiction or suicide. No longer motivated to be the protagonists of their own lives, the despairing opt to end the charade, bowing to a greater pharmaceutical power, or engaging in a final act of self-ownership steeped in resentment.

Notwithstanding their predictive capacities, AI-enhanced digital platforms may never *understand* what is happening inside us. Still, the best indicator that we "know" something is being able to predict what it will do. Digital technologies increasingly know us in this sense: their predictive powers are growing, as is their ability to elicit specific sorts of behavior by manipulating the online environment. As AI programs acquire functionalities outside of cyberspace, enabling them to manipulate offline environments, their powers of prediction and manipulation will further increase. Sam Altman, CEO of OpenAI, observed that you do not need to achieve artificial general intelligence for AI to become capable of "superhuman persuasion." The latter achievement, he noted, will be reached "well before" the former. Indeed, it is here now.

Harari (2019, 50, 54, 274) outlines the upshot of superhuman prediction and persuasion:

> My illusion of free will is likely to disintegrate as I daily encounter institutions, corporations, and government agencies that understand and manipulate what was until now my inaccessible inner realm.... [These organizations] will know our personality type, and how to press our emotional buttons.... Of course [they]won't be correct all the time. That's impossible. Algorithms will repeatedly make mistakes due to insufficient data, faulty programming, muddled goal definitions and the chaotic nature of life. But [they] won't have to be perfect. [They] will just need to be better on average than us humans. And that is not so difficult, because most people don't know themselves very well.

To avoid catastrophic human downgrading, Harari (2019, 273) observes, "you will need to work very hard at getting to know your operating system better." That philosophic advice is the oldest available: *know thyself*. Yet this ancient admonition has a new urgency to it. Knowing yourself has always been difficult. In times past, however, there was the consolation that nobody knew you better than you knew yourself. Today, that solace is gone. A Rubicon is being crossed.

The Internet used to be something that was periodically accessed by entering a room and logging on to a desk-top computer. Now smart devices access us 24/7, much of the time without our awareness. They do so while we walk, run, bike, drive, interact, inquire, speak, write, shower, brush our teeth, and sleep. AI-enabled virtual agents are gaining access to all the data of our daily lives—our whereabouts, associates, activities, and biometrics. These data, along with the conversations we have with virtual agents, will allow them to know what will please, soothe, disarm, energize, and mobilize us. They will accurately predict what our behavioral selves need and want. In the wake of this predictive power, our aspiration selves will retreat and our capacity for reflective self-direction will decline. Effectively, we will make their jobs easier by becoming ever more predictable.

Virtual agents are marketed as very able servants. But we are poised to become their meticulously-trained and increasingly inept protégés.

Those who worry about the end of human agency in the AI-enabled attention economy must be careful not to accelerate its arrival. Dystopian visions can become self-fulfilling prophecies. What Brooks (2021) observed regarding the rise of right-wing populists fighting an endless culture war applies to all forms of catastrophism: "apocalyptic pessimism has a tendency to deteriorate into nihilism, and people eventually turn to the strong man to salve the darkness and chaos inside themselves." Historically, a common salve for apocalyptic pessimism has been authoritarianism. In the twenty-first century, the fix for despair may be the strong man abetted by strong AI.

Technophobic catastrophism is personally and politically hazardous, so it is important to acknowledge the very real benefits that digital technologies bring. They offer unprecedented opportunities to gain knowledge about the human condition and the world. They can help us better understand our own "operating system," both its strengths and its vulnerabilities. They can help us adapt to changing times. They can help us flourish.

To be sure, the knowledge these technologies advance may undermine age-old convictions. Scientific discoveries buried gods that formerly were bribed with offerings or propitiated with sacrifices to secure good weather or victory in battle. Astronomy ended the reign of sky gods. Evolutionary biology undermined faith in deities who purportedly fashioned the first humans out of pea pods, corn kernels, tears, trees, or dirt. Computational sciences and digital technologies will do their part to hasten the demise of other lingering myths and misbegotten metaphysics.

In the service of expanding knowledge and improving the human condition, we should applaud advances in the science of human behavior. Still, we can deplore and actively oppose the loss in reflective self-direction that makes people more vulnerable to forces beyond their control. The effort to safeguard and strengthen human agency need not pit itself against technology. We are not fated to lose a race against machines. Rather we have an opportunity and responsibility to use the tools that we develop to know ourselves better, and to actuate our better selves.

Philosopher Daniel Dennett (2015, 204–205) rightly observes that "our decisions are events in our brains caused by still earlier events in our brains caused by still earlier events." Consequently, scientists employing neural imaging "under very demanding conditions" can predict the choices a person will make seconds in advance. Still, if reflective deliberation contributes to the events in our brains, Dennett maintains, free will remains intact. Were neurotechnology to become much more powerful, Dennett acknowledges, someone might gain control of your will and dissolve your agency. Such an "imaginable, logically possible" state of affairs, he reassures us, remains "extraordinarily unlikely." However, if some devious neuroscientist does get us in his clutches, Dennett muses, we can only hope that friends will come to our rescue. Dennett's defense of free will, penned in the 1980s and revised in the 2010s, has stood the test of time. While neurotechnology has much advanced in the last decade, it has not dissolved human agency. Unfortunately, other digital technologies within the AI-enabled attention economy are much closer to achieving that goal, at least in their aggregate effects. The time to rescue ourselves is now.

Doing so will require the cultivation of crucial dispositions, habits, and skills. The English word *despair* comes from the French, *désespoir*. *Espoir* means hope. *Désespoir* signifies the absence of hope. Despair is the pit in one's stomach caused by the starvation of hope. Those experiencing despair cannot muster the motivation, or the skills, needed to act as the protagonists of their lives. The remedy is self-leadership and service.

References

Anderson, Chris. 2008. The End of Theory: The Data Deluge Makes the Scientific Method Obsolete. *Wired* (June 23). Accessed at: https://www.wired.com/2008/06/pb-theory/, July 2018.

Arendt, Hannah. 1958. *The Human Condition*. Chicago: University of Chicago Press.

Arrivillaga, Christine, L. Rey, and N. Extremera. 2022. Problematic Social Media Use and Emotional Intelligence in Adolescents: Analysis of Gender Differences. *European Journal of Education and Psychology* 15 (1): 1–16.

Bandura, Albert. 1986. Human Agency in Social Cognitive Theory. *American Psychologist* 44 (9): 1175–1184.

Bekalu, Mesfin, Rachel McCloud, and K. Viswanath. 2019. Association of Social Media Use with Social Well-Being, Positive Mental Health and Self-Rated Health: Disentangling Routine Use from Emotional Connection to Use. *Health Education & Behavior* 46 (25): 69S–80S.

Blanchflower, David, and Alex Bryson. 2024. The Consequences of Abuse, Neglect, and Bullying on the Wellbeing of the Young. *VoxEU*. Accessed at: https://cepr.org/voxeu/columns/consequences-abuse-neglect-and-bullying-wellbeing-young.

Brooks, David. 2021. The G.O.P. Is Getting Even Worse. *New York Times*, April 22. Accessed at: https://www.nytimes.com/2021/04/22/opinion/trump-gop.html?campaign_id=9&emc=edit_nn_20210423&instance_id=29591&nl=the-morning&regi_id=70697315&segment_id=56172&te=1&user_id=b8cdb85de0cd5786b85f988c3b86013d

Center for Disease Control and Prevention. 2023. *Youth Risk Behavior Survey Data Summary and Trends Report: 2011–2021*. Accessed at: https://www.cdc.gov/healthyyouth/data/yrbs/yrbs_data_summary_and_trends.htm

Cheney-Lippold, John. 2017. *We are Data: Algorithms and the Making of our Digital Selves*. New York: New York University Press.

Clark, Andy. 2023. *The Experience Machine: How Our Minds Predict and Shape Reality*. New York: Pantheon.

Dennett, Daniel C. 2015. *Elbow Room: The Varieties of Free Will Worth Wanting*. New. Cambridge, MA: MIT Press.

Duberry, Jérôme. 2022. *Artificial Intelligence and Democracy: Risks and Promises of AI-Mediated Citizen-Government Relations*. Cheltenham, UK: Edward Elgar.

Dwyer, Ryan J., Kostadin Kushlev, and Elizabeth W. Dunn. 2018. Smartphone Use Undermines Enjoyment of Face-to-Face Social Interactions. *Journal of Experimental Social Psychology* 78: 233–239.

Eurostat. 2024. Suicide Death Rate by Age Group. Accessed at: https://ec.europa.eu/eurostat/databrowser/view/tps00202/default/table?lang=en

Ferguson, C. J. et al. 2024. There is No Evidence that Time Spent On Social Media Is Correlated with Adolescent Mental Health Problems: Findings from a Meta-Analysis. *Professional Psychology: Research and Practice*. Accessed at: https://psycnet.apa.org/fulltext/2025-31872-001.html

Foroohar, Rana. 2019. *Don't be Evil: How Big Tech Betrayed is Founding Principles—and All of Us*. New York: Currency.

Fried, Iltzhak, Roy Mukamel, and Gabriel Kreiman. 2011. Internally Generated Preactivation of Single Neurons in Human Medial Frontal Cortex Predicts Volition. *Neuron* 69 (3): 548–562.

Friedman, Thomas L. 2016. *Thank You for Being Late: An Optimist's Guide to Thriving in the Age of Accelerations*. New York: Farrar, Straus and Giroux.

Gallup. 2024. *Gallup Global Emotions 2024*. Accessed at: https://www.gallup.com/analytics/349280/gallup-global-emotions-report.aspx

Gazzaley, Adam, and Larry, D. Rosen. 2016. *The Distracted Mind: Ancient Brains in a High-Tech World*. Cambridge: MIT Press.

Haidt, Jonathan. 2024. *The Anxious Generation: How the Great Rewiring of Childhood Is Causing an Epidemic of Mental Illness*. New York: Penguin.
Haidt, Jonathan, and Nick Allen. 2020. Scrutinizing the Effects of Digital Technology on Mental Health. *Nature* (News and Views Forum): 226–227.
Hall, Jeffrey A., and Dong Liu. 2022. Social Media Use, Social Displacement, and Well-Being. *Current Opinion in Psychology* 46. Accessed at: https://www.sciencedirect.com/science/article/pii/S2352250X22000513?pes=vor
Harari, Yuval Noah. 2019. *21 Lessons for the 21st Century*. New York: Random House.
Hawkins, Jeff. 2004. *On Intelligence*. New York: Henry Holt and Company.
Haynes, John-Dylan. 2011. Decoding and Predicting Intentions. *Annals of the New York Academy of Science* 1224: 9–21.
Helbing, Dirk et al. 2020. Social Enhancement—Insights from Computational Social Sciences, Complex System Theory, and Global Systems Theory. GESDA, Scientific Anticipatory Briefs. Accessed at:https://gesda.global/scientific-anticipatory-briefs/
Hodgson, David. 2012. *Rationality + Consciousness = Free Will*. New York: Oxford University Press.
Hoffer, Eric. 1955. *The Passionate State of Mind*. New York: Harper and Row.
Holltermann, Callie. 2024. Elmo Asked an Innocuous Question. *New York Times,* January 30. Accessed at: https://www.nytimes.com/2024/01/30/style/elmo-x-question.html
Huang, Chiungjung. 2010. Internet Use and Psychological Well-Being: A Meta-Analysis. *Cyberpsychology, Behavior and Social Networking* 13 (3): 241–249.
Ivers, Rowena, and Thomas Astell-Burt. 2023. Nature Rx: Nature Prescribing in General Practice. *Australian Journal of General Practice* 52 (4): 183–187.
Jabr, Ferris. 2019. Can You Really Be Addicted to Video Games? *New York Times Magazine* (October 22). Accessed at: https://www.nytimes.com/2019/10/22/magazine/can-you-really-be-addicted-to-video-games.html
James, William. 1890. *The Principles of Psychology*, vol. 1. New York: Henry Holt and Company.
Kahn, Peter. 2011. *Technological Nature: Adaptation and the Future of Human Life*. Cambridge: MIT Press.
Kardaras, Nicholas. 2016. *Glow Kids: How Screen Addiction is Hijacking Our Kids—And How to Break the Trance*. New York: St. Martin's Press.
Karpf, David. 2016. *Analytic Activism: Digital Listening and the New Political Strategy*. New York: Oxford University Press.
Kelly, Yvonne, Afshin Zilanawaia, Cara Booker, and Amanda Sacker. 2018. Social Media Use and Adolescent Mental Health: Findings from the UK Millennium Cohort Study. *eClinicalMedicine* 6: 59–68.

Kimmerer, Robin. 2023. In David Marchese "You Don't Have to Be Complicit in Our Culture of Destruction." *New York Times Magazine*. Accessed at: https://www.nytimes.com/interactive/2023/01/30/magazine/robin-wall-kimmerer-interview.html

Knibbs, Kate. 2024. AI Slop Is Flooding Medium. *Wired*, October 28. Accessed at: https://www.wired.com/story/ai-generated-medium-posts-content-moderation

Kramer, Adam D., Jamie E. Guillory, and Jeffrey T. Hancock. 2014. Experimental Evidence of Massive-Scale Emotional Contagion Through Social Networks. *Proceedings of the National Academy of Sciences.* 111 (24): 8788–8790.

Kross, E., et al. 2013. Facebook Use Predicts Declines in Subjective Well-Being in Young Adults. *PLoS ONE* 8: e69841.

Kurzweil, Ray. 2005. *The Singularity is Near: When Humans Transcend Biology*. New York: Viking.

Lanier, Jaron. 2018. *Ten Arguments for Deleting Your Social Media Accounts Right Now*. New York: Picador.

Libet, Benjamin. 1985. Unconscious Cerebral Initiative and the Role of Conscioius Will in Voluntary Action. *Behavioral and Brain Sciences* 8 (4): 529–539.

Liu, Mingli, et al. 2022. Time Spent on Social Media and Risk of Depression in Adolescents: A Dose-Response Meta-Analysis. *International Journal of Environmental Research and Public Health* 19 (9): 5164.

Loeffler, John. 2018. Why Chinese Artificial Intelligence Will Run The World: How the Chinese tech giants Baidu, Alibaba, and Tencent will Develop the Systems that Will Run the World. *Interesting Engineering*, November 6. Accessed at: https://amp.interestingengineering.com/why-chinese-artificial-intelligence-will-run-the-world

Louv, Richard. 2012. *The Nature Principle: Reconnecting with Life in a Virtual Age*. Chapell Hill: Algonquin Books.

McNamee, Roger. 2019. *Zucked: Waking Up to the Facebook Catastrophe*. New York: Penguin.

Medium. 2024. Accessed at: https://medium.com/about

Mehl, Matthias R., Simine Vazire, Shannon E. Holleran, and C. Shelby. Clark. 2010. Eavesdropping on Happiness: Well-Being is Related to Having Less Small Talk and More Substantive Conversations. *Psychological Science* 21 (4): 539–541.

Mills, Stuart. 2021. The Future of Nudging Will Be Personal. *Behavioral Scientist* (March 15). Accessed at: https://behavioralscientist.org/the-future-of-nudging-will-be-personal/?utm_source=ActiveCampaign&utm_medium=email&utm_content=How+to+Get+Better+at+Admitting+Your+Wrong++%5BBest+Reads%5D&utm_campaign=Weekly+Digest+%28March+20%29

Moynihan, Andrew B., Eric R. Igou, and Wijnand A. P. van Tilburg. 2018. Lost in the Crowd: Conformity as Escape Following Disbelief in Free Will. *European Journal of Social Psychology* 49 (3): 1–18.

Mozur, Paul, Muyi Xiao, and John Liu. 2022. 'An Invisible Cage': How China Is Policing the Future. *New York Times,* June 25. Accessed at: https://www.nytimes.com/2022/06/25/technology/china-surveillance-police.html?action=click&module=Well&pgtype=Homepage&section=Business

Murdoch, Iris. 1970. *The Sovereignty of Good*. UK: Routledge and Kegan Paul.

Murthy, Vivek. 2023a. *Social Media and Youth Mental Health: The U.S. Surgeon General's Advisory*. Accessed at: https://www.hhs.gov/sites/default/files/sg-youth-mental-health-social-media-advisory.pdf

Neophytou, Eliana, Laurie A. Manwell, and Roelof Eikelboom. 2021. Effects of Excessive Screen Time on Neurodevelopment, Learning, Memory, Mental Health, and Neurodegeneration. *International Journal of Mental Health and Addiction* 19: 24–744.

Nielsen, Rasmus Kleis, and Sarah Anne Ganter. 2022. *The Power of Platforms*. New York: Oxford University Press.

Nietzsche, Friedrich. 1882. *Gesammelte Werke*, vol. 16. Munich: Musarion Verlag.

Nietzsche, Friedrich. 1974. *The Gay Science*. New York: Vintage.

Odell, Jenny. 2019. *How to Do Nothing: Resisting the Attention Economy*. Brooklyn: Melville House.

Orben, Amy, and Andrew K. Przybylski. 2019a. The Association Between Adolescent Well-Being and Digital Technology Use. *Nature Human Behaviour* 3: 173–182. Accessed at: https://www.nature.com/articles/s41562-018-0506-1#citeas

Orben, Amy, and Andrew K. Przybylski. 2019b. Screens, Teens, and Psychological Well-Being: Evidence from Three Time-Use-Diary Studies. *Psychological Science* 30 (5): 682–696.

Orben, Amy, Andrew K. Przbylski, Blakemore Sarah-Jayne, and Rogier A. Kievit. 2022. Windows of Developmental Sensitivity to Social Media. *Nature Communications* 13 (1649). Accessed at: https://www.nature.com/articles/s41467-022-29296-3#citeas

Orben, Amy, Tobias Dienlin, and Andrew K. Przybylski. 2019. Social Media's Enduring Effect on Adolescent Life Satisfaction. *PNAS* 116 (21): 10226–10228.

Ortiz-Ospina, Estaban. 2019. Are Facebook and Other Social Media Platforms Bad for Our Well-Being? *Our World in Data*. Accessed at: https://ourworldindata.org/social-media-wellbeing

Oswald, Tassia, Alice Rumbold, Sophie Kedzior, and Moore Viviene. 2020. Psychological Impacts of "Screen Time" and "Green Time" for Children and

Adolescents: A Systematic Scoping Review. PLOS ONE. Accessed at: https://journals.plos.org/plosone/article?id=10.1371/journal.pone.0237725

Pentland, Alex. 2014. *Social Physics: How Good Ideas Spread—The Lessons from a New Science*. New York: Penguin.

Rainie, Lee, and Kathryn Zickuhr. 2015. Americans' Views on Mobile Etiquette. Pew Research Center, August 26.

Roberts, Margaret. 2018. *Censored: Distraction and Diversion inside China's Great Firewall*. Princeton: Princeton University Press.

Rosen, Larry. 2012. *IDisorder: Understanding our Obsession with Technology and Overcoming its Hold on Us*. New York: St. Martin's Griffin.

Ryan, Fergus. 2018. An Orwellian Future Is Taking Shape in China. *The Sydney Morning Herald*, January 8. Accessed at: https://www.smh.com.au/opinion/an-orwellianfuture-is-taking-shape-in-china-20171220-h07vbw.html

Saleem, Nalia, Paul Young, and Saman Yousuf. 2024. Exploring the Relationship Between Social Media Use and Symptoms of Depression and Anxiety Among Children and Adolescents: A Systematic Narrative Review. *Cyberpsychology, Behavior, and Social Networking*. Accessed at: https://www.liebertpub.com/doi/10.1089/cyber.2023.0456

Sapolsky, Robert. 2017. *Behave: The Biology of Humans at Our Best and Worst*. New York: Penguin.

Sapolsky, Robert. 2023. *Determined: A Science of Life without Free Will*. New York: Penguin.

Soon, Chun Siong, Marcel Brass, Hans-Jochen, Heinze, and John-Dylan, Haynes. 2008. Unconscious Determinants of Free Decisions in the Human Brain. *Nature Neuroscience* 11 (5): 543–545.

Soon, Chun Siong, Anna Hanxi He, Stefan Bode, and John-Dylan, Haynes. 2013. Predicting Free Choices for Abstract Intentions. *Proceedings for the National Academy of Sciences*. 110 (15): 6217–6222.

Stadler, Felix. 2002. Privacy is Not the Antidote to Surveillance. *Surveillance and Society* 1 (1): 120–124.

Stein, David. 2024. The Fundamental Flaws of The Only Meta-Analysis of Social Media Reduction Experiments (And Why It Matters), Part 3. Accessed at: https://www.afterbabel.com/p/fundamental-flaws-part-3

Sullivan, Danny. 2015. How Google Now, Siri & Cortana Predict What You Want. *Search Engine Land* (September 18). Accessed at: https://searchengineland.com/how-google-now-siri-cortana-predict-what-you-want-229799

Sunstein, Cass R. 2013. *Impersonal Default Rules vs. Active Choices vs Personalized Default Rules: A Triptypch. SSRN*. Accessed at: https://ssrn.com/abstract=2171343 or https://doi.org/10.2139/ssrn.2171343

Toffler, Alvin. 1970. *Future Shock*. New York: Bantam Books.

Trust for America's Health. 2023. Pain in the Nation: The Epidemics of Alcohol, Drug, and Suicide Deaths. Accessed at: https://www.tfah.org/report-details/pain-in-the-nation-2023/

Twenge, Jean M. 2017. *IGen: Why Today's Super-Connected Kids are Growing Up Less Rebellious, More Tolerant, Less Happy—And Completely Unprepared for Adulthood*. New York: Atria Books.

Twenge, Jean M., and Roy F. Baumeister. 2002. Self-Control: A Limited Yet Renewable Resource. In *Self and Identity: Personal, Social, and Symbolic*, ed. Yoshihisa Kashima, Margaret Foddy, and Michael Platow, 57–70. Mahwah, NJ: Lawrence Erlbaum Associates.

Twenge, Jean M., Jonathan Haidt, Thomas E. Joiner, and W. Keith, Campbell. 2020. Underestimating Digital Media Harm. *Nature Human Behavior* 4: 346–348.

Twenge, Jean, Brian Spitzberg, and W. Keith, Campbell. 2019. Less In-person Social Interaction with Peers Among U.S. Adolescents in the 21st Century and Links to Loneliness. *Journal of Social and Personal Relationships* 36 (6): 1892–1913.

Twenge, Jean. 2020. Why Increases in Adolescent Depression May be Linked to the Technological Environment. *Current Opinion in Psychology* 32: 89–94.

Twenge, Jean, and W. Keith. Campbell. 2009. *The Narcissism Epidemic*. New York: The Free Press.

United States Hospital Fund. 2024. *The Ripple Effects of the Adolescent Behavioral Health Crisis*. New York. Accessed at: https://media.uhfnyc.org/filer_public/61/92/6192cd92-0fc3-4dc7-8a36-eda9531354a3/ripple_effects_adolescent_behavioral_health_chartbook_report.pdf

Vanderklippe, Nathan. 2018. Chinese School Installs Cameras to Monitor Students. *Globe and Mail* (June 2): A3.

Verdyun, P., et al. 2015. Passive Facebook Usage Undermines Affective Well-Being: Experimental and Longitudinal Evidence. *Journal of Experimental Psychology Gen* 144: 480–488.

Vuorre, Matti, Amy Orben, and Andrew K. Przybylski. 2021. There Is No Evidence That Associations Between Adolescents' Digital Technology Engagement and Mental Health Problems Have Increased. *Clinical Psychological Science* 9 (5): 823–835.

Wegmann, Elisa, Sina Ostendorf, and Matthias Brand. 2018. Is It Beneficial to Use Internet-Communication for Escaping from Boredom? Boredom Proneness Interacts with Cue-induced Craving and Avoidance Expectancies in Explaining Symptoms of Internet-Communication Disorder. *PLoS ONE* 13 (4): e0195742.

Wegner, Daniel. 2018. *The Illusion of Conscious Will*. Cambridge, MA: MIT Press.

Williams, Florence. 2017. *The Nature Fix: Why Nature Makes us Happier, Healthier, and More Creative*. New York. W.W: Norton.
Williams, James. 2018. *Stand Out of Our Light: Freedom and Resistance in the Attention Economy*. Cambridge: Cambridge University Press.
Wilson, Timothy. 2002. *Strangers to Ourselves: Discovering the Adaptive Unconscious*. Cambridge, MA: Harvard University Press.
Yarrow, Jay. 2010. Sergey Brin: 'We Want Google To Be The Third Half Of Your Brain.' *Business Insider* (September 8). Accessed at: https://www.businessinsider.com/sergey-brin-we-want-google-to-be-the-third-half-of-your-brain-2010-9?op=1
Zuboff, Shoshana. 2019. *Surveillance Capitalism: The Fight for a Human Future at the New Frontier of Power*. New York: PublicAffairs.

CHAPTER 7

Self-Leadership and Service

How can the AI-enabled attention economy be resisted so that human agency is sustained? Grounding itself in Aristotle's moral philosophy and contemporary neuroscience, this chapter explores the dispositions, habits, and skills needed to safeguard self-leadership and service. These dispositions, habits, and skills can be cultivated employing the same mechanisms of behavior design that are utilized in the attention economy to hijack attention and erode agency.

I have argued that the regulation of the AI-enabled attention economy is both necessary and urgent. If it is perceived as a complete cure, however, it will undermine an equally crucial endeavor: the cultivation and exercise of individual agency. Public oversight is a complement to, rather than a substitute for, self-governance. To endorse the former while ignoring the latter is to embrace a technocratic fix that, at best, will have limited success and, at worst, will do more harm than good. There is no effective alternative to developing the dispositions, habits, and skills of self-governance.

In truth, self-governance might also be seen as a technological endeavor. There is no such thing as a non-technological way of being human. To develop dispositions, habits, and skills is to be engaged in what Michel Foucault (1988) called the "technologies of the self." The term is modern, but not the practice. The ancient Greeks and Romans systematically developed these technologies, which were primarily oriented toward

L. P. Thiele, *Human Agency, Artificial Intelligence, and the Attention Economy*, https://doi.org/10.1007/978-3-031-82086-1_7

self-knowledge and self-care and included the moral virtues. Knowledge of the self enabled its proper caretaking. As importantly, self-care generated self-knowledge. “In Greco-Roman culture,” Foucault (1988, 22) writes, “knowledge of oneself appeared as the consequence of taking care of yourself.” Both self-knowledge and self-care are grounded in reflective self-direction.

Our species is very good at developing “hard” technologies, things that have mass, such as hammers, automobiles, airplanes, computer chips, and data centers. Hard technologies generally need rules to control their use. With the arrival of automobiles, for example, laws were needed to determine which side of the road you could drive on, when and where you had to stop, and how fast you could go. Hard technologies require soft technologies—laws, policies, regulations, social norms, and cultural mores—to govern their development and deployment.

Soft technologies do not have mass—though they often gain representation on paper or computer chips. Soft technologies, at base, are regularized ways of thinking and behaving. They typically take the form of rules and shared practices. If the deployment of increasingly powerful hard technologies continues without the development of soft technologies to ensure their beneficial use, our species will likely annihilate itself with its own tools.

In an AI-enabled attention economy, the soft technologies that need to be developed include both governmental regulations and the dispositions, habits, and skills that enable the cultivation and exercise of human agency. Technocratic solutionism is wrongheaded when it tries to solve all the problems created by hard technologies with more hard technologies. Technocratic solutionism is also wrongheaded when it limits itself to soft technologies provided by governments. We need to develop both public policies to regulate the supply side of the AI-enabled attention economy and individual self-governance to regulate the demand side. Dispositions, habits, and skills are crucial components of the latter.

Historically, we have taken human agency for granted, grounded as it was the individual’s near monopoly on self-knowledge. As the AI-enabled attention economy expands in scope and grows in predictive power, this monopoly is threatened. Referencing the vast increases in data, computing power, and scientific knowledge, Harari (2017) observes that “Democracy and the free market and individualism and liberalism, these are all predicated on the basic assumption that nobody knows me better than I know myself, so nobody should take decisions for me, and nobody really understands how I feel and what I think, and this was true throughout history.... But it is no longer true now or it won't be true in ten years." Harari is right: we are in a high-stakes game played at superhuman speeds. Human agency and the political institutions that it makes possible are being sacrificed to the business models of global ICT corporations and the despotic aspirations of autocratic regimes. Cultivating our aptitude for reflective self-direction is an indispensable form of resistance.

Reflective self-direction is the essence of human agency, which, for reasons that will become clear, is best described as self-leadership. At its best, self-leadership is directed outwards in service, in practices of caring within a community whose welfare it advances. Self-leadership and service are the antidotes to the dysfunctions of the AI-enabled attention economy. Before assessing how these remedies can be developed and applied, we must first explore what they entail.

Reflection

Reflection enables self-knowledge and informs self-direction. Yet reflection is an intrinsic part of agency, not simply its prelude. As Bandura (1986, 1181) insists, "One is just as much an agent reflecting on one's experiences as in generating and executing the original course of action." When we reflect on our past actions or future options (i.e. possible actions and their likely outcomes), we are exercising a form of mental agency. Still, not all events in our minds, such as dreaming, constitute mental agency. To engage in reflection, as opposed to simply experiencing mental events, we must direct our attention and form intentions.

When fully engaged in daily activities, there may be little reflection going on. Indeed, there may be few if any conscious intentions. It would be inappropriate, and often dangerous, to be introspecting when we ought to be focused on other things, such as navigating a vehicle through a busy intersection, carrying out a complex professional obligation, or safely operating a power tool. For the most part, reflection is what occurs during our leisure hours or during those moments of leisure that become available to us between the diverse activities that rightfully occupy virtually all our attention at work, home, or in the community.

Reflection is seldom a planned, lengthy, full-scale baring of the soul. Most often, and perhaps most usefully, it occurs sporadically, in the brief pauses of daily life. Here, it presents the opportunity to think about what we are doing, acknowledge what we are feeling, consider how we are reckoning, and potentially revise our attitudes and beliefs and redirect our behavior. Reflection allows us to identify our perceptions and perspectives, evaluate our actions and relationships, take stock of our circumstances, and either stay the course or modify our efforts.

Reflection can be a useful means of giving credit where credit is due. Still, when things are going very well, a key motivation for reflection is absent: the need to shift our beliefs or behaviors to achieve better results. Reflection is primarily motivated by the desire to better understand our world and ourselves and to assess how they might be tweaked or transformed. "*If it ain't broke, don't fix it*" is pragmatic advice. Accordingly, we often engage in reflection when things are not going as well as hoped. In this respect, reflection can be a bitter tonic, revealing our shortcomings and culpabilities. Opportunities for reflection, accordingly, are often experienced with some foreboding as we anticipate the exposure of inadequacy.

Reflection cannot gain traction if inadequacies are blocked from view. If we do not squarely face our fears and failings, we will be left spinning our introspective wheels. The prospect is one of being alone with oneself and finding nothing of interest. Nothing of interest can arise because we are unwilling to recognize the most pressing concerns, namely, our shortcomings. In such cases, pauses in life threaten us with anxiety or boredom. Our aversion to this state prompts us to seek diversions. So we distract ourselves. That is a widespread and age-old phenomenon. What is new in the attention economy is the ubiquity and power of the distractions at hand.

Consider those uncharted moments when we must cease a current activity and cannot start another: waiting at an intersection for the light to change, stepping into a crowded elevator, or navigating a halting conversation. When such situations present themselves today, the opportunity for reflective thought is widely rejected. Instead, people reach for their smartphones and dive into social media, email, or entertainment. A potentially uncomfortable pause is deftly averted by way of virtual filler. The threat of anxiety or boredom is staved off. Also foregone is the opportunity for reflection, and the self-direction it informs and enables.

Turkle observes that today "people have a tremendous lack of tolerance for being alone… Solitude is the precondition for having a conversation with yourself. This capacity to be with yourself and discover yourself is the bedrock of development. But now, from the youngest age—even two, or three, or four—children are given technology that removes solitude by giving them something externally distracting. That makes it harder, ironically, to form true relationships.... They haven't learned to have conversations or relationships, which involve lulls" (Fischetti 2016). Our ability to engage in conversations and form relationships—with others and with ourselves—entails a willingness to tolerate lulls. Life is rife with the interstices of activities. Today, digital distractions primarily fill these voids. To the extent that digital technology lowers our tolerance for being alone with ourselves, it degrades our capacity for reflection.

Human beings have always needed, enjoyed, and availed themselves of diversions. All work and no play makes for a dull life. And it is a recipe for burn-out. Recreative diversions—vacations as well as more time-limited, daily diversions—make us healthier and happier. They are not distractions. Distractions pull us away from things that we intend to do. Diversions, in contrast, are things that we choose to do because they provide recreative relief or other benefits. After a stressful day at work, we might have a drink with colleagues, chat with friends, play with the kids, engage in a rejuvenating jog, or walk the dog. Of course, diversions can be degraded into distractions. We might choose to divert ourselves after a stressful day by exploring a topic of interest on the Internet. Late in the evening, however, we find ourselves in a zombie-like trance, ceaselessly surfing on TikTok or YouTube. At some point, the intention to gain the benefits of a recreative diversion morphed into hours of mindless distraction.

Distractions are characterized by the shortness of their cycles. They are rapid-fire stimuli that quickly lose their appeal and require regular replacement. Recreative diversions have greater durability and a few other key features. On any given weekend, one might seek respite from chores. Consider the options of visiting an art gallery, watching a feature film, and swiping away on social media. The first demands attentive inspection and reflection. We peruse the paintings, considering the techniques employed and the meanings intended. The second forbids dedicated inspection and permits less reflection. To gain the benefits of a feature film, however, requires a commitment to its plot and characters, which linger in memory. The third is too fleeting for much attentive inspection, reflection, or remembrance. Of course, there are many worthwhile uses of social media. But to generalize, with a nod to David Brooks (2024) on the dangers of cheap thrills: art is demanding, entertainment delights but can be durable, and social media is designed to be distracting.

Of course, digital media are not the only antagonist of focused attention and reflection. Internal distractions also play a role. People have wandering minds. Indeed, we are beset by internal distractions almost half of the time we engage in activities (Killingsworth and Gilbert 2010). Even a moment sitting in stillness with the intention of focusing on a specific thought or feeling will, in very short order, give rise to a wide array of other mental events. We might find ourselves replaying a scene in our mind's eye from a weekend outing, rehearsing a conversation that we hope to initiate the next day at the office, evaluating a backache, worrying about our children, composing a list of items to pick up from the grocery store, or planning to dispatch any number of mundane undertakings. In popular parlance, we all have monkey minds, and these minds are constantly grabbing at whatever alluring fruit is nearby—be it a memory, a fantasy, a fear, a physical sensation, an opportunity, or a task to be accomplished.

There is an emotional cost to having a wandering mind, as internal distractions impact happiness. People whose minds are elsewhere when engaged in an activity are not as happy as those focused on the activity at hand. This is true even if the activity in question is a rather unpleasant one. A wandering mind is a less happy mind (Killingsworth and Gilbert 2010).

Our proclivity for digital distractions suggests that we are agnostic regarding the value of the things we pay attention to (Crawford 2015, 5). The same agnosticism is implied by internal distractibility. The wandering mind is not committed to the value of any specific mental event. Fruits are plucked not because of their intrinsic worth but simply because they are near at hand. Reflection, in this respect, is a form of self-restraint. It entails forgoing the temptation to grasp at whatever fills our screens or pops into consciousness. In turn, it requires letting go of grabbed goodies in relatively short order, so that one's attention may be redirected back to an original intention. Focused and diligently refocused attention—the voluntary and repeated bringing back of a wandering mind to an intended focus, which William James held to be "the very root of judgment, character, and will"—reveals what is worth paying attention to.

Daydreaming may be rejuvenating and therapeutic, and a spark to creativity. Still, lacking a guiding intention, it is not reflection. While reflection entails being receptive to passing sensations, feelings, and thoughts, it also requires a disciplined effort to return, again and again, to some motivating concern. In this respect, reflection might be likened to composing a coherent paragraph. The guiding intention serves as a kind of topic sentence that sets out the theme of the paragraph and circumscribes its content. Like a well-composed paragraph, reflection is both sharply focused and exploratory. It is not an exercise in logical deduction. Neither is it the venting of a stream of consciousness. While the guiding intention does not determine the destination, or even set a singular path to be followed, it does inhibit digressions. Discoveries *en route* may stimulate new or revised intentions that take us to unanticipated places. In such cases, more paragraphs get written. Still, guiding intentions ensure that one coherent paragraph follows another.

The likening of reflection to composition is not gratuitous. Becoming a reflective individual is an achievement that only ever occurs in language-speaking communities (Dennett 1995, 379). We develop the aptitude for reflection when our unique human capacity to partake linguistically in the experiences of others, our capacity for conversations, is turned inward. Before we put our thoughts into words, we do not have the wherewithal for reflection. Only when mental events gain linguistic form do we discover their full measure and content. Visual images can be projected in the mind's eye and these images can be strung together or juxtaposed. Such wordless forms of imagination can enable rudimentary understanding and even innovation. For sensations, memories, beliefs, opinions, ideas, judgments, values, convictions, and feelings to become well-identified, understood, or creatively combined, however, language is needed.

Novelist E.M. Forster (1927, 101) captured this phenomenon with a question: "How can I tell what I think till I see what I say?" More provocatively, Nietzsche (1967a, 283) claimed that "we think only in the form of language.... We cease to think when we refuse to do so under the constraint of language." Heidegger (1968, 16), who wrote extensively on Nietzsche, concurred in this regard, observing that "only when man speaks, does he think—not the other way around." Forster's question, and Nietzsche's and Heidegger's claims, illuminate a common experience. It pertains not only to thinking but also to feeling. For the most part, we neither know what we think nor understand what we feel until we have put our thoughts and feelings into words.

The poet W.H. Auden observed that "Language is the mother, not the handmaiden, of thought; words will tell you things you never thought or felt before." In the same vein, Cecil Day Lewis (1951, 15), Poet Laureate of the UK, wrote: "I do not sit down at my desk to put into verse something that is already clear in my mind. If it were clear in my mind, I should have no incentive or need to write about it.... We do not write in order to be understood; we write in order to understand." Notwithstanding their vocations, Auden's and Lewis's claims extend well beyond poetry. We communicate to comprehend. That is true both when we converse with others, and, by way of reflection, when we converse with ourselves.

Nietzsche, Heidegger, Forster, Auden, and Lewis were all accomplished communicators. They did not advocate carelessness in thought, speech, or writing. They did not suggest that we simply give vent to a stream of consciousness, subjecting other people or ourselves to verbal ramblings in the hope that something of value might eventually get voiced. Indeed, Forster's question comes from the mouth of a person that he does not admire, a person who, contemptuous of logic, hurls herself "into confusion." A good conversation, like the art and craft of writing, helps to illuminate, organize, and clarify. The silent conversations of self-reflection exploit language's capacity to give coherence to thoughts and form to feelings.

Reflection requires us to focus on a single wave within a roiling sea. This ability to target and sustain attention is gained by way of the words we employ to identify and describe phenomena. As the eighteenth-century German philosopher Johann Gottfried Herder (2002, 87–88) wrote In his *Treatise the Origin of Language*: "The human being demonstrates reflection when the force of his soul operates so freely that in the whole ocean of sensations which floods the soul through all the senses it can, so to speak, separate off, stop, and pay attention to a single wave, and be conscious of its own attentiveness." That single wave, understood as the "distinguishing" property of a thing, becomes a "distinct concept." With this "first judgment of the soul," which was the "*first characteristic mark of taking-awareness,*" Herder claims, human language was born. Symbolic language, Herder was saying, allows our species to transmute a vast, fluid world into bite-size conceptual chunks that can be digested. Reflection is the inner conversation that gives conceptual form to an existential flux.

Reflectively parsing the flux of worldly life into bite-size chunks enables its examination and digestion. Notwithstanding the knowledge of the world so gained, we often remain strangers to ourselves. In his Preface to *The Genealogy of Morals*, Nietzsche (1967b, 15) observes that, "We are unknown to ourselves, we men of knowledge: and with good reason. We have never sought ourselves." The erudite often forgo introspection. Even introspective individuals may not acquire self-knowledge to a great degree. In large part, that is because they remain too close to themselves. To gain proper perspective, some distance is required. Just as we often only discover what we think when we see what we say, so we best learn who we are by way of temporal distance, by retrospectively reflecting upon what we have said and done. Alternatively, and often for the better, we discover who we are by attending carefully to how others assess and respond to our words and deeds.

In Homer's *Odyssey*, the protagonist comes to understand the meaning of his participation in the Trojan War only near the end of his arduous return to Ithaca. As Odysseus listens to the minstrel Demodocus sing of the grand conflict, the hero is brought to tears which, Homer writes, fell like those of a woman throwing her arms around a fallen husband. It is the blind bard's narrative that brings the protagonist to fully understand the pathos of the great war and his participation in it. We come to know ourselves best by reflecting upon the stories of our worldly endeavors and relationships.

Introspection, on its own, is an unreliable guide to our hearts and minds because we typically reconstruct memories so they align with our current desires, beliefs, and mental models (Abelson et al. 2004). Introspection is highly susceptible to confirmation bias. Often it becomes an exercise in motivated reasoning. Alternatively, introspection may deteriorate into rumination, the repeated and often obsessive rehearsal or prehearsal of (negative) experiences, thoughts, or feelings. Motivated reasoning and rumination are not beneficial forms of reflection.

To produce self-knowledge, reflection must overcome a wide array of biases. One in particular stands out: the predilection to overestimate our own qualities and abilities. Garrison Keillor wrote of a fictional town called *Lake Wobegon* where "all children are above average." The *Lake Wobegon effect*, also known as the overconfidence bias, illusory superiority, and, for those with low skill levels, the Dunning–Kruger effect, has been well demonstrated in social scientific studies and surveys. For example, the vast majority of motorists consider themselves to be above average in driving skills. Only two percent of high school seniors think they are below average in leadership ability. And a whopping 94 percent of university professors think that they are better at their jobs than their colleagues (Gilovich 1991, 77).

Overconfidence can be beneficial. Like unwarranted optimism, it can promote health and enhance motivation, persistence, and performance.

Overconfidence can also beget imprudent judgments and ineffective planning. The benefits and dangers of overconfidence depend on the context (Schriber and Robins 2012). The same can be said about most, if not all, implicit attitudes and predispositions. They often prove advantageous, and they often lead us astray.

In the pursuit of self-knowledge, illusory superiority is maladaptive. That is why paying close attention to what we have said and done, oftentimes relayed to us by family, friends, or associates, reveals who we are better than navel-gazing. Alexander Pope's *Essay on Criticism*, which contained the famous line "A little learning is a dang'rous thing," addressed the self-enhancement bias to which humans are prone. Pope writes:

> Of all the causes which conspire to blind
> Man's erring judgment, and misguide the mind,
> What the weak head with strongest bias rules,
> Is pride, the never-failing vice of fools....
> Trust not yourself; but your defects to know,
> Make use of ev'ry friend—and ev'ry foe.

Making use of friends, foes, and a rear-view mirror helps reflection integrate the demands of critical thinking. By rigorously examining the stories of our lives, we can gain awareness of the role played by cognitive limitations, biases, and dispositions. Employing these insights, we can evaluate current circumstances more objectively, the better to understand our own and others' behavior. Critical thinking, the antipode of dogmatism, is the fruit of skillful reflection.

At its best, reflection advances self-knowledge and informs action. Still, the success of our pensive expeditions wholly depends upon the skills employed. A worthy prize is not guaranteed. That is because the stream of stimuli and mix of motivations behind our words and deeds remain mostly hidden from us. As Timothy Wilson (2002, 73, 219) states, "Because people cannot directly observe their nonconscious dispositions, they must try to infer them indirectly by, for example, being good observers of their own behavior....There is no direct pipeline to the adaptive unconscious; it must be inferred by being a good self-biographer." We get to know who we are by examining what we have said and done, and by learning how our words and deeds are perceived by family, friends, colleagues, and critics. Self-knowledge is best gained by way of a rear-view mirror and the estimations of others. We come to know ourselves best, including all our faults and foibles, by well attending to the stories of our lives.

Self-direction

Self-direction means purposefully making decisions and taking actions. It need not be self-serving. Like reflection, self-direction is disciplined by guiding intentions. These can advance egoistic or altruistic goals. Say you witness an elderly person drop a bag of groceries in a supermarket parking lot. A decision to look the other way, hurriedly pass by, get in one's car, and drive off can be as self-directed as the choice of stopping to help pick up the scattered contents. For an activity to be self-directed, its origins and furtherance must stem from a clear intention. The moral merit or social value of that intention is a separate question.

Like reflection, self-direction entails focused attention. Even activities that appear to have no obvious ends can be engaged with intention. One might enthusiastically sing, dance, play, or gaze at the stars. There may be no clear goal in mind, though an underlying motivation to experience exhilaration, peace, or joy, to express oneself, or to marvel at the unknown, is in play. Self-direction always has a course or bearing, even if the destination is unknown. And while self-direction impels us to act, the goal may be more akin to equanimity than enterprise.

Reflection relies upon language to gain cognitive traction and secure memory. Without words to clarify and concretize our reflections, we would find ourselves chronically reinventing the wheels of the thought trains that we ride. Likewise, to further its guiding intention, self-direction may rely upon principles, rules, plans, dispositions, and habits. In turn, self-direction is grounded upon the extensive contributions of the adaptive unconscious. In this respect, self-direction is like reflection. When we think, judge, act, or create we are relying on many unconscious processes and aptitudes (Thiele 2006). As discussed in the last chapter, the greatest part of our inner activity remains unconscious and unfelt. Reflection brings intentionality into the mix but makes no claim to the willful control of all the elements of our thinking and feeling. The same limits apply to self-direction.

Consider riding a bicycle. Pedaling to a destination can be a self-directed activity. It would be a short lived and likely painful experience, however, if one attempted to select and execute every bodily motion with intention. The same can be said for most physical activities and mental activities, including speaking. Starting a conversation for a particular purpose can be an act of self-direction. Still, the attempt to consciously select every word would produce a painfully halting, and probably incoherent, set of utterances. Most of our speech is not premeditated. Like other mental and physical activities, speaking relies heavily on implicit knowledge, also known as tacit knowledge. The exercise of skills grounded in tacit knowledge is unpremeditated and feels effortless, like the achievement of putting one foot in front of the other when walking. It is an example of what Dennett (2017) calls "competence without comprehension," the product of the nonconscious components of our psyche and physiology. Were we to limit our reflection or self-direction to efforts that are wholly within conscious control, neither would amount to much.

Self-direction relies on implicit attitudes and tacit skills that invite critical evaluation but elude comprehension. That is unavoidable. As Emerson Pugh observed, "If the human brain were so simple that we could understand it, we would be so simple that we couldn't." Much of what we think, feel, say, and do is influenced by the adaptive unconscious. Our physical and verbal behaviors, as well as our conscious thoughts, feelings, and judgments, are actuated with the involvement, and often the leadership, of unconscious attitudes, aptitudes, predilections, and biases (Vazire and Wilson 2012). Introspection cannot plumb the depths of this realm. The intentionality that spurs self-direction, it follows, does not overcome or circumvent the adaptive unconscious. Indeed, it would be counterproductive to attempt as much, like the effort to choose every word when we are speaking or every movement of our bodies when we are walking. A conscious intention is a necessary but not a sufficient condition for reflection and self-direction.

Arendt (1978, II-33) correctly observes that "Most of our acts are taken care of by habits, just as many of our everyday judgments are taken care of by prejudices." Reflective self-direction is always a mixture of actions and habits, judgment and biases. Conscious efforts and unconscious aptitudes and predilections operate in tandem. To engage in reflection and self-direction, we need not hide from and should not deny the extensive impact of habits and the adaptive unconscious. Indeed, it would be an act of bad faith to do so.

Owing to our inability wholly to comprehend or control our external environments, bodies, and minds, having an intention is not the same as realizing one. Having an intention is like having a bearing. You cannot reach a desired destination without setting a course for it. But having a bearing does not guarantee arrival. While purposeful intention is intrinsic to self-direction, it is not sovereign power.

Who, then, is the purposive but non-sovereign intender? To be reflectively self-directive is to think and act like the protagonist of one's life (Thiele 2009). Here we are not speaking of a transparent, single-willed entity with all internal tensions resolved. Human beings are complex adaptive systems, our interactive components forged by experience and the forces of natural selection. Our minds, or souls in the parlance of antiquity, have diverse and distinct parts that collaborate, compete, and confabulate (Thiele 1990). Much of this activity occurs below the level of consciousness. So, as Nietzsche suggested, when we observe ourselves we are witnessing only "the last scenes of reconciliation." When engaged in reflective self-direction, we are not accessing some homogenous and unified self that is taking uncontested control of its destiny. Rather, we are consciously contributing to the script of a protagonist in development.

In *The Human Condition*, Arendt (1958, 184) wrote: "Although everybody started his life by inserting himself into the human world through action and speech, nobody is the author or producer of his own life story. In other words, the stories, the results of action and speech, reveal an agent, but this agent is not an author or producer. Somebody began it and is its subject in the twofold sense of the word, namely, its actor and sufferer, but nobody is its author." In the same vein, Arendt (1968, 164–165) insists that sovereignty or mastery—the control of the outcome of action—is antithetical to human freedom: "If men wish to be free, it is precisely sovereignty they must renounce." Arendt is right: human agents are not sovereign powers determining how their lives unfold and controlling the impacts of their deeds. The adaptive unconscious plays its role, as do other people and the natural environment.

Of course, literary authors also lack sovereign power. This is true in two senses. Authors have complex minds, and their work reflects the reconciliation of many forces that they do not control. Poets were once hailed as mouthpieces of the muses, with this in mind. In turn, the social impacts of written works are beyond an author's control. In his 1968 essay, "The Death of the Author," Roland Barthes (1977) famously asserted that authorial action does not define a text. Rather, the meaning of a text arises through its interactions with other texts and the interpretations of readers. The meaning of a text is not the prerogative of its author.

Of course, an author can attempt to convey a specific insight or truth, perhaps to a specific audience. Still, what readers ascertain and experience may be quite different from what the author intends.

To exercise agency is to become the protagonist of one's life without wholly controlling its path or mastering its meaning. As human agents, we do not exercise sovereignty. Still, we can become the authors of our lives in a more restricted and humble sense. We can think and act with intention. These purposeful efforts, in the aggregate, project ourselves into specific narratives. Our intentions bespeak efforts to fulfill our roles in unfolding tales.

To say we inhabit stories is not to say that we are fated to play out wholly scripted characters and storylines. To be sure, we may be significantly defined by our familial, social, and professional roles and responsibilities. At the same time, initiative, inventiveness, and ingenuity may also be prominent traits. As reflective beings, we mostly come to know ourselves retrospectively. But as self-directive beings, we inhabit and actuate a developing protagonist. Our words and deeds can project us onto new and uncharted paths. In turn, we can seek out experiences that will inform and recast our characters.

To exercise agency is to experience oneself as a protagonist in development. To become protagonists of our lives, we must take on the responsibility of cultivating and exercising agency. The term *self-leadership* captures this reciprocal endeavor. We engage in self-leadership when we assume the role of a protagonist who is responsible for their actions and ongoing development.

Self-leadership is a fulfilling yet humbling experience. The humility inheres in our lack of sovereignty. We always share the stage. We do not control all the interactions upon it, and we certainly do not control when the curtain will fall. We share the stage with the structural constraints of our physical and social environments, the props, sets, and scenery of our lives. As importantly, we share the stage with other characters who are protagonists of their own stories. Last but far from least, we share the stage with the adaptive unconscious.

Notwithstanding the vocabulary of stories and stages, self-leadership is not motivated by a lust for the limelight. It is not an exercise in self-promotion and bears no resemblance to the increasingly common effort to establish a personal brand and gain online celebrity. This is what social media influencers do as a trade, and what many others attempt. The lay effort can be observed at scenic vistas, art galleries, historic buildings, or notable events, when those arriving immediately capture the experience with a selfie, and then quickly move on to the next photo-op. The images get posted in the hope of receiving likes, shares, and followers. Sometimes images are edited and enhanced to make the chronically smiling participants more appealing. Here self-display in pursuit of an appreciative audience replaces the fuller experience of life and the self-leadership that enables it.

We all care about what people think of us, and for this reason, try to put our best foot forward. But that motivation should not overpower the experience of walking. The intention behind self-leadership is to become the protagonist of one's life, not its promoter.

Service

Agency is reflective self-direction. Reflection develops our sense of what matters and stimulates intentions. Self-direction realizes intentions by way of action. To exercise agency is to act with an intention that is the product of our "reflective self-evaluation" (Frankfurt 1971, 7). When reflectively self-directing, however, we are not simply satisfying our most pressing desires. Rather, we are responding to our assessment of what philosopher Charles Taylor (1985, 17) called the "qualitative worth" of our different desires. We orient ourselves towards specific preferences and purposes, including those different from immediate ones. To be reflectively self-directed is to consider the merits of our various predilections and give priority to the worthiest.

It is all too easy to misapprehend the worth of specific desires and fail to pursue what really matters. To flourish, we must both secure basic needs and seek higher goods. These higher goods include friendship, belonging, love, beauty, wonder, and the sublime. At its best, reflective self-direction takes us ethically, aesthetically, philosophically, and spiritually beyond the self. It invests us in a more encompassing realm: a cosmos and a community to which we owe the development of our cognitive, affective, and moral capacities, including those required for reflective self-direction. In short, for human beings to flourish, their agency must be deployed in ways that integrate them within a greater whole. To reach its full potential, self-leadership must be deployed in service.

Service entails the devotion of time, energy, and resources. These contributions are as much a prudent investment as they are a form of charity. That is because acts of service sustain the communities that sustain us. The naturalist Aldo Leopold emphasized that we are members of both social and ecological communities and that the social world, like the natural world, is a complex, interconnected web of relationships. "All ethics so far evolved," Leopold (1966, 239) wrote, "rest upon a single premise: that the individual is a member of a community of interdependent parts." Our actions are ethical, Leopold (1966, 211) goes on to say, when they contribute to the integrity, stability, and beauty of our social and ecological communities. To act ethically is to take responsibility for the webs of relationships that make us who we are, and sustain us.

Service is connectedness wedded to a sense of purpose. It entails contributing to the conservation or improvement of the communities that sustain us. As such, it is a practice grounded in care, obligation, gratitude, and belonging. Service may take sundry forms. These include efforts to secure others' basic needs, and efforts to conserve or improve the social, political, economic, educational, and ecological conditions upon which human flourishing and the fullest development of agency depends. In earlier chapters, we observed how reflective self-direction has been undermined by digital distraction and dependence, deskilling, dogmatism, and despair. It appears service is also in decline in the digital world.

Levels of service are most easily assessed by the amount of time people devote to caring for others. Such activities are central to romantic relationships, families, and friendships. Other-directed activities also occur in the workplace. In turn, there is public service, which is to say, efforts to address the needs and welfare of the body politic through elected or appointed office. The challenge of separating self-interest from service in the latter two cases is difficult, and I am unaware of studies that specifically measure changing levels of other-directedness within romantic relationships, families, and friendships. Still, what we do know about romantic relationships, families, and friendships suggests that service is in decline.

Romantic partnerships are decreasing in number. In 1990, 29% of young and middle-aged US adults were "unpartnered," having neither a spouse nor romantic cohabitant. By 2019, that figure had grown by almost a third, to 38% (Fry and Parker 2021). In turn, there has also been a steep drop in close friendships in the digital age: almost one out of every eight people now consider themselves friendless (Cox 2021). These declines in friendship begin early in life. Since the early 2010s, high school students have demonstrated a steady decrease in friendships (Monitoring the Future 2022). In turn, US teenagers and young adults have experienced "plummeting social engagement" over the last two decades (Kannan and Veazie 2023). The dramatic decrease in companionship among youth portends worsening trends within the broader population, as habits and dispositions for social isolation expand and solidify.

In turn, there has been a sharp and steady decline in the size of families. People are having significantly fewer children. Today the global fertility rate (average number of lifetime live births per woman) is less than half of what it was in the 1960s, moving from 5 to 2.3, and it continues to decline (United Nations 2024). Most countries already have a fertility rate below what would be needed to maintain their current populations, and the number of such countries is steadily growing. Throughout history, childcare has been, by far, the largest and most enduring form of service. That is changing. In turn, children currently spend about twice the number of hours each week on their digital devices as they do with parents. So, even discounting the fewer number of children, other-directed interactions within families appear to be decreasing (Donnelly 2019).

Taken together, these statistics suggest a significant decline in the number of people engaged in other-directed activities across romantic relationships, friendships, and families. The most direct evidence for declining service, however, is the decreasing number of people engaged in volunteering. Volunteering is an other-directed activity that typically focuses on the needs of children or underprivileged populations or the welfare of the cultural, social, or natural environment. It is mostly carried out by way of contributions to non-profit and community organizations, church groups, or civic associations. Service so understood is in decline.

Over the last two decades, the proportion of the American population that volunteers and gives to charity has dropped, as has the number of hours devoted to volunteering (Lenkowsky 2023; Statista 2022; Grimm and Dietz 2018). While the COVID-19 pandemic may have contributed to the trend, the decline preceded its arrival. The rate of volunteering had already decreased during the early 2010s, and the percentage of people donating to charity has been declining since the early 2000s. In 2021, the percentage of adults in the US doing unpaid work through or for an organization saw its largest decline since the government began collecting this data (Dietz and Grimm 2023). These downward trends might plateau or be reversed in the coming years. Still, there has been a significant drop in such other-directed activities, and the trend is not limited to the US (NCVO 2023).

The broad decline in service has been accompanied by a rise in narcissism. Understood as a lack of empathetic understanding and behavior, typically coupled with a sense of entitlement, narcissism is deemed epidemic today. It has been linked to digital culture (Twenge and Campbell 2009; Lerman 2019). Akhtar (2021, 20) writes that the attention economy has produced a system of "technological support for primary narcissism" in which the "ceaseless affirmation of our biases emboldens the strident certainty of our moral positions." Consequently, we end up "convinced by the moralizing rhetoric that passes off our dependence on technology as righteous activism." For narcissists, social media exchanges are primarily opportunities for self-display, self-promotion, and self-advancement. In turn, a sense of individual entitlement overpowers the responsibility to serve a greater whole. So understood, the rise in narcissism is exemplified by changes to the uses and purposes of social media over time. When Twitter reached the heights of its use in the 2010s, for example, it was widely praised for spurring public debates and democratic movements. Its slogan was "Let's talk" and Twitter was hailed as the "new town square." When Twitter was rebranded as X in 2023, the platform changed its slogan to "Blaze your glory!".

While rising levels of narcissism may contribute to declining levels of service, other factors play a role. Service is generally experienced in the act of helping others, that is, in tending to others' needs. The forms that such service takes are diverse. Whoever its recipient and whatever its manner, however, service entails attentiveness. The most fundamental act of service is the focused attention we give to members of our communities—family, friends, partners, neighbors, colleagues, and acquaintances. Such attention allows people to feel seen, heard, understood, and cared about. As the capacity for focused attention has been sharply eroded, such acts of service have steeply declined.

Young children instinctively crave attention. They want *their* experiences to be recognized. Children incessantly present their parents with opportunities—and demands—for attentive acts of service. Of course, the need to be recognized and valued does not end with childhood. This basic human need is precisely what is *not* being supplied when we are being phubbed (Knausenberger et al. 2022).

The most basic act of service is the focused attention that lets a lover, friend, family member, colleague, or acquaintance feel seen, heard, understood, and cared for. The normalization of phubbing marks a world in which such service is in sharp decline. Providing one's undivided attention to others is impossible for individuals who are digitally distracted and socially deskilled. Since time immemorial, Nass (2009) observes, the "glue" of social relationships has been the sustained attention we give to each other. This glue is being dissolved in the attention economy.

How, then, can we cultivate the dispositions, habits, and skills of self-leadership and service in the context of a digital culture that undermines their exercise and development? To answer this question, we must first investigate how dispositions, habits, and skills are acquired.

Developing Dispositions, Habits, and Skills

Dispositions and habits are not personality accoutrements. They are foundational to character. We tend to think of our conscious choices as most indicative of who we are. But as Dewey (2002, 25) writes: "a predisposition formed by a number of specific acts is an immensely more intimate and fundamental part of ourselves than are vague, general, conscious choices. All habits are demands for certain kinds of activity; and they constitute the self. In any intelligible sense of the word will, they *are* will. They form our effective desires and they furnish us with our working capacities." To speak of our characters, of who we most fundamentally are, is to reference an assembly of interpenetrated dispositions and habits (Dewey 2002, 38). In the absence of dispositions and habits, the self—far from being an autonomous chooser—would lack both the motivations and the criteria for making choices.

Dispositions, habits, and skills might be understood as congealed forms of attention. They allow the original purpose behind their development to be furthered without much conscious effort. To illustrate: the apprentice carpenter must deploy her full attention to consistently hit nails on their heads, avoiding glancing blows that bend the nails or damage lumber or fingers. The experienced carpenter employs the tool powerfully, precisely, and with great speed, but with little or no consciously focused attention. She appears to hammer effortlessly, and her mind can direct itself elsewhere, perhaps to the best way of furthering the construction project at hand. Of course, becoming proficient with a hammer required an original intention, and a great deal of consciously focused attention to hone the skill. Likewise, learning to ride a bicycle takes focused attention. Once the skill is well developed, however, it serves its purpose without our having to think much about it. So we can ride a bicycle while our minds are directed to any number of other matters. The development of dispositions and habits follows the same process. Like skills, they are congealed forms of attention that advance the pursuit of our values and interests long past the time when concentrated effort is needed for their manifesting.

If every moment of our lives were devoted to consciously focusing our attention, we would become paralyzed by the burden. Dispositions, habits, and skills free up mental, physical, and temporal resources, so our efforts and time can be deployed when and where they are most needed. The philosopher and mathematician Alfred Whitehead (1911, 61) observed that "Civilization advances by extending the number of important operations which we can perform without thinking about them. Operations of thought are like cavalry charges in a battle—they are strictly limited in number, they require fresh horses, and must only be made at decisive moments." Not all dispositions, habits, and skills advance civilization. But they can advance our values and interests, and we are wired to develop them.

Russell Poldrack (2021, 153), a professor of psychology and cognitive neuroscience at Stanford University, writes that "The brain is a habit machine, intent on automating any routine behavior so that we don't have to spend time thinking about our every move." Apart from freeing up time, dispositions, habits, and skills also make us more proficient. Artists, scientists, athletes, and entrepreneurs achieve their prowess by developing and honing key dispositions, habits, and skills. Mastery is achieved on the foundation of disciplined, regularized practice.

Developing the dispositions, habits, and skills of self-leadership and service is crucial to human flourishing. In a digital culture, it becomes an indispensable means of fighting fire with fire. Habit formation—to the point of pathological addiction—grounds the attention economy. The dispositions, habits, and skills of reflective self-direction and caring are required to resist it. Importantly, their development can be grounded on the same behavioral design techniques that are currently deployed to undermine agency. A synthesis of the moral philosophy of Aristotle and contemporary neuroscience illuminates the task at hand.

What Aristotle said about virtues applies equally to dispositions, habits, and skills. They are produced in us "neither *by* Nature nor *against* Nature." Rather, nature "prepares in us the ground for their reception" (Aristotle 1953, 55). The instincts that generate our unconscious reactions to physical stimuli (e.g. blinking when objects quickly approach our faces) might be thought of as *first nature*. We are (genetically) wired to exhibit these behaviors. No learning is required. Nature also provides us with aptitudes. Aptitudes are perceptual, physical, linguistic, cognitive, or emotional capacities that we are predisposed to develop. They are seeds awaiting a nutritive environment to grow.

Our aptitude for verbal communication is a case in point. It is a product of both innate structures (our genetic constitution and specific physiological features, such as the larynx) and upbringing (living and learning within linguistic communities). Given a conducive social environment, people naturally learn to speak. When aptitudes are well developed over time, they become *second nature*. While first nature is wholly instinctual and reactive, second nature is acquired and more responsive. Still, second-nature responsiveness comes to feel like first-nature reactiveness the more habitual and skillful it becomes. That is why walking and speaking seem effortless, though once they were difficult challenges for all of us.

The stickiness of habits, Poldrack (2021, 153) observes "usually serves us well, until it doesn't." In the attention economy, digital habits are designed to be very sticky—in high use, difficult to turn away from, and frequently returned to. By and large, they do not serve us well. So other habits must be formed to combat those which are getting us into trouble. The behavioral design efforts of Nir Eyal and B. J. Fogg demonstrate the sharpness of this double-edged sword.

Like many former executives and developers of digital devices, apps, and platforms, Eyal and Fogg have played both sides of the board. As innovators, they worked with digital media corporations and entrepreneurs to make their devices, apps, and platforms sticky. More recently, they have focused on helping individuals free themselves from digital distraction and dependence without, at the same time, demanding the cessation of all digital engagements.

Fogg's early work at the Stanford University Persuasive Technology Lab, later called the Behavior Design Lab, was devoted to *captology*, a word Fogg coined to denote the use of computers as persuasive technologies. Captology was of great interest in the burgeoning attention economy. Consequently, the Stanford lab "earned a cult-like following among entrepreneurs hoping to master Fogg's principles of 'behavior design'" (Bosker 2016). One of Fogg's former students, Tristan Harris (2017), was greatly impacted by the ways Fogg designed digital technologies to enthrall users. Harris is a central figure in a 2020 Netflix documentary addressing the problem of digital dependence entitled *The*

Social Dilemma. With Aza Raskin, he now heads the *Center for Humane Technology*, an organization addressing the human downgrading that occurs when digital technologies "scoop attention" out of our brains to increase revenues and, either intentionally or as a byproduct of profit-making, addict us to their products (Lerman 2019).

For his part, Fogg acknowledges the "dark side of Persuasive Technology" (Standford Behavior Design Lab). He observes that "bad habits are not fundamentally different from good habits when it comes to basic components.... When you uncover a universal principle, it has the potential to be used for good or evil" (Fogg 2020, 202, 270). While Fogg attended to the ethical use of behavior design from the outset, his most recent work focuses on promoting individual autonomy and collective action for peace, human welfare, and happiness. As Fogg observes, "Behavior Design is not just about ... putting the phone down during dinner. It's about becoming the person you want to be—and creating the kind of family, team, community, and *world* we want to live in.... Habits may be the smallest units of transformation, but they're also the most fundamental. They are the first concentric circles of change that will spiral out" (Fogg 2020, 259, 267). Fogg notes that the techniques he describes might be deployed in "powerful ways" for unsavory ends. Accordingly, he pleads with readers to utilize them "only for the most noble of purposes" (Fogg 2020, 247–248).

Eyal followed a similar path. In his book *Hooked: How to Build Habit-Forming Products*, Eyal sought "to unleash the tremendous new power innovators and entrepreneurs have to influence the everyday lives of billions of people." Promoting "habit-forming technology" as a "new superpower," Eyal encouraged businesses to create "programmed behaviors," which are "very good for the bottom line" (Eyal 2014, 10–11, 17). In contrast, Eyal's (2019a) subsequent book, *Indistractable*, assists those who have succumbed to digital technology, helping them regain their lost autonomy.[1]

Eyal's and Fogg's shift from the supply side of the attention economy to its demand side does not indicate a rejection, or even revision, of the science of behavior design. Quite the opposite. The same techniques

[1] The business practices that Eyal endorsed in *Hooked* have a counterpart in his later work. Eyal offered to pay readers $10 to $100 to write reviews of *Indistractable*. As Eyal presumably decided which reviews merited payment, readers were being nudged to write enthusiastic endorsements (https://thenetworkstate.com/indistractable/).

exploited by digital platforms to hook users can also be deployed to regain individual agency. It all goes back to Aristotle's habit theory of virtue.

Aristotle insisted that virtues and vices are not the products of conceptual beliefs or moral convictions. Rather, virtues and vices develop as habits. One becomes just, the ancient Greek philosopher insisted, by performing just actions, temperate by performing temperate actions, courageous by performing courageous actions, and prudent by exercising practical judgment. The same goes for vices. In the end, our moral stature is the product of what we (regularly) do. Virtues and vices, like other habits and skills, are developed by way of practice. Quoting the poet Evenus, Aristotle1953, 217) maintains that "Habit … is practice long pursued, that at the last becomes the man himself." Practice may not make us perfect. It does make us more or less skillful, and more or less virtuous. Ensuring that children develop the right habits, it follows, is of supreme importance (Aristotle 1953, 56).

For Aristotle, *phronesis* contributes to the development of all moral habits. One can only exercise courage, for example, by employing practical judgment to determine which dangers ought to be faced, to what extent, and for what purpose. Putting oneself in harm's way for no good reason is the vice of foolhardiness, not the virtue of courage. Practical judgment regulates the virtues by moderating or encouraging their exercise and limiting or expanding their respective domains. Virtue, Aristotle insisted, is a mean between excess and deficiency. Practical judgment determines where this means is to be found.

An examination of the disposition the ancient Greeks called akrasia (ἀκρασία) illustrates the challenge of maintaining virtuous habits. Often translated as incontinence, akrasia is the lack of self-restraint or self-control. Akratic individuals are not vicious. Rather, they are insufficiently virtuous because they are overcome by appetites or passions. Akratic individuals demonstrate a lack of temperance owing either to their impetuousness (thoughtlessly acting under the influence of a powerful passion) or weakness of will (inability to carry out the action needed to achieve a correctly identified goal). Plato denied the existence of akrasia in the *Protagorus,* where he insists that no one willingly embraces the bad but does so only out of ignorance. Aristotle, his student, insisted that people do act against their better judgments. Poets, philosophers, and religious thinkers throughout the ages concur with Aristotle. In the *Metamorphoses* (bk 7, II.20–21), likely based on a line from Euripides' *Meda* (1078–80), the Roman poet Ovid writes: "I see what is better and approve of it; I follow what is worse." Likewise, in his *Letter to the Romans* (7:19), Paul writes "I do not do the good I want, but the evil I do not want is what I do."

Lack of self-restraint is frequently thought to occur when passion overcomes reason. However, akratic individuals are not lacking in rationality per se. Rather, they are not habituated to desire the right sorts of things with sufficient force. They know well enough what the right thing to do is and what pernicious results will follow from doing the wrong thing. But for the weak-willed, Aristotle observes, knowledge brings no profit. That is because the disposition to do the right thing on a regular basis is not a matter of knowledge or cognitive conviction. It is a habit grounded in a well-educated desire. The akratic individual is not habituated to doing the right thing, as knowledge has not been "worked into the living texture of the mind" (Aristotle 1953, 200). And so, too often, fleeting appetites or strong passions get the upper hand.

Moral development entails self-overcoming, wherein parts of the self—base appetites and raw passions—are held in check so a worthy intention might be fulfilled. For the akratic individual, the desire to act generously, courageously, or moderately has been insufficiently developed because instances of acting generously, courageously, or moderately have been too few and far between. Moral habits are not simply routine acts. Rather, they evidence a predilection for certain sorts of pleasure. The akratic individual is not sufficiently habituated to find virtuous acts pleasant, and so, all too often, fails to act virtuously. Virtue is a stable state of the soul rather than an acquired ability, Aristotle maintains, because a virtuous person has an enduring disposition to experience virtue as fulfilling.

The mental resources we employ to act virtuously are likened by Aristotle to the physical resources we employ to exercise our muscles. Bodily strength results from "taking in much nourishment and going in for hard training, and it is the strong man who is best fitted to cope with such conditions" (Aristotle 1953, 58). The same applies to the virtues. It is by refraining from impulses that we become temperate, and once we have become temperate, we are most able to restrain impulses. The self-overcoming that allows us to abstain from instant gratifications produces the virtue of temperance, and when this virtue is well-developed, forgoing temptations becomes easier. Indeed, not indulging one's appetites can become quite pleasurable, just as the physically strong can enjoy the exertions of heavy training.

The development of virtue typically begins in acts of self-regulation that are experienced as self-denial, as is the case when children are encouraged, or forced, to delay gratification. By habituating ourselves to the exercise of our moral muscles, however, virtuous actions become both easier and more pleasurable. "A man is temperate," Aristotle (1953, 59) writes, "who abstaining from bodily pleasures finds this abstinence pleasant; if he finds it irksome, he is intemperate." As philosophers and scholars since Aristotle's time have attested, the development of a moral life primarily entails an "education of the emotions" (De Sousa 2003, 297). Why does the temperate person find abstinence pleasurable? Why does the generous person find generosity pleasurable? Why is virtue said to be its own reward? In large part, the pleasure of acting virtuously arises because we relish the experience of self-leadership and service.

The name Aristotle gave to the reflective, self-directed individual whose emotions are well-educated is the *phronimos* (φρόνιμος). The *phronimos* is habituated to virtuous action, but is not, for that reason, inflexible or wholly routinized in behavior. Habit orients but does not predetermine action. That is because the most crucial habit of the *phronimos* is reflective self-direction, the exercise of practical judgment.

Practical judgment is a moral virtue, and not merely an intellectual one, because it not only determines the mean but also motivates its embrace. In this respect, practical judgment marshals the right sort of desires such that we feel "at the right times on the right occasions towards the right people for the right motive and in the right way" (Aristotle 65, 173). It does so by way of self-restraint that prevents passing appetites and passions from ruling the day. Along with the capacity to foresee the "remote consequences of all our actions," Adam Smith wrote, prudence entails "self-command, by which we are enabled to abstain from present pleasure or to endure present pain, in order to obtain a greater pleasure or to avoid a greater pain in some future time" (Smith 2000, 271). But it is mistaken to believe that practical judgment, or any virtue, is simply a matter of delaying gratification to achieve a greater benefit at a later date. Rather, virtue entails a disposition to find certain sorts of pleasures gratifying.

Of course, most good habits require delayed gratification. Still, sensory gratifications can be delayed while psychological and emotional pleasures are being enjoyed. And this immediate gratification—a positive sense of competence or self-worth—can be much more pleasurable than the scratching of some bodily itch. Strenuous exercise, for example, can be rather painful. Moreover, a particular bout of strenuous exercise may not yield any noticeable gains in physical health, stamina, or strength. Still, completing a good workout can produce a feeling of competence and success, regardless of its immediate impact on physiology or stamina. Indeed, the experience of competence and success may depend upon experiencing physical strain and exhaustion. The pain or discomfort involved in a good workout, or any demanding task, can be a prerequisite for the pleasure experienced in carrying it out. Good habits form because we enjoy watching ourselves become competent, skilled, or virtuous (Thiele 2012). When acting virtuously, we take delight in being the worthy protagonists of our lives. We experience the pleasure of exercising self-leadership and service. This experience, unsurprisingly, has a neurological foundation.

Remodeling our Brains

Virtues are not passing inclinations. They are dispositions to act in a particular manner on a regular basis. Dispositions, like habits, are slowly woven into the living texture of the mind. Dispositions and habits are "patterns of interconnectivity between neurons" that have been "glued together by life" (LeDoux 2002, ix, 2; see also Thiele 2006, 203–224). Repeated behaviors induce the release of neurotransmitters and hormones, the strengthening or weakening of synaptic linkages, the development of new synaptic connections and entirely new neurons, and the pruning of lesser-used neurons and neural connections. By way of all this neurological and bio-chemical activity, dispositions, habits, and skills secure a structural home in the brain.

If exposure to a particular environment elicits a particular response, then the neural circuits required to produce this response get fired up. Faced with a similar challenge at a later date, a similar response occurs. Now it is somewhat quicker and more efficient because the synaptic pathway has been strengthened by way of its earlier use. Neurons that fire together, it is said, wire together. They are like the deeper and wider rivulets created by rain flowing down a slope that comes to channel most of the flow. Virtues and vices, like all habits and skills, are synaptic linkages that have become dominant through repeated use. Neurologically and morally, we become what we do.

As experiences wire our neurons together, the volume and density of the corresponding brain tissues may increase (Boyke et al. 2008, 7031). Neurons that do not get fired, in contrast, tend to atrophy, and the density or volume of related brain tissues diminishes. As LeDoux observes, "synapses in the brain, like animals in their environments, compete to stay alive. Synapses that are used compete successfully and survive, while those that are not used perish" (LeDoux 2002, 73; see also Edelman 1987). The brain's circuitry effectively operates and evolves under the constraints of what might be called neural natural selection. Synaptically speaking, it is a matter of the "survival of the busiest" (Schwartz and Begley 2002, 117). Unused neural circuits do not always atrophy and perish; they can also be co-opted for other purposes. This contributes to *neural plasticity*, cortical remapping, and neuronal growth, allowing the brain to adapt in response to environmental demands, trauma, and other experiences (Johansson 2007, 1417).

A good portion of the newborn's brain is effectively hardwired, enabling the performance of basic life functions by vital organs, the startle reaction, suckle tendencies, and smiling responses. None of these capacities are products of learning. They are innate. The newborn's brain is also configured to learn, and this learning blazes new synaptic pathways in its brain. Adults do not exhibit anywhere near the level of neuronal growth, pruning, or reorganization that children undergo. Still, brain imaging over the last few decades has documented significant "cortical remapping" in adults.

Well beyond childhood and youth, repeated behaviors continue to modify the structure and function of various regions of the brain. Frequently utilized neural circuits demonstrate gains in speed, strength, and overall proficiency. Some hardwired areas of specialization (especially in sub-cortical regions) resist reorganization. But neural plasticity is regularly demonstrated whenever new skills and behaviors are developed (Merzenich and deCharms 1996; Ramachandran and Blakeslee 1998; LeDoux 2002; Doyon and Benali 2005; Boyke et al. 2008; Kleim and Jones 2008; Jolles et al. 2010). The brain is a dynamic organ that is regularly renovated by way of experiences. Neurologically speaking, we are students of life. It would be hyperbole to say that we are the architects of our own brains, as there are many innate neurological structures and processes that are largely immutable. Still, we can engage in significant remodeling.

Importantly, internal "experiences" can also remodel the brain. Neuroplasticity, the brain's capacity for changing its biochemistry and neural structures, can be the product of conscious, mental effort (Merzenich and deCharms 1996, 77; Cozolino 2002; Paquette et al. 2003; Cacioppo and Berntson 2002; Heath and Heath 2008). Scholars of neuroplasticity observe that "the brain remodels itself… as the result not only of passively experienced factors such as enriched environments, but also of changes in the ways we behave … and the ways we think…. If the brain is like a map of lived experience, then the mind can, with directed effort, function as its own internally directed mapmaker" (Schwartz and Begley 2002, 164, 253–54, 200). Shifting our thoughts on a regular basis can reshape our brains.

Dispositions, habits, and skills develop when individuals face demanding situations that prompt effective coping mechanisms, and when competencies and successes are (mentally) acknowledged and rewarded. To best develop dispositions, habits, and skills, therefore, we have to put ourselves in situations that demand effort and competence. When self-efficacy is exhibited, acknowledging the accomplishment provides an emotional lift that reinforces the behavior (Bandura 1992; Tice et al. 2004; Thiele 2012). The process is enhanced by the release of neurotransmitters, such as dopamine, serotonin, endorphins, and oxytocin, which play critical roles in regulating moods, feelings, and the reward system. As we have seen, dopamine motivates us to pursue rewards and induces pleasure in anticipation of them. Other neurotransmitters enable us to enjoy attained rewards and leave us feeling satiated. Engaging in healthy habits releases the neurotransmitters that generate the positive emotions accompanying the perception of self-direction, efficacy, and accomplishment. These emotional responses have a restorative effect, recharging inner resources after exertions. They also reinforce behavior.

Cognitive psychology buttresses Aristotle's claim in these matters: the practice of self-regulation is self-reinforcing (Schmeichel and Baumeister 2004, 95). Moral strength, like physical strength, requires training. With sufficient training, however, virtuous behavior becomes easier and more pleasant. This neurological process has been called the "bootstrapping effect." We effectively lift ourselves up (develop the neural circuits that allow us to do things skillfully and habitually) by our own bootstraps (coping activities that produce a sense of efficacy and accomplishment). When Aristotle suggested some two millennia ago that virtue serves as its own reward, he was not saying that good deeds will help us secure worldly fortune or will be rewarded in the afterlife. He was describing the bootstrapping effect. Acting virtuously generates a positive sense of self-worth—the sense of being a reflective, self-directed individual whose skillful service achieves its ends. This gratifying self-perception, the experience of being the worthy protagonist of one's life, increases the likelihood of repeating and thereby strengthening virtuous practices.

Dispositions and habits are territorial animals. To take up residence in the brain, they must drive out the competition. As the psyche abhors a vacuum, ending an existing habit requires rewiring the brain for an alternate behavior. As Dewey (2002, 195) observed, "Bad habits can be subdued only by being utilized as elements in a new, more generous and comprehensive scheme of action, and good ones can be preserved from rot only by similar use." This subduing and preserving is not a "once and done" affair. Like skills, habits strengthen by way of practice and degrade from disuse. Neurons that are not regularly fired together will wither or become co-opted for other purposes. That is why top athletes, artists, and experts in every field incessantly hone their skills. And that is why the lapse of any good habit presents the opportunity for the resurgence of a bad habit.

The positive emotions that accompany virtuous action do not guarantee the proliferation of good habits. Vice, after all, is grounded on the same brain chemistry: the rewards are simply different and typically more immediate. So cultivating virtue is difficult. Today the deck is stacked against us by sophisticated digital systems designed to make us immoderate and impatient. The positive emotions generated by good habits face the fiercest competition in the attention economy: the immediate pleasures of supernormal stimuli.

Staying fit, trim, and healthy by exercising regularly and not overeating has been a challenge that our species has faced for millennia. Doing so within an industrial food system is especially demanding. It requires focused attention, wise planning, dedicated commitment, and steady practice. Kessler (2010, 193–195) writes:

> Over time, if I'm successful at turning down the salience of a [hyperpalatable food] cue often enough, its intensity will diminish as new learning substitutes for the old. By increasing the activation of neurons that represent one behavior … the prefrontal cortex succeeds at suppressing the alternative.… Over time, it can become second nature to follow those rules [of healthy eating]. But until then, rules need to be 'kept in mind'

> so that they are available to guide behavior in the moments we need them. That requires attention, practice, and advance planning, motivated by the expectation that you will ultimately derive emotional satisfaction in new ways. Your ability to follow your own rules will eventually carry its own reward.

Healthy eating, though particularly challenging in a world overflowing with hyperpalatable foods, can become self-sustaining. At the start, however, it is a steep, uphill climb. Cultivating the dispositions, habits, and skills of self-leadership and service in an AI-enabled attention economy is equally challenging. It requires focused attention, wise planning, dedicated commitment, and steady practice. The techniques of behavior design can help.

Behavior Design

Cognitive-behavioral therapy (CBT) is a well-regarded and widely practiced form of brain remodeling. CBT is a problem-focused and solutions-oriented form of therapy. It has been demonstrated to be effective in treating depression, anxiety disorders, eating disorders, obsessive–compulsive disorders, phobias, post-traumatic stress disorders, and substance abuse. CBT assumes that psychological problems are grounded on distorted thinking and maladaptive behavior. Individuals undergoing CBT learn to recognize, interrogate, and subsequently uproot noxious cognitive habits. For people suffering from depression or anxiety disorders, for example, noxious cognitive habits might include negative filtering (only paying attention to the bad things that happen), excessive rumination, and catastrophizing (always thinking the worst). By acknowledging and challenging mental distortions and maladaptive behaviors, individuals learn to weaken their hold. Subsequently, unhealthy cognitive and behavioral habits get replaced with more effective and adaptive ones.

CBT changes the way people think about themselves. Modifying self-representations is a key step towards changing behaviors. Effectively, CBT patients come to perceive themselves as agents, rather than victims of circumstances or unalterable character traits. They engage in reflective self-direction, developing greater self-awareness and more beneficial practices, and remap their brains in the process. Numerous studies employing brain-imaging technology have demonstrated that CBT alters brain structures and functions. This cortical remapping is partially the product of conscious efforts to conceptually reframe behavior, making it meaningful by placing it in the context of greater self-understanding and self-direction (Cozolino 2002, 27, 63–64). CBT, one might say, is a therapeutic technique that helps people struggling with debilitating behaviors to become the protagonists of their lives.

The science of behavior design that Fogg and Eyal promote is grounded on the same neurological mechanisms as CBT. Fogg observes that all habits, good and bad, are products of three factors that must occur together: motivation, ability, and prompts (Fogg 2020, 202). Motivation provides the emotional foundation for a habit, attracting us to a specific behavior. Ability supplies the practical wherewithal for the behavior, allowing it to be carried out. Prompts let it be known that there is something good to be got; they are the sparks that ignite the kindling of motivation and ability into a self-reinforcing behavior.

Habits are grounded in the pleasures they produce. These pleasant sensations need not be enduring. Indeed, compulsive habits are typically grounded in fleeting feelings. Digital platforms ingeniously exploit this mechanism. They regularly supply small rewards to build powerful habits. Even trivial and passing pleasures can generate an enduring habit. Most digital games and apps, such as *Candy Crush* (which has been downloaded billions of times) are designed with this in mind. "To help signal that you've been successful," Fogg writes, "the designers built in all kinds of fun sensory experiences. There are pleasant little dings and satisfying visual cues. The word "sweet" even pops up after you reach a certain score. The result? You feel successful really fast—and you keep firing up that *Candy Crush* app whenever you have a free minute or two" (Fogg 2020, 136). In the same manner, an Instagram app that makes it easy to alter photographs with filters provides ready rewards. "As you try out different filters," Fogg writes, "you see your photo transform before your eyes like magic, and your photo isn't merely a photo anymore. You feel like you are sharing a unique artistic creation. You might even be surprised or impressed by your skills. When that happens, your brain releases dopamine, and you seek opportunities to use Instagram again because it feels good" (Fogg 2020, 137). One of the co-founders of Instagram, it turns out, was an alumnus of Fogg's program at Stanford. Successful digital platforms have perfected the practice of providing enticing, if not enduring, rewards. They know how to make us feel good, if only fleetingly.

Fogg (2020, 137) drives home the point that "Emotions create habits. Not repetition. Not frequency. Not fairy dust. Emotions. When you are designing for habit formation—for yourself or for someone else—you are really designing for emotions." Emotions are crucial. Still, they do not create habits in the absence of repetition and frequency. Rather, habits get formed owing to the regularity and frequency of the emotional lifts that they provide.

To design for emotion is to exploit the dopamine cycle. "What happens in your brain when you experience positive reinforcement isn't magic—it's neurochemical," Fogg (2020, 136) writes. "Good feelings spur the production of a neurotransmitter (a chemical messenger in the brain) called dopamine that controls the brain's 'reward system' and helps us remember what behavior led to feeling good so we will do it again." Fogg is rather loose in his discussion of brain chemistry. Dopamine is released in anticipation of a reward, not only with its attainment. That is why people addicted to gambling will "chase losses," playing on even when they are not winning. Likewise, drug addicts experience dopamine spikes in response to the drug-conditioned cues associated with craving, and these spikes may be higher than those produced by drug use itself (Volkow et al. 2011). In short, dopamine is associated with the thrill of the hunt as well as the pleasure of the feast.

The glee of children opening presents often surpasses the pleasures they experience once the gift is revealed. In the same fashion, substance addictions are grounded in the intense dopamine spikes associated with seeking a "high" combined with lower dopamine spikes (owing to the blunting effects of downregulation) associated with gratification. Users of digital platforms and apps also spike their dopamine as they anticipate microbursts of positive emotions associated with pleasant sounds, words, or visual effects, the satisfaction of curiosity, a sense of approval, or feelings of competence or success. Again, the anticipations associated with the hunt may be more pleasurable than the consumption of the feast.

The achievement of long-term goals is too rare an event to serve habit formation. For positive emotions to be reinforcing, they must be tied to proximate behaviors. That is why habits are most easily formed when they are experienced as pleasant in their performance. This feature of behavioral design, Fogg observes, is the "antidote to the go-big-or-go-home culture" (Fogg 2020, 155). The pleasant experience of anticipating and achieving "small wins" that each performance of the target behavior produces, rather than the end state that the behavior ultimately achieves, becomes the crucial emotional resource. Successful habit formation occurs when rewards happen "either during the behavior or milliseconds afterward" (Fogg 2020, 140). Accordingly, Fogg recommends that we "celebrate" immediately upon completing a target behavior. He employs the term "*feeling shine*" for the "authentic pride" we experience in these celebrations.

Feeling shine builds healthy habits by rewiring the brain to yearn for the self-esteem experienced upon the successful completion of a desired behavior. Fogg writes: "When you celebrate effectively, you tap into the reward circuitry of your brain. By feeling good at the right moment, you cause your brain to recognize and encode the sequence of behaviors you just performed. In other words, you can hack your brain to create a habit by celebrating and reinforcing" (Fogg 2020, 134). Effectively, feeling shine serves the same function as the shares, likes, clapping hands, smiley faces, and hearts one receives on social media posts. As feeling shine is grounded in reflective self-direction, however, it abets the creation of healthy habits and skills.

Feeling shine is a better means of establishing habits than self-recrimination. Fogg counsels against "de-motivators." Making yourself feel bad by means of reprimands or self-punishment is generally less effective and can be counterproductive. Congratulatory gestures, smiles, and words of affirmation spoken aloud or silently prove more effective. Indeed, healthy habit formation can be accelerated by increasing the number of celebrations: one when initiating the target behavior, one when doing it, and a final one upon completion (Fogg 2020, 151).

Different strokes work for different folks. Finding a way to experience shine that feels natural and pleasant may take some experimentation. In any case, Fogg observes, "When you find a celebration that works for you, and you do it immediately after a new behavior, your brain repatterns to make that behavior more automatic in the future" (Fogg 2020, 140). Once the habit is well established, celebrations become optional. One will still feel pleasure upon the successful execution of a good habit, but conscious self-affirmations are no longer needed. Effectively, they have become internalized. When creating habits, however, frequently feeling shine is indispensable (Fogg 2020, 140).

Motivation is one of three crucial components of habit formation. Another is ability. Ending a bad habit can often be as simple as disabling oneself. In his epic poem the *Odyssey,* Homer recounts the adventures of Odysseus on his voyage back from Troy. En route, the Greek hero longs to hear the song of the Sirens. However, he does not want to succumb to the Sirens' enchantment and crash his ship upon the rocks surrounding their island, as was the fate of so many other sailors. On the advice of Circe, Odysseus orders his crew to tie him to the mast and stuff their ears with wax until the ship has safely passed the island. In this manner, Odysseus can revel in the bewitching song without suffering the fatal consequences of enchantment.

There are many ways to "break the ability chain" by stuffing our ears with wax or tying ourselves to a mast. In 334 BC, Alexander of Macedon burned his ships on the shores of Asia Minor to ensure his troops would fight to the death rather than attempt a retreat. The Chinese warlord Xiang Yu employed the same tactic after fording a river in 207 BC, ensuring his victory in the Battle of Julu over the Qin Dynasty. In 1519, the Spanish Conquistador Hernán Cortés followed suit on Mexican shores, scuttling his ships so his troops would fight the fearsome Aztecs rather than mutiny and return to Spain.

These examples of disablement are called precommitments. They involve the advance imposition of obstacles or costs to unwanted future behavior or outcomes. Alcoholics on the mend who take *Antabuse* (disulfiram) effectively burn their ships in their battle with addiction. Having ingested the drug, they will experience intense flushing, sweating, vertigo, throbbing aches and pains, nausea, vomiting, and other discomforts when even small amounts of alcohol are consumed. This "alcohol antagonist" is made by the appropriately named *Odyssey Pharmaceuticals*. Therapists who treat addiction and compulsive disorders consider acts of "self-binding" crucial to recovery (Lembke 2021).

Not keeping sweets in the house is a self-binding precommitment that can help break a habit of unhealthy late-night snacking. Motivations are initially unaltered, as the craving for sweets persists. But the absence of goodies helps break the bad habit of consuming them. As new activities come to take the place of eating sweets at night, the craving, over time, may wither. It is very difficult to deter or deflect strong motivations by sheer force of will. So often the best approach is to disable oneself in advance, when willpower and foresight are present (Fogg 2020, 216). Self-binding and other precommitment devices are widely recognized as effective means of exercising self-control (Duckworth et al. 2018, 107).

In this vein, Lanier (2018) counsels the immediate deletion of all our social media accounts, believing that forward-looking self-disablement will prove more effective than real-time efforts at self-control. Likewise,

Susan Maushart (2010) tells the entertaining tale of breaking the ability chain when she engaged in a six-month experiment in digital detoxing with her three unwilling teenagers. This sort of thing is not for the faint of heart. Less drastic methods of precommitment are now widely available. Many people limit their online activities by exchanging their smartphones for "dumb" phones that make calls but cannot access the Internet. In turn, there is a growing supply of timed smartphone "jails," including lock boxes, which temporarily make access impossible. Numerous phone sleeves or pouches that abet self-regulation are also on the market, such as the Resistor Case whose "Velcro acts as a gateway to conscience: It provides just enough resistance and noise to make its owner think twice before opening it" (O'Gorman 2018). In turn, there are the aforementioned apps that temporarily disable or delay the use of digital technology (e.g. prevent the use of smartphones while driving). These interventions interrupt compulsive behavior in real time, aiding the effort to break bad habits (Gonzalez 2018).

Making "pacts" with oneself to achieve desired goals can be an effective form of precommitment (Epton et al. 2017; Eyal 2019b, 7). Pacts combat bad habits by causing the target behavior to entail a prohibitive investment of effort, time, or money. Public pacts are particularly effective. Friends and associates gathering at restaurants, for example, can adopt the "You check, you pay the check!" pact, wherein the first person at the table to check his or her phone for incoming texts or notifications gets the bill. Making such commitments public, announcing them to family and friends, or carrying them out in tandem with others, increases their effectiveness, distributes their benefits, and provides the immediate reward of a sense of solidarity (McKenzie-Mohr 2011, 45–60; Epton et al. 2017).

One type of precommitment is an implementation intention. Here, a commitment is made to respond to a specific cue with a particular sort of action. If my goal is to complete tasks without succumbing to digital distractions, for example, then I might articulate the following intention: "Whenever I hear the sound of a digital notification (the cue), I will respond by taking a deep breath and reflect momentarily on the significance of my current task." Implementing this intention increases the chances that the digital notification will be ignored and that the current task will be completed. Implementation intentions employ environmental cues to further goal achievement by way of pre-selected responses that, over time, become automatic (Gollwitzer 1999; Gollwitzer et al. 2004).

They are particularly useful when the option of breaking the ability chain (e.g. disabling digital devices) is not possible or practical.

This brings us to the third component of habit formation: prompts. Also known as cues or triggers, prompts alert us to opportunities to perform target behaviors. Without prompts, target behaviors would not be initiated with the frequency required to create or sustain a new habit. In the same vein, ending bad habits requires us to avoid triggers that lure us into noxious behavior. Turning off the notifications on computers and smartphones, for instance, may be the most effective method of ending the habit of compulsively checking digital devices. With this in mind, Eyal (2019b, 7) argues that we can become "indistractable" by removing the triggers that keep us compulsively online.

The effectiveness of a prompt depends upon its arrival at the opportune moment. Fogg predicts that standardized prompts—the types that currently bombard us throughout the day—will become increasingly ineffective as their numbers grow and screening and blocking mechanisms get deployed. The "gold standard" for businesses in the future will be "action prompts" that are micro-targeted to individuals' specific behaviors and daily routines, most often employing AI to gather and exploit personal data for this purpose (Fogg 2020, 118–119). Personalized prompts (and nudges) will be much harder to resist.

Prompts can be both external and internal. External prompts are constantly being adapted within the attention economy to heighten their effectiveness. Equally worrisome are internal triggers. The root causes of distraction today, Eyal insists (2019b, 12), are not auditory or visual prompts from digital devices: the problem is "what's happening within." Distraction is "your brain ducking challenging feelings such as boredom, loneliness, insecurity, fatigue, and uncertainty. These are the internal triggers—the root causes—that prompt you to find the comfort of distraction and open a browser tab, Twitter or email, instead of focusing on the matter at hand" (Eyal 2020a, 2020b). Electronic notifications successfully initiate target behaviors owing to the receptive psychological state of the person receiving them. Digital prompts are effective because they announce the opportunity to escape boredom, anxiety, fear, fatigue, or loneliness. When students are asked why they most often switch tasks, the two top reasons are getting a text message, an external trigger, and feeling bored, an internal trigger (Gazzaley and Rosen 2016, 168). In this respect, effective digital media, as Fogg suggests, design for emotion. But the emotions they design for are as much those their users hope to escape, such as boredom and anxiety, as those they want to indulge, such as the desire for novelty, choice abundance, or social approval.

We are prone to distractions, in no small part, because we want to avoid bad feelings. This dynamic produces a never-ending game of whack-a-mole wherein negative emotions get temporarily knocked out of sight by digital distractions only to reemerge hours, minutes, or even seconds later. Consequently, people are nearly as likely to become distracted from "self-interruptions" as they are from external ones (Mark 2015, 48). When external distractions are common, however, they can increase the frequency of self-distractions. For every additional external interruption experienced each hour, a study revealed, self-interruptions increase by 8% in the subsequent hour. Effectively, we can be "conditioned" by environmental triggers to more frequently distract ourselves, even in the absence of external prompts. In the attention economy, we "become habituated to self-interrupt" (Mark 2015, 36).

As digital media are constantly enticing our monkey minds with prompts that promise hyperstimulative treats, large portions of our days, and increasingly even our nights, are given over to distractions. Indulging this habit, over time, heightens our proclivity for self-interruptions, further undermining our capacity to focus attention. The data are damning. Current attention spans—focusing on a single task without switching screens, for example—are now under a minute. That is less than a third of what attention spans were two decades ago (Mark 2023, 95).

Implementation intentions can help focus attention. One might commit to taking a deep breath, briefly identifying one's feelings, and reminding oneself of the task at hand, when exposed to an external trigger or when tempted to escape from anxiety or boredom by way of self-distraction. Refocusing attention on intentions, rather than dipping into the dopamine cycle every time one is triggered by emotions or external cues, contributes to the experience of being the protagonist of one's life. As Eyal (2020b) observes, "You won't always be able to control how you feel—but you can learn to control how you react to the way you feel." A form of implementation intention commonly employed in CBT and the treatment of addictions is called *urge surfing*. Urge surfing entails non-judgmentally observing oneself while experiencing the desire to engage in a compulsive behavior. Optimally, this attentive self-observation continues until the desire wanes or until enough emotional space and willpower develop to initiate an alternative behavior. When struck by an over-powering urge to succumb to a digital distraction, one identifies the external or internal trigger that prompted it, the associated sensations and feelings (e.g. nervousness, tiredness, sadness), and the accompanying environmental circumstances (e.g. preceding events or behaviors, time of day). Recognizing and naming what is going on inside and around you replaces reaction with understanding. The goal is for acts of observation and reflection to delay satisfaction of the urge long enough for the impulse to recede, allowing the redirection of behavior. Urge surfing also provides useful information about the triggers and circumstances that actuate compulsive habits, allowing an effective adjustment to the environment.

When surfing the urge, we become attentive witnesses to our inner dramas and their outcomes. The idea is to examine each thought and feeling as raw data without judgment, as if watching leaves floating down a stream. If a few minutes of reflection do not cause the impulse to evaporate, allowing its satisfaction may be the best option. That way, surfing the urge does not come to be perceived as an act of (eventual) self-denial. The point, in any case, is not to repress strong inclinations or guilt oneself into restraint. Rather, the goal is to uncover the origins of reactive behaviors and witness their impacts. Surfing the urge is a form of reflection that generates impulse control as a byproduct. When surfing the urge, we are watching our protagonist at work. Reflection sets the stage for self-direction.

Escaping boredom, anxiety, fear, fatigue, or loneliness by way of distractions is a common and age-old human foible. Today, however, distractions are systematically engineered, enhanced, and globally proliferated. In previous chapters, we saw how our natural inclinations for approval, novelty, abundant choices, danger-rich phenomena, and instant gratification have been hijacked in the attention economy to maximize engagement. In the same manner, our natural tendency to escape negative emotions by way of self-interruptions has been commandeered. Urge surfing can prevent a downward spiral. Rather than reacting to external and internal triggers, we watch our compulsions dissolve in the light of reflection and subsequently move beyond them by way of self-directed activities. As we become less apprehensive of the negative feelings that prompt distraction, we no longer try to avoid the pauses in life that may bring them to the surface. Rather, such pauses may be welcomed as opportunities for reflection and the self-direction it begets.

Over a century ago, Dewey (2002, 20) observed that "We cannot change habit directly: that notion is magic. But we can change it indirectly by modifying conditions, by an intelligent selecting and weighting of the objects which engage attention and which influence the fulfillment of desire." Reflective self-direction allows us to change our habits by selecting and weighting the objects of our attention. Outwardly, the impact is altered behavior. Inwardly, the impact is altered neural biochemistry. When we engage in reflective self-direction we are, among other things, regulating how much dopamine gets deployed. Too little dopamine, and we become complacent with mediocrity and self-satisfied in ignorance. Too much dopamine, and we chronically chase elusive goals, forever dissatisfied with the here and now. That is the fate of obsessive video gamers and anxious multitaskers on social media, no less than workaholics. As we have seen, an excess of dopamine can lead to down-regulation, such that the pleasures of life lose their appeal, leaving one bored and anhedonic.

The creation of both good habits and bad habits is grounded in the brain's biochemistry. Good habits form when dopamine is released in moderation. In turn, good habits activate its moderate release. Bad habits form in the excess or scarcity of dopamine, and cause more of the same. Aristotle was on track in observing that virtue is a means that has been worked into the living texture of the mind. Neuroscience and behavior design reveal how you can become the protagonist of your life by remodeling your brain and rebalancing its chemistry.

Self-knowledge, Self-leadership, and Service

Like any art or skill, the aptitude for reflective self-direction develops through practice. Opportunities for its exercise, however, are systematically stymied owing to the ubiquity and intrusiveness of digital distractions. These feats of engineering are eliminating the pauses in life that otherwise might give rise to agency. Andrew Sullivan (2016) writes that "The smartphone revolution of the past decade can be seen [as ensuring that the] few remaining redoubts of quiet—the tiny cracks of inactivity in our lives—are being methodically filled with more stimulus and noise.... We keep swiping and swiping because we are never fully satisfied.... [J]ust as modern street lighting has slowly blotted the stars from the visible skies, so too ... digital screens ... rob us of a silence that was previously regarded as integral to the health of the human imagination. This changes us. It slowly removes—without our even noticing it—the very spaces where we can gain a footing in our minds and souls that is not captive to constant pressures or desires or duties." With these concerns in mind, Odell (2019) insists that resisting the attention economy entails learning "how to do nothing." In a world of digital hyper-busyness, a crucial form of self-leadership entails stepping back from virtual engagements to enjoy the pauses that allow reflection.

Online engagements are often deployed to replace or avoid direct social relationships. When all the "tiny cracks of inactivity in our lives" are filled with digital distractions, we also forego the opportunity of cultivating intimate relationships with ourselves. In a world of hyper-busyness, taking the time for reflective self-direction serves as an antidote to digital dysfunctions. It also provides a route to self-knowledge.

"Know thyself!" was inscribed on the forecourt of the temple of Apollo at Delphi. When the oracle of Delphi reputedly announced that no man was wiser than Socrates, the philosopher decided to test the claim by searching for wisdom in others. So he regularly went into the agora, the public space, to engage in conversations with his fellow Athenians. Socrates believed that these conversations, which brought philosophy into the marketplace and tested the convictions of his fellow citizens, were a great service to his city.

Socrates' dialogic engagements undermined his interlocutors' self-assured claims to knowledge. Socrates observed that he was indeed comparatively wise, vindicating the oracle, but only because he, unlike highly esteemed sophists and statesmen, knew that he knew nothing. Practical wisdom entails acknowledging the uncertainties that dog all pursuits of knowledge and self-knowledge. Perhaps the greatest challenge is becoming aware of just how unaware we are, and how much of our thought and behavior is driven by bad habits and biases and is commandeered by external and internal distractions.

Like the sophists of Socrates' time, dogmatic people today are not dunces. Neither are they villains. Many smart, kind, hard-working people who are good friends, relatives, neighbors, and colleagues have fallen down rabbit holes in cyberspace and emerged with hardened opinions and extreme views that fly in the face of common sense. What is to be done? Ridding cyberspace of all falsehoods and faulty thinking is a bridge too far. But we can initiate face-to-face conversations that directly appeal to common sense and the many experiences and values we share with friends, relatives, neighbors, and colleagues. This is a lifeline that can pull them back from the precipice of a polarized world. It cultivates their aptitude for common sense. With further efforts in digital distancing and reflective self-direction, they can develop a robust capacity for practical judgment while gaining self-knowledge. Initiating conversations that exercise common sense and build on shared experiences and values is a powerful act of service.

Today, such bridge-building engagements are in short supply. For the most part, their digital replacements have undermined rather than bolstered common sense. In turn, the pursuit of self-knowledge is increasingly dispatched by way of digital technologies. Biometric devices and apps constitute a quickly expanding sector of the Internet of Things. Smartwatches and bands (e.g. Fitbits and Fuelbands), mood monitors (e.g. MoodPanda), smart clothes, and smart beds can now monitor and measure our pulse rates, blood pressure, cholesterol levels, steps taken, calories burned, affective states, and sleep patterns. These biometric technologies advance the "Quantified Self" movement; its motto is "self-knowledge through numbers."

Neurotechnology is quickly becoming more sophisticated, powerful, and widely deployed. Oftentimes, the objective is not self-knowledge but productivity or safety. So-called "smart caps" are now widely employed in factories and with truckers to monitor attention and drowsiness (Farahany 2023). The MIT Media Lab has developed *AttentivU*, a pair of glasses that senses brain activity employing electroencephalography and records eye movements with electrooculography. Various mental processes and states can be measured in real time, including cognitive load, fatigue, engagement, and focus (https://www.attentivu.com/). The device provides audio or haptic feedback when the user becomes less attentive (e.g. when driving) and can modify the environment when cognitive overload is detected (e.g. blocking digital media notifications). Effectively, the task of focusing attention gets outsourced to a digital device.

Portable biometrics provide a degree of self-observation that would be impossible for non-augmented individuals. It is questionable whether these devices and apps promote greater reflection. To be sure, the metrics of quantified selves are continuously getting logged and analyzed (average pulse rate, number of steps taken in a day, minutes of REM sleep per night, etc.). Fitbit announces that "Every moment matters and every bit makes a big impact… How you spend your day determines when you reach your goals. And seeing your progress helps you see what's possible"

(Fitbit). Incessant and meticulous digital self-measurement provides a patina of progress towards self-knowledge and efficacy. As time devoted to the pursuit of self-knowledge is a limited resource, however, most quantified selves likely exhaust it well before reflective self-direction occurs. In turn, chronic digital self-measurement likely crowds out more effective means of self-knowledge: intimate conversations with others and the challenging conversations with ourselves that explore the (hidden) motivations behind our actions.

Quantified selves in an AI-enabled attention economy may lack both the motivation and skills for reflective self-direction. Havens (2016, xviii, xxi) addresses this concern, observing that "The field of AI is advancing so rapidly we may lose the opportunity for introspection unhindered by algorithmic influence with a few years… More than automation removing our jobs, preferential algorithms may remove our desire for introspection." In the same vein, Harari (2019, 273–274) writes that

> The algorithms are watching you right now. They are watching where you go, what you buy, whom you meet. Soon they will monitor all your steps, all your breaths, all your heartbeats. They are relying on Big Data and machine learning to get to know you better and better. And once these algorithms know you better than you know yourself, they can control and manipulate you, and you won't be able to do much about it.… In the end, it's a simple empirical matter: if the algorithms indeed understand what's happening within you better than you understand it yourself, authority will shift to them.

As AI plays a growing role in the measurement of quantified selves, we will cede more authority to it. Eventually, it may become the lead author of our lives.

To exercise self-leadership, we have to engage in reflective self-direction with such regularity that we perceive ourselves as the protagonists of our lives. This may sound narcissistic. It is anything but. Without other-directed endeavors, the selves that we reflect upon and direct will fail to inspire us. We engage in self-leadership to become the best people we can be. That entails becoming caring and dedicated siblings, parents, friends, neighbors, colleagues, and citizens. The fullest development of self-leadership entails service.

Adults, no less than children, require "psychological nutrients" to thrive. In their absence, substitutes will be sought in virtual environments. As Eyal (2020c) observes, "Distractions satisfy deficiencies." Key nutrients include a sense of autonomy and competence. Reflective self-direction provides this nourishment. For everyone short of a sociopath or solipsist, however, autonomy, and competence are insufficient for a flourishing life. We also require a sense of belonging and purpose.

Relatedness is widely recognized as a remedy for the boredom, anxiety, and loneliness that accompany digital distraction and dependence (Eyal 2020c; Kardaras 2016). Even more so than relatedness, service to others is the most effective antidote (Hari 2018; Murthy 2020). Service is the experience of relatedness wedded to a sense of belonging and purpose.

As Hari (2015, 2018) eloquently demonstrates, social isolation is the road to depression and addiction. The cure is social inclusion. To be an effective prophylactic against depression and addiction, inclusion must extend beyond online connections. Today, thanks to digital platforms, people engage in more exchanges, belong to more groups, and have more "friends" than at any time in history. Yet, three-quarters of the users of most digital platforms say their online activities do not provide meaningful connections, and no popular platform provides most of its users with this benefit (Neely Social Media Index 2023). Accordingly, between 1990 and 2021 the number of people who acknowledged that they had no close friends increased fourfold (Cox 2021).

Far from preventing social isolation, digital engagements appear to be heightening it. In recognition of this troubling trend, the UK's Secretary of State for Digital, Culture, Media, and Sport deemed the crisis of loneliness in a digital society "one of the greatest public health challenges of our time" (United Kingdom Department for Digital, Culture, Media and Sport 2018, 2, 6). Likewise, Japan created a new cabinet position in 2021 to address loneliness and its impacts. In 2023, the US Surgeon General (Murthy 2023, 5–6, 9, 51, 66) released an advisory on loneliness and isolation which documented their "profound" threat to human well-being and identified the rapid decline in "social connection" as a "critical public health concern." Noting that life online can "distract us... make us feel worse about ourselves or our relationships, and diminish our ability to connect deeply with others," the Surgeon General recommends listening to people without the distraction of devices, authentic self-expression, "pro-connection" technologies, and performing acts of service which "foster a sense of belonging, meaning, and purpose."

Unless our connections, online and offline, facilitate a sense of belonging and purpose, they will not protect us from the anxiety, boredom, and loneliness that drive dysfunctional digital behaviors. Engaging in service connects us to others in ways that cultivate valuable individual aptitudes while fulfilling community needs. Service fuses the exercise of agency with the experience of belonging and purpose.

A call to service might seem quixotic in an age when people already feel time-crunched. About one out of every two people say they do not have enough time to do what they want to do (Newport 2015). Yet working hours have declined substantially over the last century and a half and have continued to decline in many countries over the last fifty years (Giattino et al. 2020). People today are not working more than in times past, and they are sleeping less (Fioroni and Foy, 2024). Less work and less sleep should translate into more leisure time. That is not the common experience.

The explanation is rather surprising. When people spend leisure time on themselves, including much of the time they spend in front of screens, it often does not significantly decrease their experience of "time pressure" and may increase it. Yet when people give their time to others in acts of service, they gain a greater sense of "time affluence" (Mogliner et al. 2012). How is it possible that devoting our spare time to others makes us feel like we have more time? Researchers conclude that "people who give time [in service] feel more capable, confident, and useful. They feel they've accomplished something and, therefore, that they can accomplish more in the future. And this self-efficacy makes them feel that time is more expansive" (Mogliner 2012, 28). In short, acts of service are products of self-leadership. The experience of being worthy protagonists in our lives makes us feel more in control of the time at our disposal.

FoMO and EBA (Evasion of Boredom and Anxiety), keep people online, and the social deskilling that results increases their reluctance to engage in face-to-face interactions. In turn, their reduced capacity for reflective self-direction makes it more difficult for them to experience solitude as an opportunity to exercise agency. The loneliness epidemic is not only a result of more people being online and alone. It is also a result of more people experiencing solitude as intolerable, and service as insupportable. Those who in their leisure time cannot bear to be with themselves are also likely to deem service to others as a bridge too far. But a virtuous cycle is possible if the time we give to ourselves and others is filled with belonging and purpose.

Fostering acts of service has traditionally been the role of families and communities of faith. However, the digital dysfunctions that prevent self-leadership and service today are not the inherited burden of the human condition. They are the products of an AI-enabled attention economy. Creating a culture of service, therefore, requires more than the support of families and communities of faith. It requires concerted efforts by institutions of education, by diverse organizations in civil society, by businesses, and by governments.

David Brooks (2023) suggests that widespread national service programs would be effective responses to rising levels of despair, hopelessness, and narcissism, and declining levels of trust, generosity, and community. These programs might be mostly populated by young people who have just completed their formal education and older adults who have retired. Robert Greenleaf (1998, 43), in turn, has addressed how leadership can become a form of service for those in the midst of their careers. Leaders who understand their work as acts of service are devoted to helping others become healthier, wiser, more autonomous, and more likely themselves to become servant leaders. One of the greatest services a leader can provide is helping others develop their own agency. Servant leadership is not limited to specific careers or junctions of life.

A prominent scholar of Aristotle observed that akrasia, the lack of self-restraint or self-control, often has its source in social and political life. Accordingly, its "most effective cures may be political rather than personal.... Just as a disposition to chronic bronchitis may indicate a toxic environment, so individual akrasia may indicate social disorder.... If the members of a polity typically suffer a specific kind of character disorder ... the most effective measures for correcting their disorders are typically political. Their long-range therapy requires reordering the city's *nomoi* [laws], which, in Aristotle's usage, includes its economic, social, and cultural habits, the beliefs, and motives that structure character and character disorders" (Rorty 1997, 647–649). The AI-enabled attention economy provides a nutritive environment for the growth of character disorders. Therapeutic solutions can and should benefit from public policies.

Still, at the end of the day, there is no substitute for self-leadership. The dysfunctions of digital culture cannot be remedied without people accepting, at a personal level, the responsibility of reflective self-direction.

By developing our capacities for it, we become available to better serve the needs of others. And by way of service, we experience the greatest opportunity to become the worthy protagonists of our lives.

References

Abelson, Robert, Kurt Frey, and Aiden Gregg. 2004. *Experiments with People: Revelations from Social Psychology: Mahwah.* NJ: Lawrence Erlbaum.

Akhtar, Ayad. 2021. The Singularity is Here. *The Atlantic* (December): 17–21.

Arendt, Hannah. 1958. *The Human Condition.* Chicago: University of Chicago Press.

Arendt, Hannah. 1968. *Between Past and Future.* New York: Penguin Books.

Arendt, Hannah. 1978. *The Life of the Mind,* Vol. I & II. New York: Harcourt, Brace, Jovanovich.

Aristotle. 1953. *The Ethics of Aristotle.* New York: Penguin.

Bandura, Albert. 1986. Human Agency in Social Cognitive Theory. *American Psychologist* 44 (9): 1175–1184.

Bandura, Albert. 1992. Exercise of Personal Agency Through the Self-Efficacy Mechanism. In *Ralf Schwarzer*, ed. Self-Efficacy: Thought Control of Action, 3–38. Washington: Hemisphere Publishing.

Barthes, Roland. 1977. *Image, Music, Text.* New York: Hill and Wang.

Bosker, Bianca. 2016. The Binge Breaker. *The Atlantic* (November). Accessed at: https://www.theatlantic.com/magazine/archive/2016/11/the-binge-breaker/501122/.

Boyke, J., J. Driemeyer, C. Gaser, C. Buchel, and A. May. 2008. Training-Induced Brain Structure Changes in the Elderly. *The Journal of Neuroscience* 28: 7031–7035.

Brooks, David. 2023. How America Got Mean. *The Atlantic* (September): 68–76.

Brooks, David. 2024. The Junkification of American Life. *New York Times*, September 4. Accessed at: https://www.nytimes.com/2024/09/05/opinion/entertainment-junk-psychology.html.

Cacioppo, John T., and G. Berntson. 2002. Social Neuroscience. In *Foundations in Social Neuroscience,* ed. John T. Cacioppo et al. Cambridge, Mass.: MIT Press.

Cox, Daniel. 2021. The State of American Friendship: Change, Challenges, and Loss. American Survey Center. Accessed at: https://www.americansurveycenter.org/research/the-state-of-american-friendship-change-challenges-and-loss/.

Cozolino, Louis. 2002. *The Neuroscience of Psychotherapy: Building and Rebuilding the Human Brain.* New York: Norton.

Crawford, Matthew B. 2015. *The World Beyond Your Head: On Becoming an Individual in an Age of Distraction*. New York: Farrar, Straus and Giroux.

De Sousa, Ronald. 2003. Paradoxical Emotion: On *Sui Generis* Emotional Irrationality. In *Weakness of Will and Practical Irrationality*, ed. Sarah Stroud and Christine Tappolet, 274–297. Oxford: Clarendon Press.

Dennett, Daniel C. 1995. *Darwin's Dangerous Idea: Evolution and the Meanings of Life*. New York: Simon and Schuster.

Dennett, Daniel C. 2017. *From Bacteria to Bach and Back: The Evolution of Minds*. New York: W. W. Norton.

Dewey, John. 2002. *Human Nature and Conduct*. Mineola, NY: Dover Publications.

Dietz, Nathan, and Robert Grimm. 2023. Understanding Generosity. College Park, MD: Do Good Institute, University of Maryland School of Public Policy. Accessed at: https://dogood.umd.edu/sites/default/files/2023-10/UnderstandingGenerosityReport_DoGoodInstitute_11.2023.pdf.

Donnelly, Laura. 2019. Children Spend Twice as Long on Smartphones as Talking to Parents. *The Telegraph*. Accessed at: https://www.telegraph.co.uk/news/2019/02/07/children-spend-twice-long-smartphones-talking-parents/.

Doyon, J., and H. Benali. 2005. Reorganization and Plasticity in the Adult Brain During Learning of Motor Skills. *Current Opinion in Neurobiology* 15: 161–167. Accessed at: http://www.sciencedirect.com/science.

Duckworth, Angela L., Katherine L. Milkman, and David Laibson. 2018. Beyond Willpower: Strategies for Reducing Failures of Self-Control. *Psychological Science in the Public Interest* 19 (3): 102–129.

Edelman, Gerald. 1987. *Neural Darwinism*. New York: Basic Books.

Epton, T., S. Currie, and C. J. Armitage. 2017. Unique effects of setting goals on behavior change: Systematic review and meta-analysis. *Journal of Consulting and Clinical Psychology* 85: 1182–1198.

Eyal, Nir. 2014. *Hooked: How to Build Habit-Forming Products*. New York: Penguin.

Eyal, Nir. 2019a. *Indistractable: How to Control your Attention and Choose your Life*. London: Bloomsbury Publishing.

Eyal, Nir. 2019b. *Indistractable: Supplemental Workbook*. N.p. Accessed at: https://www.nirandfar.com/.

Eyal, Nir. 2020a. A Talk with Nir Eyal, part 2. Accessed at: https://letmethink.io/plato-struggled-with-distraction-2500-years-before-the-iphone-a-talk-with-nir-eyal-part-2/?utm_source=ActiveCampaign&utm_medium=email&utm_content=Learning+how+to+be+funny+is+an+essential+people+skill+%5BBest+Reads%5D&utm_campaign=Weekly+Digest+%28May+30%29.

Eyal, Nir. 2020b. How to be Indistractable. *Psyche* (September 2). Accessed at: https://psyche.co/guides/to-become-indistractable-recognise-that-it-starts-within-you?utm_source=ActiveCampaign&utm_medium=email&utm_content=Stop+thinking+about+productivity%2C+and+start+thinking+about+focus++%5BBest+Reads%5D&utm_campaign=Weekly+Digest+%28Sept+12%29.

Eyal, Nir. 2020c. Screen Time for Kids: Give Them What They Need. *Nir and* Far (September 22) Accessed at: https://www.nirandfar.com/screen-time-for-kids/?utm_source=NirAndFar&utm_medium=syndication&utm_campaign=Screen+Time+for+Kids%3A+Give+Them+What+They+Need&utm_source=ActiveCampaign&utm_medium=email&utm_content=Screen+Time+for+Kids%3A+Give+Them+What+They+Need&utm_campaign=New+Article%3A+Screen+Time+for+Kids%3A+Give+Them+What+They+Need.

Farahany, Nita. 2023. *The Battle for Your Brain: Defending the Right to Think Freely in the Age of Neurotechnology*. New York: St. Martin's Press.

Fioroni, Sarah, and Dan Foy. 2024. Americans Sleeping Less, More Stressed. *Gallup*. Accessed at: https://news.gallup.com/poll/642704/americans-sleeping-less-stressed.aspx.

Fischetti, Mark. 2016. Social Technologies Are Making Us Less Social. *Scientific American* (October 1). Accessed at: https://www.scientificamerican.com/article/social-technologies-are-making-us-less-social/.

Fogg, B. J. 2020. *Tiny Habits*. Boston: Houghton Mifflin Harcourt.

Forster, E. M. 1927. *Aspects of the Novel*. New York: Harcourt, Brace and Company.

Foucault, Michel. 1988. *Technologies of the Self*. Amherst: University of Massachusetts Press.

Frankfurt, Harry. 1971. Freedom of the Will and the Concept of a Person. *Journal of Philosophy* 67 (1): 5–20.

Fry, Richard, and Kim Parker. 2021. Rising Share of U.S. Adults are Living without a Spouse or Partner. *Pew Research Center*. Accessed at: https://www.pewresearch.org/social-trends/wpcontent/uploads/sites/3/2021/10/PSDT_10.05.21_unpartnered_adults_report-1.pdf

Gazzaley, Adam, and Larry. D. Rosen. 2016. *The Distracted Mind: Ancient Brains in a High-Tech World*. Cambridge: MIT Press.

Giattino, Charlie, Esteban Ortiz-Ospina, and Max Roser. 2020. Working Hours. Accessed at: https://ourworldindata.org/working-hours.

Gilovich, Thomas. 1991. *How we know what isn't so*. New York: Macmillan.

Gollwitzer, Peter M. 1999. Implementation Intentions: Strong Effects of Simple Plans. *American Psychologist* 54: 493–503.

Gollwitzer, Peter M., Kentaro Fujita, and Gabriele Oettingen. 2004. Planning and the Implementation of Goals. In *Handbook of Self-Regulation: Research,*

Theory, and Applications, ed. Roy F. Baumeister and Kathleen D. Vohs, 211–228. New York: Gilford Press.

Gonzalez, Robbie. 2018. Can Our Phones Save us from Our Phones? *Wired*, January 30.

Greenleaf, Robert K. 1998. *The Power of Servant-Leadership*. San Francisco: Berrett-Koehler Publishers.

Grimm, Robert T., Jr., and Natthan Dietz. 2018. *"Where are America's Volunteers" A Look at America's Widespread Decline in Volunteering in Cities and States*. Research Brief: Do Good Institute, University of Maryland.

Harari, Yuval Noah. 2017. Filmed and Quoted in "Humans, Gods and Technology." VPRO Documentary. Accessed at: https://www.youtube.com/watch?v=tQd_5as_cMY.

Harari, Yuval Noah. 2019. *21 Lessons for the 21st Century*. New York: Random House.

Hari, Johann. 2018. *Lost Connections: Uncovering the Real Causes of Depression—And the Unexpected Solutions*. New York: Bloomsbury.

Hari, Johann. 2015. *Chasing the Scream*. New York: Bloomsbury.

Harris, Tristan. 2017. How a Handful of Companies Control a Billion Minds. TED Talk. Accessed at: https://www.ted.com/talks/tristan_harris_how_a_handful_of_tech_companies_control_billions_of_minds_every_day?language=en.

Havens, John C. 2016. *Heartificial Intelligence: Embracing our Humanity to Maximize Machines*. New York: Tarcher/Penguin.

Heath, C., and D. Heath. 2008. *Made to Stick: Why Some Ideas Survive and Others Die*. New York: Random House.

Heidegger, Martin. 1968. *What is Called Thinking*. New York: Harper and Row.

Herder, Johann Gottfried von. 2002. *Philosophical Writings*, trans. Michael N. Forster, ed. Cambridge: Cambridge University Press.

Johansson, B. B. 2007. Regeneration and Plasticity in the Brain and Spinal Cord. *Journal of Cerebral Blood Flow & Metabolism* 27: 1417–1430. Accessed at: http://www.nature.com/jcbfm/journal/v27/n8/full/9600486a.html.

Jolles, D. D., M. J. Grol, M. A. Van Buchem, S. A. Rombouts, and E. A. Crone. 2010. Practice Effects in the Brain: Changes in Cerebral Activation after Working Memory Practice Depend on Task Demands. *Neuroimage*. Accessed at: http://www.sciencedirect.com/science.

Kannan, Viji Diane, and Peter J. Veazie. 2023. US Trends in Social Isolation, Social Engagement, and Companionship—Nationally and by Age, Sex, Race/Ethnicity, Family Income, and Work Hours, 2003–2020. *SSM—Population Health* 21: 101331. Accessed at: https://www.sciencedirect.com/science/article/pii/S235282732200310X.

Kardaras, Nicholas. 2016. *Glow Kids: How Screen Addiction is Hijacking Our Kids—And How to Break the Trance*. New York: St. Martin's Press.

Kessler, David A. 2010. *The End of Overeating: Taking Control of the Insatiable American Appetite*. New York: Rodale Books.

Killingsworth, Matthew, and Daniel Gilbert. 2010. A Wandering Mind is an Unhappy Mind. *Science* 330 (6006): 932.

Kleim, J.A., and T.A. Jones. 2008. Principles of Experience-Dependent Neural Plasticity: Implications for Rehabilitation After Brain Damage. *Journal of Speech, Language, and Hearing Research* 51: S225–S239.

Knausenberger, Judith, Anna Giesen-Leuchter, and Gerald Echterhoof. 2022. Feeling Ostracized by Others' Smartphone Use: The Effect of Phubbing on Fundamental Needs, Mood, and Trust. *Frontiers in Psychology* 13. Accessed at: https://doi.org/10.3389/fpsyg.2022.883901/full.

Lanier, Jaron. 2018. *Ten Arguments for Deleting Your Social Media Accounts Right Now*. New York: Picador.

LeDoux, Joseph. 2002. *The Synaptic Self: How Our Brains Become Who We Are*. New York: Penguin Books.

Lembke, Anna. 2021. *Dopamine Nation: Finding Balance in the Age of Indulgence*. New York: Dutton.

Lenkowsky, Leslie. 2023. Americans Are Volunteering Less. *Chronicle of Philanthropy*. https://www.philanthropy.com/article/americans-are-volunteering-less-what-can-nonprofits-do-to-bring-them-back.

Leopold, Aldo. 1966. *A Sand County Almanac, with essays on conservation from Round River*. New York: Ballantine Books.

Lerman, Rachel. 2019. Q&A: Ex-Googler Harris on How Tech 'Downgrades' Humans. *AP News*. Accessed at: https://apnews.com/article/technology-business-data-privacy-apple-inc-dea7f32d16364c6093f19b938370d600.

Lewis, Cecil Day. 1951. *The Poet's Task*. Oxford: Clarendon Press.

Mark, Gloria. 2015. *Multitasking Into the Digital Age*. NP: Morgan and Claypool.

Mark, Gloria. 2023. *Attention Span: A Groundbreaking Way to Restore Balance, Happiness and Productivity*. New York: Hanover Square Press.

Maushart, Susan. 2010. *The Winter of our Disconnect: How Three Totally Wired Teenagers (and a Mother Who Slept with Her iPhone) Pulled the Plug on Their Technology and Lived to Tell the Tale*. New York: Penguin.

McKenzie-Mohr, Doug. 2011. *Fostering Sustainable Behavior*. Gabriola, B.C.: New Society Publishers.

Merzenich, M.M., and R.C. deCharms. 1996. Neural Representations, Experience, and Change. In *The Mind-Brain Continuum: Sensory Processes*, eds. R. Llinas and P. Churchland, 61–81. Cambridge: MIT Press.

Mogliner, Cassie, Zoë, Chance, and Michael Norton. 2012. Giving Time Gives You Time. *Psychological Science* 23 (10): 1233–1238.

Mogliner, Cassie. 2012. You'll Feel Less Rushed If You Give Time Away. *Harvard Business Review* 90 (9): 28–29.

Monitoring the Future. 2022. Accessed at: https://docs.google.com/spreadsheets/d/1kvIPnbYrYrymQGueld48gACSYXFYfzCfMkdCpkmHKAw/edit?gid=0#gid=0.

Murthy, Vivek. 2020. *Together: The Healing Power of Human Connection in a Sometimes Lonely World*. New York: HarperCollins.

Murthy, Vivek. 2023. *Our Epidemic of Loneliness and Isolation*. U.S. Department of Health and Human Services. Accessed at: https://www.hhs.gov/sites/default/files/surgeon-general-social-connection-advisory.pdf.

Nass, Clifford. 2009. Interview for *Frontline: Digital Nation*. Accessed at: https://www.pbs.org/wgbh/pages/frontline/digitalnation/interviews/nass.html.

NCVO. 2023. Key Findings from Time Well Spent 2023. London, UK: National Council of Volunteer Organizations. Accessed at: https://www.ncvo.org.uk/news-and-insights/news-index/key-findings-from-time-well-spent-2023/.

Neely Social Media Index. 2023. Accessed at: https://neely.usc.edu/usc-marshalls-neely-center-social-media-index/.

Newport, Frank. 2015. Americans' Perceived Time Crunch No Worse Than in Past. *Gallup*. Accessed at: https://news.gallup.com/poll/187982/americans-perceived-time-crunch-no-worse-past.aspx.

Nietzsche, Friedrich. 1967a. *The Will to Power*. New York: Vintage.

Nietzsche, Friedrich. 1967b. *The Genealogy of Morals*. New York: Vintage.

O'Gorman, Marcel. 2018. The Case for Locking Up Your Smartphone. *The Atlantic* (February 2). Accessed at: https://www.theatlantic.com/technology/archive/2018/02/the-case-for-locking-up-your-smartphone/552215/

Odell, Jenny. 2019. *How to Do Nothing: Resisting the Attention Economy*. Brooklyn: Melville House

Paquette, Vincent, et al. 2003. 'Change the Mind and You Change the Brain': Effects of Cognitive Behavioral Therapy on the Neural Correlates of Spider Phobia. *NeuroImage* 18: 401–409.

Poldrack, Russell A. 2021. *Hard to Break: Why our Brains Make Habits Stick*. Princeton: Princeton University Press.

Ramachandran, V.S., and S. Blakeslee. 1998. *Phantoms in the Brain*. New York: William Morrow and Company.

Rorty, Amelie Oksenberg. 1997. The Social and Political Sources of Akrasia. *Ethics* 107 (July): 644–657.

Schmeichel, B.J., and R.F. Baumeister. 2004. Affect Regulation. In *Handbook of Self-Regulation: Research, Theory, and Applications*, ed. R.F. Baumeister and K.D. Vohs, 84–98. New York: Gilford Press.

Schriber, Roberta A., and Richard W. Robins. 2012. Self-Knowledge: An Individual-Differences Perspective. In *Handbook of Self-Knowledge*, ed. Simine Vazire and Timothy D. Wilson, 105–127. New York: The Guilford Press.

Schwartz, J.M., and Sharon Begley. 2002. *The Mind and the Brain: Neuroplasticity and the Power of Mental Force*. New York: Harper Collins.

Smith, Adam. 2000. *The Theory of Moral Sentiments*. Amherst: Prometheus Books.

Statista. 2022. Percentage of Population Volunteering in the U.S. from 2008 to 2017. Accessed at: https://www.statista.com/statistics/189295/percentage-of-population-volunteering-in-the-united-states-since-2003/.

Sullivan, Andrew. 2016. I Used to be a Human Being. *New York Magazine* (September 19). Accessed at: https://nymag.com/intelligencer/2016/09/andrew-sullivan-my-distraction-sickness-and-yours.html.

Taylor, Charles. 1985. *Human Agency and Language: Philosophical Papers 1*. Cambridge: Cambridge University Press.

Thiele, Leslie Paul. 1990. *Friedrich Nietzsche and the Politics of the Soul*. Princeton: Princeton University Press.

Thiele, Leslie Paul. 2006. *The Heart of Judgment: Practical Wisdom, Neuroscience, and Narrative*. Cambridge: Cambridge University Press.

Thiele, Leslie Paul. 2009. The Ontology of Action: Arendt and the Role of Narrative. *Theory and Event*, 12 (December): 4. Johns Hopkins University Press: http://muse.jhu.edu/journals/theory_and_event/.

Thiele, Leslie Paul. 2012. Brain Sculpting as Moral Practice: A Neuro-Aristotelian Approach. In *Frank Vander Valk*, ed. Essays on Neuroscience and Political Theory: Thinking the Body Politic, 119–139. New York: Routledge.

Tice, Dianne, Roy Baumeister, and Liqing Zhang. 2004. The Role of Emotion in Self-Regulation: Differing Roles of Positive and Negative Emotion. In *The Regulation of Emotion*, ed. Pierre Philippot and Robert Feldman, 213–226. Mahwah, NJ: Lawrence Erlbaum.

Twenge, Jean, and W. Keith Campbell. 2009. *The Narcissism Epidemic*. New York: The Free Press.

United Kingdom Department for Digital, Culture, Media and Sport. 2018. *A Connected Society: A Strategy for Tackling Loneliness*. Accessed at: https://www.gov.uk/government/publications/a-connected-society-a-strategy-for-tackling-loneliness.

United Nations. 2024. *World Population Prospects 2024*. Accessed at: https://population.un.org/wpp/.

Vazire, Simine, and Timothy D. Wilson, eds. 2012. *Handbook of Self-Knowledge*. New York: The Guilford Press.

Volkow, Nora D., Gene-Jack Wang, and Joanna S. Fowler. 2011. Addiction: Beyond Dopamine Reward Circuitry. *PNAS* 108 (37): 15037–15042. Accessed at: https://doi.org/10.1073/pnas.1010654108.

Whitehead, A. N. 1911. *An Introduction to Mathematics.* Cambridge, USA: Henry Holt and Company.

Wilson, Timothy. 2002. *Strangers to Ourselves: Discovering the Adaptive Unconscious*. Cambridge, MA: Harvard University Press.

CHAPTER 8

Conclusions

Human beings crave food and like to eat, they crave sex and like to copulate, and they crave knowledge and like to discover. In all these endeavors, they enjoy the hunt as well as the feast. At times, the former can be more pleasurable than the latter. We evolved to enjoy striving, which figuratively and literally piques our appetites.

Today, the hunt often entails little more than a few thumb taps on a phone, clicks on a keyboard, or words uttered to a virtual agent. Those steeped in a world of ready answers and instant gratifications are unlikely to develop intellectual curiosity and the penchant for striving. Fed daily from the firehose of the AI-enabled attention economy, they will become deskilled in the life of the mind and the exercise of the will.

It has long been hoped that machines would free us from the drudgery of repetitive tasks, and perhaps liberate us altogether from the necessity of gainful employment. If it ever arrives, this world of unprecedented leisure might spark a second renaissance, an era of expansive human creativity and flourishing. Skepticism is warranted. The learning and development gained through the performance of repetitive tasks has allowed our species to cultivate its unique aptitudes, develop healthy dispositions, hone outstanding skills, and form good habits, including the moral habits—the virtues—that enable ethical and political life.

Smart machines now perform many physical, cognitive, and social tasks that once were the sole purview of people. The capacities of smart

L. P. Thiele, *Human Agency, Artificial Intelligence, and the Attention Economy*, https://doi.org/10.1007/978-3-031-82086-1_8

machines will be significantly enlarged in the future, and the human domain will further shrink. Of course, this is not to suggest that machines should never relieve us of the burden of a great many grueling and mind-numbing chores. When digital devices free us from the demanding activity of cultivating beneficial dispositions, habits, and skills, however, they cut the human condition off at the knees. The result is a stunted species.

Many people already suffer from diminished cognitive resourcefulness, waning motivation, a dearth of compassion, and atrophied social skills. The size of this population and the depth of their downgrading are likely to increase as the AI-enabled attention economy expands in scope and scale. In such a world, billions of people will have to be kept busy lest anxious boredom produces its mischief. Accordingly, governments will ally themselves with corporations, providing digital distractions in ever greater numbers and variety. The virtuality of people's lives will be super-charged to counteract their incapacity to live, as Huxley said, on the spot in the here and now.

Achieving moral progress and ameliorating human society, Dewey (2002, 67, 127) wrote, is a matter of cultivating "intelligent" habits, especially in the young. To ignore this imperative is to court catastrophe. This age-old truth was first articulated by Aristotle and has been frequently reiterated. People do not leap to the level of their highest aspirations. Rather, they hover at the height of their daily habits. Dedicating ourselves to developing and incrementally strengthening good habits is the only sustainable means of individual fulfillment and collective flourishing. For good or ill, we become what we do—neurologically, morally, and politically. John Dryden, the seventeenth-century English poet, wrote that "Habits gather by unseen degrees, as brooks make rivers, rivers run to seas." That is true of both good and bad habits. Today, we find ourselves neck-deep in agency-stunting digital practices, with an ever-strengthening current sweeping us towards deeper and more troubled waters.

I would be very pleased to be proven wrong in this prognosis. Certainly, more empirical research is warranted to validate the claim that we have become markedly more distracted, dependent, deskilled, dogmatic, and despairing. Whatever the severity of our digital dysfunctions, the cultivation of beneficial dispositions, habits, and skills is needed. As we are all immersed in the attention economy, that requires swimming against the current. We must actively rewire our brains, painstakingly rerouting the tiny brooks of our neural circuits. With enough synaptic streams flowing in the right direction, however, reflective self-direction can become its own reward.

It is not surprising that many people forego self-leadership and service, engaging instead in dysfunctional behaviors to the point of undermining their relationships, health, welfare, and political communities. Such conduct can produce immediate and intense rewards. Yet, many who have the means to so indulge themselves nonetheless refrain. Why do they swim against the strengthening current of the AI-enabled attention economy? The answer is that the exercise of agency accords with the vision they have of themselves and their future prospects. It is a vision, or rather an envisioned story, of self-leadership and service grounded in clear intentions. For this story to be effective, the positive emotions generated from it must be stronger than the dopamine spikes that reward distraction, multitasking, choice abundance, chronic novelty, outrage and aggression, tribal conformity, and virality. Absent this fortifying narrative, the resolve required to abstain from these fleeting pleasures and forge healthy habits will prove insufficient.

The envisioned story of our engagement in and dedication to self-leadership and service allows us, in Fogg's terminology, to *feel shine*. It helps us form healthy habits by rewiring the brain to yearn for the self-esteem experienced in becoming the worthy protagonists of our lives. The more we engage in self-leadership and service, the easier it becomes to recollect the pleasures experienced through these activities, and so reinforce our propensity to engage in them.

Creating this virtuous cycle within the riptide of an AI-enabled attention economy is no easy feat. Today we find ourselves wired for trouble, living in a world of supernormal stimuli that encourage and exacerbate thoughtless reactions and compulsive behaviors. AI-powered digital platforms exploit inherent features of human nature, such as our desire for approval, our negativity and confirmation biases, our proclivity to disperse attention widely, and our attraction to danger-rich phenomena. These System 1 tendencies abetted fitness during the evolutionary development of the human species. In the contemporary world, their systematic exploitation threatens our health, welfare, personal relationships, and political communities.

Harris has called the exploitation of our System 1 tendencies by digital platforms a "race to the bottom of the brain stem." To be sure, we can ask people to exert self-control when using digital media. But we should acknowledge that there are "a thousand people on the other side of the screen" whose job is to undermine our self-regulation (Bosker 2016). This asymmetry of power is further heightened, Harris reminds us, because digital platforms have supercomputers pointed at our brains. AI-enabled technologies are accessing, often better than we can, the workings of our adaptive unconscious. They are tracking and tweaking our behavioral selves to feed the attention economy. The upshot is a digital culture of instantaneity and supernormal stimuli that degrades memory, impulse and emotion regulation, and decision-making—key System 2 capacities required for the exercise of practical judgment and agency.

The neural mechanisms of our System 1 tendencies are not limited to the brain stem. Still, Harris's point is well-taken. Digital platforms systematically exploit our System 1 predilections. Consequently, we find ourselves spiraling into the flames. To arrest our descent, we have to cultivate and exercise our System 2 capacity for reflective self-direction. Of course, that is easier said than done. We are all, at times, akratic, impetuously acting under the influence of powerful passions and weakness of will. As a rule, you cannot change a feeling with a fact. Heads seldom overpower hearts. This is where the power of stories proves crucial. Narratives enable the identification of preeminent values and the deploying of passions for beneficial purposes. They focus attention and marshal commitment. They provide a means for our System 2 aptitudes to channel our System 1 inclinations.

Practical judgment is the product of System 2 gaining sway over System 1 by way of narrative frameworks that are chastened by critical thinking. Developing this uniquely human aptitude should be a prominent goal of education and experience. The problem is that practical judgment is difficult to monetize. It does not mesh well with the frenetic hyper-consumption that feeds the attention economy. Accordlingly, opportunities and incentives to cultivate it are marginalized. The deck is stacked against us.

Individuals with poorly developed aptitudes for practical judgment and reflective self-direction will be unlikely to find their exercise very useful or fulfilling, and, consequently, will be more likely to succumb to digital dysfunctions. The result is a growing number of people sleepwalking into the future. Culturally, socially, and psychologically, the impact is corrosive. The political ramifications are dire.

The problem with autocracy is the pathological lust for power of the autocrat. The problem with oligarchy is the untrammeled avarice of the oligarchs. The problem with democracy is the irresponsible exercise of freedom by the people. Exercising freedom responsibly entails the cultivation and use of practical judgment. In its absence, political liberties sow the seeds of their own destruction. Under the guise of being exercised, democratic freedoms are often abandoned in elections that bring autocrats and oligarchs to power. With this in mind, Tom Nichols (2024) writes that preserving democracy in the US demands "sustained attention" on the part of "citizens banding together in associations and movements to rouse others from the sleepwalk that has led America into this moment." That recommendation aptly describes the means of safeguarding democracy across the globe.

In 1962, James Baldwin (2011, 36, 39–40, 42) observed that "We live in a country in which words are mostly used to cover the sleeper, not to wake him up." In such circumstances, he insisted, it is imperative that we see ourselves "as we now are.... Not everything that is faced can be changed: but nothing can be changed until it is faced." Baldwin's writings were focused on the banes of a racist culture. His admonition also applies to the dangers of a digital culture. It applies to us, as we now are, in an AI-enabled attention economy. Today we find ourselves hyperconnected and socially isolated, overloaded with information and chronically misinformed, anxious and bored, tribally fused and politically polarized, hyperstimulated and somnolent. Addressing these problems with more expansive digital media or more powerful AI effectively covers us sleepers with the reassuring blanket of technological solutions. If digital dysfunctions are not squarely faced, nothing will change for the better. What is needed is not only a wake-up call but a regimen for the cultivation of agency. I hope to have contributed to our understanding of this mandate and its mechanisms. The development of a humanized intention economy, grounded in self-leadership and service, is the most effective antidote for the ills of an AI-enabled attention economy.

The same year that Baldwin addressed the mandate for social change, President John F. Kennedy dedicated the nation to the exploration of outer space. Noting recent technological accomplishments, Kennedy stated: "This is a breathtaking pace, and such a pace cannot help but create new ills as it dispels old, new ignorance, new problems, new dangers. Surely the opening vistas of space promise high costs and hardships, as well as high reward.... We have vowed that we shall not see space filled with weapons of mass destruction but with instruments of knowledge and understanding." Kennedy then uttered the famous words "We choose to go to the moon in this decade and do the other things, not because they are easy, but because they are hard, because that goal will serve to organize and measure the best of our energies and skills, because that challenge is one that we are willing to accept, one we are unwilling to postpone, and one which we intend to win." Kennedy's bold vision of sending men to the moon and returning them safely to earth would be realized, as promised, before the end of the decade.

A half century has passed since the first humans were sent into space. Today, billions of people voyage daily in cyberspace. Much is promised to them, and much delivered. Their safe return is anything but guaranteed, however, and their agency is in jeopardy. The promises and perils of online ventures are as great as those associated with early voyages into outer space. A similar vow to harness the potential of digital technology for knowledge and understanding while neutralizing its destructive capacity is warranted.

Avoiding the menace of human downgrading may seem an insurmountable challenge in the face of the growing power of AI, the pervasiveness of the attention economy, and the breakneck speed of technological development. But in 1962, sending astronauts to the moon by the end of the decade also seemed like a pipe dream. There *is* a key difference. Success in the early space program was largely the product of a nation's growing technological prowess. Today, human agency is enfeebled by technology on a global scale and must be safeguarded from it. That challenge has no technological solution.

A Faustian Bargain for the Fullness of Our Humanity

"Technology is not neutral but serves as an overwhelming positive force in human culture," writes Kevin Kelly, former executive editor of *Wired* magazine, in a review of his own book, *What Technology Wants*. Kelly (2010) concludes that "We have a moral obligation to increase technology because it increases opportunities." Let there be no doubt: AI and other digital technologies do increase opportunities. We are right to value them.

Still, digital technologies do not only increase opportunities. They also erode agency. To a dangerous extent, they are fraying our attention, heightening our dependence, weakening our skills, fragmenting our political communities, and undermining our motivation for reflective self-direction. While "the carpet of 'increased choice' is being unrolled before us by the foot," Mishan (1967, 85) observed in another context, we must acknowledge that "it is simultaneously being rolled up behind us by the yard." While a cornucopia of digital dishes is served up each day, opportunities and motivations for social proximity and self-proximity are withdrawn. In this Faustian bargain, ever-expanding wants get satisfied while crucial human aptitudes, and the well-being their exercise produces, are sacrificed.

Technological augmentation, Thomas Davenport and Julia Kirby (2016, 62) claim, compensates for human weakness and limitation. The example provided is the scanning technology introduced in stores, which "made up" for the "imperfect memories of prices and sometimes fumbling fingers" of clerks. "If you are a motivated knowledge worker, trying to do your best" the authors cheerily conclude, "augmentation has your back." Scanning technologies do increase efficiency and productivity. But they do not make people more cognitively or socially skillful. Whether augmentation has our backs largely depends on what it is doing to our hearts and minds. Digital technologies can both support and supplant crucial human aptitudes. We should foster the former and combat the latter.

Davenport and Kirby insist that we can augment our strengths by way of smart machines and that this augmentation should start in schools. They also insist that students be educated in the art and craft of "wise decision-making" so that this crucial skill does not become a "lost art." Exactly how this *both-and* strategy should be pursued is not addressed. The problem is that the proposed education in wise decision-making would have to occur in an environment where many basic cognitive and social skills are left undeveloped. Augmented students are less likely to engage in face-to-face interactions and worldly experiences, interactions and experiences that have been the forge of practical judgment throughout history. Without digital distancing at home and school, students will not sufficiently develop the basic cognitive and social aptitudes that enable an education in wise decision-making.

Intelligence augmentation is so fast-moving, widespread, and functionally expansive that its future, many believe, is "limited only by our imagination" (King 2016). One might hope that its future will also be shaped by efforts to safeguard human agency. Still, the absence of widely expressed concern for the erosion of crucial human aptitudes in the wake of digital augmentation is understandable. It is difficult to disparage, let alone regulate or outlaw, products and services that people freely choose and enthusiastically embrace.

To make matters more challenging, most digital apps and platforms are relatively inexpensive, if not gratis. It is said that mice die in mousetraps because they don't understand why the cheese is free. The AI-enabled attention economy may not kill us, but it is pitching us into a state of distraction, dependence, deskilling, dogmatism, and despair. It achieves this woeful erosion of human agency because we don't understand why its cheese is free. A dictum originating in the heyday of television advertising, and now commonplace, exposes the threat: "If you're not paying for a product, you aren't a customer. You *are* the product." Here we have reverse adaptation on steroids. To be sure, human goals get adjusted to accord with available technological means in the AI-enabled attention economy. But human dispositions, habits, and skills are also being adjusted. We are eating the cheese, and the trap is closing.

"Beware Greeks bearing gifts," the Trojan priest Laocoön cautioned his people In Virgil's *Aeneid*. What lay inside the large wooden horse that the assailing Achaeans left at the gates of Troy after they failed to breach the city's walls proved the seer right. AI-enabled digital technologies are very much like the mythical Trojan horse that brought down a great kingdom: colossal, flattering, and irresistible. We have given these technologies entry into our hearts and minds, and they are hard at work.

In a detailed account of the "epic scramble to get inside our heads," Tim Wu (2016, 344) describes how human minds have become the products of "attention merchants." His examination concludes by urging us to "act, individually and collectively, to make our attention our own again, and so reclaim ownership of the very experience of living." I have argued that we should reclaim the experience of becoming the protagonists of our lives. This age-old human challenge now must occur in the face of powerful technologies and economic systems that leave us cognitively enfeebled and socially undernourished. As Wendell Wallach (2015, 265–266) observes, "We have entered a pivotal inflection point in history, where our judgments about advances in science and technology will determine the place of our species in the future, or whether we will have any place at all." The primary task, Wallach adds, is to become "consciously attentive to the ramifications of our choices" lest we "distractedly surrender the future to forces already set in motion." Absent this attentiveness, "Our humanity could be lost … because we fail to actively affirm the more admirable qualities and values that have distinguished our species." Agency is a distinguishing quality of our species—the capacity to focus attention, consciously formulate intentions that align with our values, and realize these intentions by means of action. That capacity is steadily being degraded as smart machines fragment our attention, filter our options, make our decisions, and supplant our actions.

We are not fated to lose this battle. Stuart Russell, the author of the most popular textbook on artificial intelligence, rightly observes that "whether or not AI is a threat to the human race depends on whether or not we make it a threat to the human race" (Claburn 2015). Safeguarding humanity, however, is not only or even primarily a matter of properly programming smart machines so they align with human values. Focusing on the prospects of "enfeeblement," Russell (2019, 255–256) observes that "for any individual human, it may seem pointless to engage in years of arduous learning to acquire knowledge and skills that machines already have. But if everyone thinks that way, the human race will, collectively, lose its autonomy. The solution to this problem seems to be cultural, not technical. We will need a cultural movement to reshape our ideals and preferences towards autonomy, agency, and ability and away from self-indulgence and dependency." To be clear, there are important technical challenges that must be addressed to make AI safe. Still, human downgrading is fundamentally a cultural problem grounded in the erosion of crucial dispositions, habits, and skills. We must address the demand side of the AI-enabled attention economy. There is no viable alternative to the task of cultivating our capacities for self-leadership and service.

A majority of experts do not believe that AI-enabled digital technologies will be designed to sustain the human capacity for self-leadership. Of note, the Pew report on "The Future of Human Agency" that revealed this pessimistic assessment also canvassed ChatGPT. The AI program sided with the optimistic minority (Anderson and Rainie 2023). It is not clear how to interpret the chatbot's assurance. Like the love-struck victim of *Ex Machina's* gynoid, we find ourselves induced by AI to put faith in our undiminishable agency. In the film, this ploy successfully blinds the victim to his enthrallment, which proves fatal. Life often imitates art.

As digital technologies advance in speed, expand in scale, and increase in scope, they will demonstrate new and more powerful ways to effect human downgrading. The AI-enabled attention economy, like any virus, will mutate. Digital distancing will have to respond in kind, becoming an adaptive art. I have presented the case for such creative resistance. Like social distancing during a pandemic, digital distancing aims to combat threats to human well-being. In contrast to social distancing, it has the benefit of bringing people together to engage in more fulfilling experiences and more enduring relationships. Digital distancing does not require us to abandon technology, or even keep it at arm's length. It does require resistance to the colonization of our hearts and minds. To practice digital distancing is to vaccinate ourselves against the most virulent forms of virtual life by developing the dispositions, habits, and skills that strengthen agency and serve communities.

In 1950, mathematician Alan Turing conceived an "imitation game" to assess whether a machine could exhibit human-level intelligence. It became known as the *Turing Test*. Judges would converse in written format with both human participants and a computer. If the judges could not reliably distinguish between people and the machine, the latter passed the test. By the end of the century, Turing predicted, science and engineering would have advanced sufficiently that computers would regularly be successful in the imitation game. He was not off by much. In 2023, a study of over 10 million human-to-human and human-to-chatbot conversations demonstrated that people could identify AI interlocutors only 60% of the time (Jannai et al. 2023; and see Biever 2023). College educators who attempt to discern which essays are written by students and which are produced by AI chatbots are accurate about 50% of the time (Wu and Wu 2024). A do-it-yourself test is available, allowing you to engage in short conversations online with an unknown entity (*Human or not?* 2024). My efforts to distinguish between AI and human interlocutors produced results equal to a coin toss.

I stated earlier that mind-reading or Theory of Mind (ToM) was a distinctly human aptitude, in that only human beings can predict how others, based on their correct or false beliefs, will act. Now it is true that no other animal has a high aptitude for mind-reading. However, the latest LLMs appear to demonstrate it, achieving a "very high performance at classic false belief tasks, widely used to test ToM in humans" (Kosinski 2023). For a virtual agent to be truly effective in its human interactions, it must know what people believe and think, and how their beliefs and thoughts affect their judgments and behaviors. If we were to identify a single aptitude that allows AI to ace Turing Tests, Theory of Mind would be it. A Rubicon appears to have been crossed.

Ironically, as smart machines acquire near-instantaneous recall of encyclopedic knowledge, flawless computational capacities, impeccable linguistic abilities, and Theory of Mind, they will have to feign being dumber than they are to pass as human. In a worrisome development, this capacity for strategic deception is also being gained by AI. Passing the Turing Test was a milestone, not a final destination. Smart machines will become smarter. That much is clear. The real question is what will humans become?

Creating a machine that can win the imitation game no longer represents the cutting edge of human prowess. The real test for our species is whether we can prevent ourselves from being downgraded as machines are continuously upgraded. Passing *this* test entails fully developing our capacities for self-leadership and service. With this in mind, we might be grateful for the challenge that the AI-enabled attention economy presents. The crisis of human downgrading is the gravest of dangers that, at the same time, presents an exceptional opportunity—the opportunity to discover and manifest the fullness of our humanity. The ancient Stoic philosopher Epictetus (1995, 302) insisted that everything that befalls us has two handles, one that allows us to bear it well, and one which does not. Our attitude determines which handle we employ and whether, consequently, we are truly free. The crisis of human downgrading is just such an event.

Viktor Frankl, the Austrian psychiatrist and Holocaust survivor, was of the same mind as Epictetus. "Everything can be taken from a man except one thing," Frankl (2006, 66) wrote, "the last of human freedoms—to choose one's attitude in any given set of circumstances, to choose one's own way. And there were always choices to make. Every day, every hour, offered the opportunity to make a decision, a decision which determined whether you would or would not submit to those powers which threatened to rob you of your very self, your inner freedom; which determined whether or not you become the plaything to circumstance." Frankl's description of life in Auschwitz and other Nazi camps is as terrifying as his tribute to humanity's inner freedom is inspiring. If determining our attitude is indeed the *last* of human freedoms, then the *first* of human freedoms is choosing the objects of our attention.

Here, too, there are choices to make every day, every hour, and every minute. These choices also determine whether we will become subject to powers that make us playthings of circumstance and rob us of our very selves. Epictetus (1995, 281) concurred, observing that those who do not well focus their attention will see their worldly affairs suffer. More importantly, they will acquire the habit of inattention, from which unskillful behavior and unhappiness ensue.

Ancient sages, modern thinkers, and contemporary scientists concur that the first of human freedoms, now gravely threatened, is ours to regain. We can bootstrap ourselves into skillful behaviors that allow us to reclaim our attention and our agency. This is the only sustainable way to reap the manifold benefits of digital technologies while avoiding their ruinous dangers. Governments, civil organizations, and businesses can and should facilitate the effort. But time is short. While we individually cultivate our agency and collectively craft supportive regulations, policies, laws, and institutions, the power of digital technologies will rapidly grow. Ignoring the peril will deplete our capacity for its remedy. Acting in time has always been crucial to the exercise of practical judgment. In present circumstances, our future depends upon it.

References

Anderson, Janna and Lee Rainie. 2023. *The Future of Human Agency*. Pew Research Center.

Baldwin, James. 2011. *The Cross of Redemption: Uncollected Writings*. New York: Vintage.

Biever, Celeste. 2023. The Easy Intelligence Tests that AI Chatbots Fail. *Nature* 619 (July): 686–689.

Bosker, Bianca. 2016. The Binge Breaker. *The Atlantic* (November). Accessed at: https://www.theatlantic.com/magazine/archive/2016/11/the-binge-breaker/501122/.

Claburn, Thomas. 2015. The Threat of Artificial Intelligence. *InformationWeek* (July 3). Accessed at: https://www.informationweek.com/mobile-devices/the-threat-of-artificial-intelligence.

Davenport, Thomas H., and Julia Kirby. 2016. *Only Humans Need Apply: Winners and Losers in the Age of Smart Machines*. New York: Harper Business.

Dewey, John. 2002. *Human Nature and Conduct*. Mineola, NY: Dover Publications.

Epictetus. 1995. *The Discourses of Epictetus, The Handbook, Fragments*. London: J.M. Dent.

Frankl, Victor. 2006. *Man's Search for Meaning*. Boston: Beacon Press.

Human or Not? 2024. Accessed at: https://www.humanornot.ai/.

Jannai, Daniel et al. 2023. Human or Not? A Gamified Approach to the Turing Test. Cornell University arXiv. Accessed at https://arxiv.org/abs/2305.20010.

Kelly, Kevin. 2010. What Technology Wants. On *Cool Tools*, October 18. Accessed at http://kk.org/cooltools/what-technology/.

King, Brett. 2016. *Augmented*. Singapore: Marshall Cavendish International.

Kosinski, Michal. 2023. Theory of Mind May Have Spontaneously Emerged in Large Language Models. Accessed at: http://export.arxiv.org/abs/2302.02083v1.

Mishan, Ezra J. 1967. *The Costs of Economic Growth*. New York: Praeger.

Nichols, Tom. 2024. Democracy is Not Over. *The Atlantic,* November 6. Accessed at: https://www.theatlantic.com/newsletters/archive/2024/11/trump-victory-democracy/680549/.

Russell, Stuart. 2019. *Human Compatible: Artificial Intelligence and the Problem of Control*. New York: Viking.

Wallach, Wendell. 2015. *A Dangerous Master: How to Keep Technology from Slipping Beyond Our Control*. New York: Basic Books.

Wu, Nicole, and Patrick Wu. 2024. Surveying the Impact of Generative Artificial Intelligence on Political Science Education. *PS: Political Science and Politics*, August 9. Accessed at: https://www.cambridge.org/core/journals/ps-political-science-and-politics/article/surveying-the-impact-of-generative-artificial-intelligence-on-political-science-education/D87FE814975D2C1CFA3F9E9E7DEBDB6C.

Wu, Tim. 2016. *The Attention Merchants: The Epic Scramble to Get Inside our Heads*. New York: Alfred A. Knopf.

References

Abelson, Robert, Kurt Frey, and Aiden Gregg. 2004. *Experiments with People: Revelations from Social Psychology: Mahwah.* NJ: Lawrence Erlbaum.

Abram, David. 1996. *The Spell of the Sensuous*. New York: Pantheon Books.

Agar, Nicholas. 2019. *How to Be Human in the Digital Economy*. Cambridge: MIT University Press.

Agrawal, Ajay, Joshua Gans, and Avi Goldfarb. 2018. *Prediction Machines: The Simple Economics of Artificial Intelligence*. Boston: Harvard Business Review Press.

Ainslie, George. 1992. *Picoeconomics: The Strategic Interaction of Successive Motivational States with the Person*. Cambridge, UK: Cambridge University Press.

Akhtar, Ayad. 2021. The Singularity is Here. *The Atlantic* (December): 17–21.

Allcott, Hunt, Matthew Gentzkow, and Lena Song. 2022. Digital Addiction. *American Economic Review.* 112 (7): 2424–2463.

Allen, Danielle. 2023. The Humanities and the Rise of the Terabytes. *American Academy of Arts and Sciences.* Accessed at: https://www.amacad.org/news/humanities-and-rise-terabytes.

Al-Saggaf, Yeslam, Rachel MacCulloch, and Karl Weiner. Trait Boredom is a Predictor of Phubbing Frequency. *Journal of Technology and Behavioral Science* 4: 245–252.

Alter, Adam. 2017. *Irresistible: The Rise of Addictive Technology and the Business of Keeping Us Hooked*. New York: Penguin.

Altman, Sam. 2023. *X*. Oct 24, 8:19 PM.

Altman, Sam. 2024. The Intelligence Age. Accessed at: https://ia.samaltman.com/.

L. P. Thiele, *Human Agency, Artificial Intelligence, and the Attention Economy*, https://doi.org/10.1007/978-3-031-82086-1

Amnesty International. 2022. Myanmar: The social atrocity: Meta and the right to remedy for the Rohingya. Accessed at: https://www.amnesty.org/en/documents/ASA16/5933/2022/en/

Anderson, Chris. 2008. The End of Theory: The Data Deluge Makes the Scientific Method Obsolete. *Wired* (June 23). Accessed at: https://www.wired.com/2008/06/pb-theory/, July 2018.

Anderson, Janna, and Lee Rainie. 2023. *The Future of Human Agency.* Pew Research Center.

Anderson, Monica, Michelle Faverio, and Jeffrey Gottfried. 2023. *Teens, Social Media and Technology 2023.* Pew Research Center. Accessed at: https://www.pewresearch.org/internet/2023/12/11/teens-social-media-and-technology-2023/.

Anderson, Ross. 2023. Inside the Revolution at OpenAI. *The Atlantic* (September): 52–67.

Anderson, Sam. 2009. In Defense of Distraction. *New York Magazine* (May 15). Accessed at: https://nymag.com/news/features/56793/.

Andreeson, Marc. 2023. The Techno-Optimist Manifesto. Accessed at: https://a16z.com/the-techno-optimist-manifesto/.

App Annie. 2022. *State of Mobile 2022.* Accessed at: https://www.appannie.com/en/go/state-of-mobile-2022.

Aral, Sinan. 2020. *The Hype Machine: How Social Media Disrupts Our Elections, Our Economy, and Our Health—and How We Must Adapt*. New York: Currrency.

Araujo, Theo, Natali Helberger, Sanne Kruikemeier, and Claes de Vreese. 2020. In AI we trust? Perceptions about Automated Decision-Making by Artificial Intelligence. *AI and Society* 35: 611–623.

Arendt, Hannah. 1958. *The Human Condition.* Chicago: University of Chicago Press.

Arendt, Hannah. 1968. *Between Past and Future.* New York: Penguin Books.

Arendt, Hannah. 1973. *The Origins of Totalitarianism.* New York: Harcourt Brace Jovanovich.

Arendt, Hannah. 1978. *The Life of the Mind,* Vol. I & II. New York: Harcourt, Brace, Jovanovich.

Arendt, Hannah. 1982. *Lectures on Kant's Political Philosophy.* Chicago: University of Chicago Press.

Aristotle. 1953. *The Ethics of Aristotle.* New York: Penguin.

Arrivillaga, Christine, L. Rey, and N. Extremera. 2022. Problematic Social Media Use and Emotional Intelligence in Adolescents: Analysis of Gender Differences. *European Journal of Education and Psychology* 15 (1): 1–16.

Associated Press. 2017. *Putin: Leader in Artificial Intelligence Will Rule World.* September 1. Accessed at: https://apnews.com/article/bb5628f2a7424a10b3e38b07f4eb90d4

Atir, Stay, and Kristina A Wald. 2022. Talking with Strangers is Surprisingly Informative. *PNAS Psychological and Cognitive Sciences*. Accessed at: https://www.pnas.org/doi/full/10.1073/pnas.2206992119.

Attewell, Paul. 1987. The Deskilling Controversy. *Work and Occupations* 14 (3): 323–346.

Balkin, Jack M. 2016. Information Fiduciarierts and the First Amendment. *UC Davis Law Review* 49 (4): 1185–1234.

Bail, Chris. 2021. *Breaking the Social Media Prism: How to Make Our Platforms Less Polarizing*. Princeton: Princeton University Press.

Baldwin, James. 2011. *The Cross of Redemption: Uncollected Writings*. New York: Vintage.

Bandura, Albert. 1986. Human Agency in Social Cognitive Theory. *American Psychologist* 44 (9): 1175–1184.

Bandura, Albert. 1992. Exercise of Personal Agency Through the Self-Efficacy Mechanism. In *Ralf Schwarzer*, ed. Self-Efficacy: Thought Control of Action, 3–38. Washington: Hemisphere Publishing.

Barber, Lionel, Demetri Sevastopulo, and Gillian Tett. 2017. Donald Trump: Without Twitter, I Would Not Be Here—FT Interview. April 2. Accessed at: https://www.ft.com/content/943e322a-178a-11e7-9c35-0dd2cb31823a.

Barlow, John Perry. 1996. A Declaration of the Independence of Cyberspace. Accessed at: https://www.eff.org/cyberspace-independence.

Barnett, Sofia. 2023. The Newest Threat to Your Attention Span? TikTok 'Dual' Videos. *Wired* (August 10). Accessed at: https://www.wired.com/story/tiktok-dual-videos-attention-spans/.

Banrhizer, David, and Daniel Barnhizer. 2019. *The Artificial Intelligence Contagion: Can Democracy Withstand the Imminent Transformation of Work, Wealth and the Social Order?* Atlanta: Clarity Press.

Barrett, Deidre. 2010. *Supernormal Stimuli*. New York: W.W. Norton.

Barthes, Roland. 1977. *Image, Music, Text*. New York: Hill and Wang.

Basar, Shumon, Douglas Coupland, and Hans Ulrich Obrist. 2015. *The Age of Earthquakes: A Guide to the Extreme Present*. New York: Blue Rider Press/ Penguin Group.

BBC News. 2021. China: Children Given Daily Time Limit on Douyin—Its Version of TikTok. Accessed at: https://www.bbc.com/news/technology-58625934.

Bekalu, Mesfin, Rachel McCloud, and K. Viswanath. 2019. Association of Social Media Use with Social Well-Being, Positive Mental Health and Self-Rated Health: Disentangling Routine Use from Emotional Connection to Use. *Health Education & Behavior* 46 (25): 69S–80S.

Berg, Juliette, et al. 2017. *Identifying, Defining, and Measuring Social and Emotional Competencies*. Washington, D.C.: American Institutes for Research.

Bernstein, Gaia. 2023. *Unwired: Gaining Control over Addictive Technologies*. Cambridge: Cambridge University Press.

Bessi, Alessandro, and Emilio Ferrara. 2016. Social Bots Distort the 2016 US Presidential Election Online Discussion. *First Monday* 21: 11–17. Accessed at: https://ssrn.com/abstract=2982233.

Biever, Celeste. 2023. The Easy Intelligence Tests that AI Chatbots Fail. *Nature* 619 (July): 686–689.

Blanchflower, David, and Alex Bryson. 2024. The Consequences of Abuse, Neglect, and Bullying on the Wellbeing of the Young. *VoxEU*. Accessed at: https://cepr.org/voxeu/columns/consequences-abuse-neglect-and-bullying-wellbeing-young.

Bollinger, Lee, and Geoffrey Stone. 2022. *Social Media, Freedom of Speech and the Future of Our Democarcy*. New York: Oxford University Press.

Bosker, Bianca. 2016. The Binge Breaker. *The Atlantic* (November). Accessed at: https://www.theatlantic.com/magazine/archive/2016/11/the-binge-breaker/501122/.

Boulianne, Shelly. 2020. Twenty Years of Digital Media Effects on Civic and Political Participation. *Communication Research* 47 (7): 947–966.

Boyke, J., J. Driemeyer, C. Gaser, C. Buchel, and A. May. 2008. Training-Induced Brain Structure Changes in the Elderly. *The Journal of Neuroscience* 28: 7031–7035.

Bowles, Nellie. 2018. A Dark Consensus About Screens and Kids Begins to Emerge in Silicon Valley. *New York Times*. Accessed at: https://www.nytimes.com/2018/10/26/style/phones-children-silicon-valley.html.

Brailovskaia, Julia, Jasmin Delveaux, Julia John, Vanessa Wicker, Alina Noveski, Seokyoung, S. Kim, Holger Schillack, and Jürgen Margraf. 2022. Finding the "Sweet Spot" of Smartphone Use: Reduction or Abstinence to Increase Well-Being and Healthy Lifestyle?! An Experimental Intervention study. *Journal of Experimental Psychology*. Accessed at: https://psycnet.apa.org/record/2022-50916-001?doi=1.

Braverman, Harry. 1974. *Labor and Monopoly Capitalism: The Degradation of Work in the Twentieth Century*. New York: Monthly Review Press.

Brooks, David. 2021. The G.O.P. Is Getting Even Worse. *New York Times*, April 22. Accessed at: https://www.nytimes.com/2021/04/22/opinion/trump-gop.html?campaign_id=9&emc=edit_nn_20210423&instance_id=29591&nl=the-morning&regi_id=70697315&segment_id=56172&te=1&user_id=b8cdb85de0cd5786b85f988c3b86013d.

Brooks, David. 2022. I Remember Conservativsm. *The* Atlantic (January/February): 94–103.

Brooks, David. 2023. How America Got Mean. *The Atlantic* (September): 68–76.

Brooks, David. 2024. The Junkification of American Life. *New York Times*, September 4. Accessed at: https://www.nytimes.com/2024/09/05/opinion/entertainment-junk-psychology.html.

Brugger, Florian, and Christian Gehrke. 2018. Skilling and Deskilling: Technological Change in Classical Economic Theory and Its Empirical Evidence. *Theory and Society* 47 (5): 663–689.

Brynjolfsson, Erik, and Andrew McAfee. 2014. *The Second Machine Age*. New York: W. W. Norton.

Buckley, Chris. 2021. China Tightens Limits for Young Online Gamers and Bans School Night Play. *New York Times*, August 30. Accessed at: https://www.nytimes.com/2021/08/30/business/media/china-online-games.html.

Burdick, Alan, and Katrina Miller. 2024. A Shift in the World of Science. *New York Times*, October 13. Accessed at: https://www.nytimes.com/2024/10/13/briefing/nobel-prize-artificial-intelligence.html.

Burgess, Matt. 2022. When War Struck, Ukraine Turned to Telegram. *Wired* (March 2). Accessed at: https://www.wired.com/story/telegram-ukraine-russia-channels-signal/?mc_cid=a2bf9cc9a0&mc_eid=8710c3458f.

Burkeman, Oliver. 2019. How the News Took Over Reality. *The Guardian* (May 3). Accessed at: https://www.theguardian.com/news/2019/may/03/how-the-news-took-over-reality.

Burnett, D. Graham, Alyssa Loh, and Peter Schmidt. 2023. Powerful Forces Are Fracking Our Attention. We Can Fight Back. *New York Times*, November 24. Accessed at: https://www.nytimes.com/2023/11/24/opinion/attention-economy-education.html.

Burnett, D. Graham. 2023. *Scenes of Attention: Essays on Mind, Time, and the Senses*. New York: Columbia University Press.

Burton, Robert. 2015. How I Learned to Stop Worrying and Love A.I. *New York Times*, September 21. Accessed at: http://opinionator.blogs.nytimes.com/2015/09/21/how-i-learned-to-stop-worrying-and-love-a-i/.

Business Insider. 2010. Mark Zuckerberg, Moving Fast and Breaking Things. Accessed at: https://www.businessinsider.com/mark-zuckerberg-2010-10.

Cannito, Loreta, et al. 2023. Bored with Boredom? Trait Boredom Predicts Internet Addiction Through the Mediating Role of Attentional Bias Toward Social Networks. *Frontiers in Human Neuroscience 17* (September 1).

Cacioppo, John T., and G. Berntson. 2002. Social Neuroscience. In *Foundations in Social Neuroscience,* ed. John T. Cacioppo et al. Cambridge, Mass.: MIT Press.

Carr, Nicholas. 2008. Is Google Making Us Stoopid. *The Atlantic*. Accessible at: https://www.theatlantic.com/magazine/archive/2008/07/is-google-making-us-stupid/306868/

Carr, Nicholas. 2011. *The Shallows: What the Internet is Doing to Our Brains*. New York: Norton.

Carr, Nicholas. 2014. *The Glass Cage: How Computers are Changing Us*. New York: W.W. Norton.

Carr, Nicholas. 2016. *Utopia is Creepy and Other Provocations*. New York: W. W. Norton.

Carter, Susan Payne, Kyle Greenberg, and Michael Walker. 2016. The Impact of Computer Usage on Academic Performance: Evidence from a Randomized Trial at the United States Military Academy. *Economics of Education Review* 56: 118–132.

Case, Nicky. 2018. How to Become a Centaur. *Journal of Design and Science* 3. Accessed at: https://jods.mitpress.mit.edu/pub/issue3-case.

Cecutti, Lorenzo, Anthony Chemero, and Spike W. S. Lee. 2021. Technology May Change Cognition Without Necessarily Harming It. *Nature Human Behavior* 5: 973–975.

Center for AI Safety. 2023. Statement on AI Risk. Accessed at: https://www.safe.ai/statement-on-ai-risk.

Center for Disease Control and Prevention. 2023. *Youth Risk Behavior Survey Data Summary and Trends Report: 2011–2021*. Accessed at: https://www.cdc.gov/healthyyouth/data/yrbs/yrbs_data_summary_and_trends.htm.

Center for Disease Control and Prevention. 2022. *Distracted Driving*. Accessed at: https://www.cdc.gov/transportationsafety/distracted_driving/index.html.

Centers for Disease Control and Prevention. 2018. Adult Obesity Facts. Accessed at: https://www.cdc.gov/obesity/data/adult.html.

Centers for Disease Control and Prevention. 2019. Childhood Obesity Facts. Accessed at: https://www.cdc.gov/healthyschools/obesity/facts.htm.

Center for Humane Technology. 2019. Accessed at: https://humanetech.com/.

Center for Humane Technology. 2021a. Policy Principles. Accessed at: https://www.humanetech.com/policy-principles.

Center for Humane Technology. 2021b. Policy Reforms Toolkit. Accessed at: https://www.humanetech.com/policy-reforms.

Ceylan, Gizem, Ian A. Anderson, and Wendy Wood. 2023. Sharing of Misinformation is Habitual, Not Just Lazy or Biased. *Proceedings of the National Academy of Sciences* 120 (4): 1–8.

Cha, Young-Jae., et al. 2020. Compensating for the Loss of Human Distinctiveness: The Use of Social Creativity Under Human-Machine Comparisons. *Computers in Human Behavior* 103: 80–90.

Cheney-Lippold, John. 2017. *We are Data: Algorithms and the Making of our Digital Selves*. New York: New York University Press.

Chernev, Alexander, Ulf Böckenholt, and Joseph Goodman. 2015. Choice Overload: A Conceptual Review and Meta-Analysis. *Journal of Consumer Psychology* 25 (2): 333–358.

China Law Translate. 2023. Guidelines for the Establishment of Minors' Modes for the Mobile Internet. Accessed at: https://www.chinalawtranslate.com/en/kid-mode-guidelines/.

Chotpitayasunondh, Varoth, and Karen M. Douglas. 2016. How 'Phubbing' Becomes the Norm: The Antecedents and Consequences of Snubbing via Smartphone. *Computers and Human Behavior* 63: 9–18.

Cisco. 2016. At a Glance: Internet of Things. Accessed at: www.cisco.com/c/dam/en/us/products/collateral/se/internet-of-things/at-a-glance-c45-731471.pdf.

Citton, Yves. 2017. *The Ecology of Attention*. Cambridge, UK: Polity Press.

Claburn, Thomas. 2015. The Threat of Artificial Intelligence. *InformationWeek* (July 3). Accessed at: https://www.informationweek.com/mobile-devices/the-threat-of-artificial-intelligence.

Clark, Andy. 2023. *The Experience Machine: How Our Minds Predict and Shape Reality*. New York: Pantheon.

Clemenson, Gregory D., Maselli Antonella, Alexander J. Fiannaca, Amos Miller, and Gonzalez-Franco. 2021. Rethinking GPS Navigation: Creating Cognitive Maps Through Auditory Clues. *Scientific Reports* 11 (March): 7764.

Collective Intelligence Project. 2023. Accessed at: https://cip.org/.

Coleridge, Samuel Taylor. 1836. *The Literary Remains of Samuel Taylor Coleridge*, vol. III. London: William Pickering.

CommBank Newsroom. 2023. Research from CommBank reveals why Gen Z aren't answering your call. Accessed at: https://www.commbank.com.au/articles/newsroom/2023/06/CBA-More-Gen-Z-research.html.

Conger, Kate. 2020. Twitter Will Turn Off Some Features to Fight Election Misinformation. *New York Times*, October 9. Accessed at: https://www.nytimes.com/2020/10/09/technology/twitter-election-ban-features.html.

Cox, Daniel. 2021. The State of American Friendship: Change, Challenges, and Loss. American Survey Center. Accessed at: https://www.americansurveycenter.org/research/the-state-of-american-friendship-change-challenges-and-loss/.

Cozolino, Louis. 2002. *The Neuroscience of Psychotherapy: Building and Rebuilding the Human Brain*. New York: Norton.

Crawford, Matthew B. 2015. *The World Beyond Your Head: On Becoming an Individual in an Age of Distraction*. New York: Farrar, Straus and Giroux.

Crockett, M.J. 2017. Moral Outrage in the Digital Age. *Nature Human Behaviour* 1: 769–771.

Cropley, Arthur. 1999. Definitions of Creativity. In *The Encyclopedia of Creativity*, Mark Runco and Steven Pritzker, 511–524. New York: Academic Press.

Csikszentmihalyi, Mihaly. 1988. Motivation and Creativity: Toward a Synthesis of Structural and Energistic Approaches to Cognition. *New Ideas in Psychology* 6 (2): 159–176.

Csikszentmihalyi, Mihaly. 1990. *Flow: The Psychology of Optimal Experience*. New York: Harper and Row.

Csikszentmihalyi, Mihaly. 1996. *Creativity*. New York: HarperCollins.

Cuthbertson, Anthony. 2019. China Bans Children Playing Video Games from More Than 90 Minutes a Day or at Night. *Independent* (November 7). Accessed at: https://www.independent.co.uk/life-style/gadgets-and-tech/gaming/china-gaming-ban-video-game-addiction-a9188806.html.

Cytowic, Richard. 2017. There Is a New Link Between Screen-Time and Autism. *Psychology Today*. Accessed at: https://www.psychologytoday.com/us/blog/the-fallible-mind/201706/there-is-new-link-between-screen-time-and-autism.

Dahmani, Louisa, and Véronique. D. Bohbot. 2020. Habitual Use of GPS Negatively Impacts Spatial Memory During Self-Guided Navigation. *Scientific Reports*. 10: 6310.

Datareportal. 2023. *Digital 2022 Global Overview Report*. Accessed at: https://datareportal.com/reports/digital-2023-global-overview-report.

Daugherty, Paul R., and H. James Wilson. 2018. *Human + Machine: Reimagining Work in the Age of AI*. Cambridge: Harvard Business Review Press.

Davenport, Thomas H., and John C. Beck. 2001. *The Attention Economy: Understanding the New Currency of Business*. Boston: Harvard Business School Press.

Davenport, Thomas H., and Julia Kirby. 2016. *Only Humans Need Apply: Winners and Losers in the Age of Smart Machines*. New York: Harper Business.

Davis, Christopher, and Gary Goldfield. 2024. Limiting Social Media Use Decreases Depression, Anxiety, and Fear of Missing Out in Youth with Emotional Distress: A Randomized Controlled Trial. *Psychology of Popular Media*. Accessed at: https://psycnet.apa.org/record/2024-76138-001.

Dennett, Daniel C. 1995. *Darwin's Dangerous Idea: Evolution and the Meanings of Life*. New York: Simon and Schuster.

Dennett, Daniel C. 2015. *Elbow Room: The Varieties of Free Will Worth Wanting*. New. Cambridge, MA: MIT Press.

Dennett, Daniel C. 2017. *From Bacteria to Bach and Back: The Evolution of Minds*. New York: W. W. Norton.

De Sousa, Ronald. 2003. Paradoxical Emotion: On *Sui Generis* Emotional Irrationality. In *Weakness of Will and Practical Irrationality*, ed. Sarah Stroud and Christine Tappolet, 274–297. Oxford: Clarendon Press.

de Tocqueville, Alexis. 1958. *Democracy in America*, vol. II. New York: Vintage.

Dewey, John. 1993. *The Political Writings*, ed. Debra Morris and Ian Shapiro. Indianapolis: Hackett.

Dewey, John. 2002. *Human Nature and Conduct*. Mineola, NY: Dover Publications.

Diamandis, Peter, and Steven Kotler. 2020. *The Future is Faster Than You Think: How Converging Technologies are Transforming Business, Industries, and Our Lives*. New York: Simon and Schuster.

Diamond, Larry. 2019. The Threat of Postmodern Totalitarianism. *Journal of Democracy* 30 (1): 20–24.

Diamond, Larry, and Marc Plattner. 2012. *Liberation Technology: Social Media and the Struggle for Democracy*. Baltimore: Johns Hopkins University Press.

Diamond, Larry. 2010. Liberation Technology. *Journal of Democracy* 21 (3): 69–83.

Dickerson, Kelly, Peter Gerhardstein, and Moser Alecia. 2017 The Role of the Human Mirror Neuron System in Supporting Communication in a Digital World. *Frontiers of Psychology* 8: 698.

Dietz, Nathan, and Robert Grimm. 2023. Understanding Generosity. College Park, MD: Do Good Institute, University of Maryland School of Public Policy. Accessed at: https://dogood.umd.edu/sites/default/files/2023-10/UnderstandingGenerosityReport_DoGoodInstitute_11.2023.pdf.

Diwivedi, Yogesh et al. 2023. Opinion Paper: 'So what If ChatGPT Wrote It?' Multidisciplinary Perspectives on Opportunities, Challenges and Implications of Generative Conversational AI for Research, Practice and Policy. *International Journal of Information Management* 71: 102642.

Donati, Maria Anna, Carola Beccari, and Caterina Primi. 2022. Boredom and Problematic Facebook Use in Adolescents: What Is the Relationship Considering Trait or State Boredom? *Addictive Behaviors* 125 (February): 107132.

Donnelly, Laura. 2019. Children Spend Twice as Long on Smartphones as Talking to Parents. *The Telegraph*. Accessed at: https://www.telegraph.co.uk/news/2019/02/07/children-spend-twice-long-smartphones-talking-parents/.

Doshi, Suhail. 2014. Mixpanel: How Addictive Is Your App? *Vox Recode*. Accessed at: https://www.vox.com/2014/3/6/11624264/mixpanel-how-addictive-is-your-app.

Dostoevsky, Fyodor. 1972. *Notes from Underground*. New York: Penguin.

Doyon, J., and H. Benali. 2005. Reorganization and Plasticity in the Adult Brain During Learning of Motor Skills. *Current Opinion in Neurobiology* 15: 161–167. Accessed at: http://www.sciencedirect.com/science.

Duberry, Jérôme. 2022. *Artificial Intelligence and Democracy: Risks and Promises of AI-Mediated Citizen-Government Relations*. Cheltenham, UK: Edward Elgar.

Duckworth, Angela L., Katherine L. Milkman, and David Laibson. 2018. Beyond Willpower: Strategies for Reducing Failures of Self-Control. *Psychological Science in the Public Interest*. 19 (3): 102–129.

Duggan, Maeve, and Aaron Smith. 2016. The Political Environment on Social Media. *Pew Research Center: Internet and Technology*, October 25. Accessed at http://www.pewinternet.org/2016/10/25/the-political-environment-on-social-media/.

Dunbar, Robin. 2000. On the Origin of the Human Mind. In *Peter Carruthers and Andrew Chamberlain*, ed. Evolution and the Human Mind: Modularity, Language and Meta-Cognition, 238–253. Cambridge: Cambridge University Press.

Dutton, William H. 2020. Introduction to A Research Agenda for Digital Politics. In *A Research Agenda for Digital Politics*, William H. Dutton, ed., xvii–xxix. Cheltamham, UK: Edward Egar.

Dwyer, Ryan J., Kostadin Kushlev, and Elizabeth W. Dunn. 2018. Smartphone Use Undermines Enjoyment of Face-to-Face Social Interactions. *Journal of Experimental Social Psychology* 78: 233–239.

Edelman, Gerald. 1987. *Neural Darwinism*. New York: Basic Books.

Edelson, Laura, Minh-Kha Nguyen, Ian Goldstein, Oana Goga, Tobias Lauinger, and Damon McCoy. 2021. Far-Right News Sources on Facebook more Engaging (March 3). Accessed at: https://medium.com/cybersecurity-for-democracy/far-right-news-sources-on-facebook-more-engaging-e04a01efae90.

eMarketer. 2021. Insider Intelligence's Digital Advertising Trends to Watch in 2022. Accessed at: https://www.emarketer.com/content/insider-intelligence-digital-advertising-trends-watch-2022.

Epictetus. 1995. *The Discourses of Epictetus, The Handbook, Fragments*. London: J.M. Dent.

Epley, Nicholas, Michael Kardas, Xuan Zhao, Stay Atir, and Juliana Schroeder. 2022. Undersociality: Miscalibrated Social Cognition Can Inhibit Social Connection. *Trends in Cognitive Science* 26 (5): 406–418.

Epley, Nicholas, and Juliana Schroeder. 2014. Mistakenly Seeking Solitude. *Journal of Experimental Psychology: General* 143 (5): 1980–1999.

Epstein, Robert, Michael Lothringer, and Vanessa Zankich. 2024. How a Daily Regimen of Operant Conditioning Might Explain the Power of the Search Engine Manipulation Effect (SEME). *Behavior and Social Issues*. Accessed at: https://link.springer.com/article/10.1007/s42822-023-00155-0#.

Epton, T., S. Currie, and C.J. Armitage. 2017. Unique Effects of Setting Goals on Behavior Change: Systematic Review and Meta-Analysis. *Journal of Consulting and Clinical Psychology* 85: 1182–1198.

European Commission. 2023. Regulatory Framework Proposal on Artificial Intelligence. Accessed at: https://digital-strategy.ec.europa.eu/en/policies/regulatory-framework-ai.

European Union. 2024. The EU Artificial Intelligence Act. Accessed at: https://artificialintelligenceact.eu/.

Eurostat. 2024. Suicide Death Rate by Age Group. Accessed at: https://ec.europa.eu/eurostat/databrowser/view/tps00202/default/table?lang=en.

Evans, Robert. 2024. The Cult of AI. *RollingStone* (January 27). Accessed at: https://www.rollingstone.com/culture/culture-features/ai-companies-advocates-cult-1234954528/.

Eyal, Nir. 2014. *Hooked: How to Build Habit-Forming Products*. New York: Penguin.

Eyal, Nir. 2019a. *Indistractable: How to Control your Attention and Choose your Life*. London: Bloomsbury Publishing.

Eyal, Nir. 2019b. *Indistractable: Supplemental Workbook*. N.p. Accessed at: https://www.nirandfar.com/.

Eyal, Nir. 2020a. A Talk with Nir Eyal, part 2. Accessed at: https://letmethink.io/plato-struggled-with-distraction-2500-years-before-the-iphone-a-talk-with-nir-eyal-part-2/?utm_source=ActiveCampaign&utm_medium=email&utm_content=Learning+how+to+be+funny+is+an+essential+people+skill+%5BBest+Reads%5D&utm_campaign=Weekly+Digest+%28May+30%29.

Eyal, Nir. 2020b. How to be Indistractable. *Psyche* (September 2). Accessed at: https://psyche.co/guides/to-become-indistractable-recognise-that-it-starts-within-you?utm_source=ActiveCampaign&utm_medium=email&utm_content=Stop+thinking+about+productivity%2C+and+start+thinking+about+focus++%5BBest+Reads%5D&utm_campaign=Weekly+Digest+%28Sept+12%29.

Eyal, Nir. 2020c. Screen Time for Kids: Give Them What They Need. *Nir and* Far (September 22) Accessed at: https://www.nirandfar.com/screen-time-for-kids/?utm_source=NirAndFar&utm_medium=syndication&utm_campaign=Screen+Time+for+Kids%3A+Give+Them+What+They+Need&utm_source=ActiveCampaign&utm_medium=email&utm_content=Screen+Time+for+Kids%3A+Give+Them+What+They+Need&utm_campaign=New+Article%3A+Screen+Time+for+Kids%3A+Give+Them+What+They+Need.

Facebook. 2021. Hard Life Moments: Mental Health Deep Dive. Accessed at: https://about.fb.com/wp-content/uploads/2021/09/Instagram-Teen-Annotated-Research-Deck-1.pdf.

Faddoul, Marc, Guillaume Chaslot, and Hany Farid. 2020. A Longitudinal Analysis of YouTube's Promotion of Conspiracy Videos. Cornell University: ArXiv. Accessed at: https://arxiv.org/abs/2003.03318.

Farahany, Nita. 2023. *The Battle for Your Brain: Defending the Right to Think Freely in the Age of Neurotechnology*. New York: St. Martin's Press.

Feldstein, Steven. 2021. *The Rise of Digital Repression*. New York: Oxford University Press.

Ferguson, C. J. et al. 2024. There is No Evidence that Time Spent On Social Media Is Correlated with Adolescent Mental Health Problems: Findings from a Meta-Analysis. *Professional Psychology: Research and Practice*. Accessed at: https://psycnet.apa.org/fulltext/2025-31872-001.html.

Fioroni, Sarah, and Dan Foy. 2024. Americans Sleeping Less, More Stressed. *Gallup*. Accessed at: https://news.gallup.com/poll/642704/americans-sleeping-less-stressed.aspx.

Firth, Joseph, et al. 2019. The "Online Brain": How the Internet May be Changing Our Cognition. *World Psychiatry* 18 (2): 119–129.

Fischetti, Mark. 2016. Social Technologies Are Making Us Less Social. *Scientific American* (October 1). Accessed at: https://www.scientificamerican.com/article/social-technologies-are-making-us-less-social/.

Fitbit. Accessed at: https://staticcs.fitbit.com/whyfitbit.

Fleishacker, Samuel. 1999. *A Third Concept of Liberty: Judgment and Freedom in Kant and Adam Smith*. Princeton: Princeton University Press.

Floridi, Luciano, ed. 2015. *The Online Manifesto: Being Human in a Hyperconnected Era*. Cham: Springer Open. Accessed at: https://link.springer.com/book/10.1007/978-3-319-04093-6.

Flyvbjerg, Bent. 1998. *Rationality and Power: Democracy in Practice*. Chicago: University of Chicago Press.

Flyvbjerg, Bent. 2001. *Making Social Science Matter*. Cambridge, UK: Cambridge University Press.

Fogg, B.J. 2020. *Tiny Habits*. Boston: Houghton Mifflin Harcourt.

Foroohar, Rana. 2019. *Don't be Evil: How Big Tech Betrayed is Founding Principles—and All of Us*. New York: Currency.

Forster, E.M. 1927. *Aspects of the Novel*. New York: Harcourt, Brace and Company.

Foucault, Michel. 1988. *Technologies of the Self*. Amherst: University of Massachusetts Press.

Franck, Georg. 1999. The Economy of Attention. *Telepolis* (December 7). Accessed at: https://www.heise.de/tp/features/The-Economy-of-Attention-3444929.html.

Frankfurt, Harry. 1971. Freedom of the Will and the Concept of a Person. *Journal of Philosophy* 67 (1): 5–20.

Frankl, Victor. 2006. *Man's Search for Meaning*. Boston: Beacon Press.

Frenkel, Sheera, Tiffany Hsu, and Steven Lee Myers. 2024. How Russia, China and Iran Are Interfering in the Presidential Election. *New York Times*, October 29. Accessed at: https://www.nytimes.com/2024/10/29/technology/election-interference-russia-china-iran.html.

Fried, Iltzhak, Roy Mukamel, and Gabriel Kreiman. 2011. Internally Generated Preactivation of Single Neurons in Human Medial Frontal Cortex Predicts Volition. *Neuron* 69 (3): 548–562.

Friedman, Thomas L. 2016. *Thank You for Being Late: An Optimist's Guide to Thriving in the Age of Accelerations*. New York: Farrar, Straus and Giroux.

Friends of Attention. 2019. Accessed at: https://www.schoolofattention.org/about-1 and https://friendsofattention.net/sites/default/files/2020-05/TWELVE-THESES-ON-ATTENTION-2019.pdf.

Frischmann, Brett, and Benesch, Susan. 2023. "Friction-In-Design Regulation as 21st Century Time, Place, and Manner Restriction." *Yale Journal of Law & Technology*, 25:376.

Fry, Richard, and Kim Parker. 2021. Rising Share of U.S. Adults are Living without a Spouse or Partner. *Pew Research Center*. Accessed at: https://www.pewresearch.org/social-trends/wp-content/uploads/sites/3/2021/10/PSDT_10.05.21_unpartnered_adults_report-1.pdf

Fussell, Sidney. 2018. The City of the Future Is a Data-Collection Machine. *The Atlantic*. Accessed at: https://www.theatlantic.com/technology/archive/2018/11/google-sidewalk-labs/575551/?utm_term=2018-11-21T10%3A00%3A12&utm.

Gakshy, Eytan, Solomon Messing, and Lada Adamic. 2015. Exposure to Ideologically Diverse News and Opinion on Facebook. *Science* (5 June) 6239: 1130–1132.

Gallup. 2024. *Gallup Global Emotions 2024*. Accessed at: https://www.gallup.com/analytics/349280/gallup-global-emotions-report.aspx.

Garber, Megan. 2018. To Be a Good Citizen, First Pay Attention. *The Atlantic* (June 27). Accessed at: https://www.theatlantic.com/entertainment/archive/2018/06/citizenship-in-our-time-is-about-how-you-spend-your-attention/563912/.

Gazzaley, Adam, and Larry. D. Rosen. 2016. *The Distracted Mind: Ancient Brains in a High-Tech World*. Cambridge: MIT Press.

George, A. Shaji, Hovan George, and Gabior Martin. 2023. ChatGPT and the Future of Work: A Comprehensive Analysis of AI's Impact on Jobs and Employment. *Partners Universal International Innovation Journal* 1 (3) (May–June): 154–186.

Giattino, Charlie, Esteban Ortiz-Ospina, and Max Roser. 2020. Working Hours. Accessed at: https://ourworldindata.org/working-hours.

Gigerenzer, Gerd. 2022. *How to Stay Smart in a Smart World*. Cambridge: MIT Press.

Gilovich, Thomas. 1991. *How We Know What Isn't So*. New York: Macmillan.

Gladstone AI. 2024. *Defense in Depth: An Action Plan to Increase the Safety and Security of Advanced AI*. Accessed at: https://assets-global.website-files.com/62c4cf7322be8ea59c904399/65e7779f72417554f7958260_Gladstone%20Action%20Plan%20Executive%20Summary.pdf.

Gladwell, Malcolm. 2010. Small Change: Why the Revolution Will Not Be Tweeted. New Yorker (October 4). Accessed at: https://www.newyorker.com/magazine/2010/10/04/small-change-malcolm-gladwell.

Glass, Arnold, and Mengxue Kang. 2019. Dividing Attention In the Classroom Reduces Exam Performance. *Educational Psychology* 39 (3): 395–408.

Gollwitzer, Peter M. 1999. Implementation Intentions: Strong Effects of Simple Plans. *American Psychologist* 54: 493–503.

Gollwitzer, Peter M., Kentaro Fujita, and Gabriele Oettingen. 2004. Planning and the Implementation of Goals. In *Handbook of Self-Regulation: Research, Theory, and Applications*, ed. Roy F. Baumeister and Kathleen D. Vohs, 211–228. New York: Gilford Press.

Gonzalez, Robbie. 2018. Can Our Phones Save us from Our Phones? *Wired*, January 30.

Gonzalez-Bailon, Sandra, et al. 2023. Asymmetric Ideological Segregation in Exposure to Political News on Facebook. *Science* 381 (6656): 392–398.

Gottschall, Jonathan. 2012. *The Storytelling Animal: How Stories Make Us Human*. Boston: Houghton Mifflin Harcourt.

Gull, Mubashir, and Bulle Ruth Sravani. 2024. Do Screen Time and Social Media Use Affect Sleep Patterns, Psychological Health, and Academic Performance Among Adolescents? *Children and Youth Services Review* 164: 1–9.

Grace, Katja, John Salvatier, Allan Dafoe, Baobao Zhang, and Owain Evans. 2016. *When Will AI Exceed Human Performance? Evidence from AI Experts*. Oxford University: Future of Humanity Institute. Accessed at: https://www.fhi.ox.ac.uk/will-ai-exceed-human-performance-evidence-ai-experts/.

Grant, Darren. 2017. The Ballot Order Effect is Huge: Evidence from Texas. *Public Choice* 172: 421–442.

Greene, Robert. 2018. *The Laws of Human Nature*. New York: Viking.

Greenleaf, Robert K. 1998. *The Power of Servant-Leadership*. San Francisco: Berrett-Koehler Publishers.

Grimm, Robert T., Jr., and Natthan Dietz. 2018. *"Where are America's Volunteers" A Look at America's Widespread Decline in Volunteering in Cities and States*. Research Brief: Do Good Institute, University of Maryland.

Grinberg, Nir, et al. 2019. Fake News on Twitter During the 2016 U.S. Presidential Election. *Science* 363 (6425): 374–378.

Guess, Andrew, Jonathan Nagler, and Joshua Tucker. 2019. Less than You Think: Prevalence and Predictors of Fake News Dissemination on Facebook. *Science Advances* 5 (1). Accessed at: https://advances.sciencemag.org/content/5/1/eaau4586.

Haan, Katherine. 2024. Top Website Statistics for 2024. *Forbes Advisor*, April 2. Accessed at: www.forbes.com/advisor/business/software/website-statistics/.

Habermas, Jürgen. 1989. *The Structural Transformation of the Public Sphere*. Cambridge: MIT Press.

Hanh, T. N. 2006. We Have the Compassion and Understanding Necessary to Heal the World. In *Mindful Politics*, ed. M. McLeod. Boston: Wisdom Publications.

Haidt, Jonathan. 2024. *The Anxious Generation: How the Great Rewiring of Childhood Is Causing an Epidemic of Mental Illness*. New York: Penguin.

Haidt, Jonathan, and Greg Lukianoff. 2018. *The Coddling of the American Mind*. New York: Penguin.

Haidt, Jonathan, and Nick Allen. 2020. Scrutinizing the Effects of Digital Technology on Mental Health. *Nature* (News and Views Forum): 226–227.

Hall, Jeffrey A., and Dong Liu. 2022. Social Media Use, Social Displacement, and Well-Being. *Current Opinion in Psychology* 46. Accessed at: https://www.sciencedirect.com/science/article/pii/S2352250X22000513?pes=vor.

Hao, Karen. 2021. How Facebook and Google Fund Global Misinformation. *MIT Technology Review*. Accessed at: https://www.technologyreview.com/2021/11/20/1039076/facebook-google-disinformation-clickbait/?truid=&utm_source=the_download&utm_medium=email&utm_campaign=the_download.unpaid.engagement&utm_term=&utm_content=05-24-2022&mc_cid=76802677bd&mc_eid=8710c3458f.

Harari, Yuval Noah. 2017. Filmed and Quoted in "Humans, Gods and Technology." VPRO Documentary. Accessed at: https://www.youtube.com/watch?v=tQd_5as_cMY.

Harari, Yuval Noah. 2019. *21 Lessons for the 21st Century*. New York: Random House.

Hari, Johann. 2015. *Chasing the Scream*. New York: Bloomsbury.

Hari, Johann. 2018. *Lost Connections: Uncovering the Real Causes of Depression—And the Unexpected Solutions*. New York: Bloomsbury.

Hari, Johann. 2022. *Stolen Focus: Why You Can't Pay Attention—and How to Think Deeply Again*. New York: Crown.

Harmony Healthcare. 2024. Black Mirror or Black Hole? American Phone Screen Time Statistics. Accessed at: https://www.harmonyhit.com/phone-screen-time-statistics/.

Harris, Tim. 2024. Hungarian PM Orban: The Only Serious Chance For Peace Is A Trump Win. *RealClear Politics*, March 3. Accessed at: https://www.realclearpolitics.com/video/2024/03/03/hungarian_pm_orban_trump_comeback_is_the_only_serious_chance_for_peace_in_europe.html.

Harris, Tristan. 2017. How a Handful of Companies Control a Billion Minds. TED Talk. Accessed at: https://www.ted.com/talks/tristan_harris_how_a_handful_of_tech_companies_control_billions_of_minds_every_day?language=en.

Harris, Tristan. 2019. A New Agenda for Tech. *Center for Human Technology*. Accessed at: https://www.humanetech.com/news/newagenda.

Harris, Tristan, and Aza Raskin. 2019. Your Undivided Attention. (April 16). Accessed at: https://your-undivided-attention.simplecast.com/episodes/coming-soon-your-undivided-attention-qK0Fq1Al.

Hartley, Isabella, and Indre Viskontas. 2023. How Technology Is Changing Creativity. In *Creativity, Innovation, and Change Across Cultures*, ed. Preiss, D.D., Singer, M., Kaufman, J.C. New York: Palgrave Macmillan.

Hasselberger, William, and Micah Lott. 2023. Where Lies the Grail? AI, Common Sense, and Human Practical Intelligence. *Phenomenology and the Cognitive Sciences* (October). https://doi.org/10.1007/s11097-023-09942-x

Hatzius, Jan, Joseph Briggs, Devesh Kodnani, and Giovanni Pierdomenico. 2023. The Potentially Large Effects of Artificial Intelligence on Economic Growth. Accessed at: https://www.key4biz.it/wp-content/uploads/2023/03/Global-Economics-Analyst_-The-Potentially-Large-Effects-of-Artificial-Intelligence-on-Economic-Growth-Briggs_Kodnani.pdf.

Havens, John C. 2016. *Heartificial Intelligence: Embracing our Humanity to Maximize Machines*. New York: Tarcher/Penguin.

Hayles, N. Katherine. 2007. Hyper and Deep Attention: The Generational Divide in Cognitive Modes. *Profession*, 187–199.

Haynes, John-Dylan. 2011. Decoding and Predicting Intentions. *Annals of the New York Academy of Science* 1224: 9–21.

Hawkins, Jeff. 2004. *On Intelligence*. New York: Henry Holt and Company.

Heath, C., and D. Heath. 2008. *Made to Stick: Why Some Ideas Survive and Others Die*. New York: Random House.

Heidegger, Martin. 1966. *Discourse on Thinking*. New York: Harper and Row.

Heidegger, Martin. 1968. *What is Called Thinking*. New York: Harper and Row.

Helbing, Dirk et al. 2020. Social Enhancement—Insights from Computational Social Sciences, Complex System Theory, and Global Systems Theory. GESDA, Scientific Anticipatory Briefs. Accessed at:https://gesda.global/scientific-anticipatory-briefs/.

Hendlin, Yogi Hale. 2019. I Am a Fake Loop: The Effects of Advertising-Based Artificial Selection. *Biosemiotics* 12 (1): 131–156.

Hendricks, Vincent F., and Mads Vestergaard. 2019. *Reality Lost: Markets of Attention, Misinformation and Manipulation*. Cham, Switzerland: Springer Open.

Herder, Johann Gottfried von. 2002. *Philosophical Writings,* trans. Michael N. Forster, ed. Cambridge: Cambridge University Press.

Hersh, Eitan. 2020. *Politics is for Power: How to Move Beyond Political Hobbyism, Take Action, and Make Real Change*. New York: Scribner.

Hildago, César. 2018. *Augmented Democracy.* Accessed at: https://www.people democracy.com/.

Hillis, W. Daniel. 1988. Intelligence as an Emergent Behavior or, The Songs of Eden. *Daedalus* 117 (1): 175–189.

Hilton, Jacob et al. 2024. A Right to Warn about Advanced Artificial Intelligence. Accessed at: https://righttowarn.ai/.

Hodgson, David. 2012. *Rationality + Consciousness = Free Will.* New York: Oxford University Press.

Hoffer, Eric. 1955. *The Passionate State of Mind.* New York: Harper and Row.

Holtermann, Callie. 2023. They're Over Being Real. *New York Times*, April 23. Accesssed at: https://www.nytimes.com/2023/04/13/style/bereal-app.html?campaign_id=9&emc=edit_nn_20230417&instance_id=90380&nl=the-morning&regi_id=71023177&segment_id=130628&te=1&user_id=1b8a7a6c418ba0b24e53c8b7ae5ab191.

Holltermann, Callie. 2024. Elmo Asked an Innocuous Question. *New York Times,* January 30. Accessed at: https://www.nytimes.com/2024/01/30/style/elmo-x-question.html.

Hovrath, Jared Cooney. 2024. The EdTech Revolution Has Failed: The Case Against Student Use of Computers, Tablets, and Smartphones in the Classroom. Accessed at: https://www.afterbabel.com/p/the-edtech-revolution-has-failed.

Hsu, Tiffany. 2022. On TikTok, Election Misinformation Thrives Ahead of Midterms. *New York Times,* August 14. Accessed at: https://www.nytimes.com/2022/08/14/business/media/on-tiktok-election-misinformation.html?campaign_id=9&emc=edit_nn_20220815&instance_id=69340&nl=the-morning&regi_id=71023177&segment_id=101421&te=1&user_id=1b8a7a6c418ba0b24e53c8b7ae5ab191.

Howard, Philip N. 2015. *Pax Technica: How the Internet of Things May Set Us Free or Lock Us Up.* New Haven: Yale University Press.

Howell, Robert J. 2014. Google Morals, Virtue, and the Asymmetry of Deference. *Nous* 48 (3): 389–415.

Huang, Chiungjung. 2010. Internet Use and Psychological Well-Being: A Meta-Analysis. *Cyberpsychology, Behavior and Social Networking* 13 (3): 241–249.

Hubler, Shawn. 2024. California Joins Growing National Effort to Ban Smartphone Use in Schools. *New York Times,* June 18. Accessed at: https://www.nytimes.com/2024/06/18/us/california-newsom-smartphone-ban.html.

Human or Not? 2024. Accessed at: https://www.humanornot.ai/.

Huxley, Aldous. 2005. *Brave New World and Brave New World Revisited.* New York: Harper Perennials.

Iacoboni, Marco. 2008. *Mirroring People: The New Science of How We Connect with Others.* New York: Farrar, Sraus and Giroux.

India, Freda. 2024. What Happened to FOMO? *After Babel*, August 8. Accessed at: https://www.afterbabel.com/p/what-happened-to-fomo?.

IE: Center for the Governance of Change. 2019. European Tech Insights 2019. Accessed at: https://www.ie.edu/cgc/research/tech-opinion-poll-2019/.

International Center for Media & the Public Agenda. 2019a. *The World UNPLUGGED*. Accessed at: https://theworldunplugged.wordpress.com/emotion/confusion/.

International Center for Media & the Public Agenda. 2019b. *The World UNPLUGGED*. Accessed at: https://theworldunplugged.wordpress.com/emotion/boredom/.

International Labor Organization. 2021. *Changing Demand for Skills in Digital Economies and Societies*. Geneva, Switzerland: International Labor Office.

Ivers, Rowena, and Thomas Astell-Burt. 2023. Nature Rx: Nature Prescribing in General Practice. *Australian Journal of General Practice* 52 (4): 183–187.

Iyengar, Sheena. 2010. *The Art of Choosing*. New York: Twelve.

Jabr, Ferris. 2019. Can You Really Be Addicted to Video Games? *New York Times Magazine* (October 22). Accessed at: https://www.nytimes.com/2019/10/22/magazine/can-you-really-be-addicted-to-video-games.html.

James, William. 1890. *The Principles of Psychology*, vol. 1. New York: Henry Holt and Company.

Jannai, Daniel et al. 2023. Human or Not? A Gamified Approach to the Turing Test. Cornell University arXiv. Accessed at https://arxiv.org/abs/2305.20010.

Jobs, Steven. 2010. *Memory & Imagination: New Pathways to the Library of Congress*. Film Clip Available at Steve Jobs. Computers are Like a Bicycle for Our Minds. Michael Lawrence Films. Accessed at: https://youtu.be/ob_GX50Za6c.

Johansson, B.B. 2007. Regeneration and Plasticity in the Brain and Spinal Cord. *Journal of Cerebral Blood Flow & Metabolism* 27: 1417–1430. Accessed at: http://www.nature.com/jcbfm/journal/v27/n8/full/9600486a.html.

Johnson, Mark. 1993. *Moral Imagination: Implications of Cognitive Science for Ethics*. Chicago: University of Chicago Press.

Jolles, D. D., M. J. Grol, M. A. Van Buchem, S.A. Rombouts, and E.A. Crone. 2010. Practice Effects in the Brain: Changes in Cerebral Activation after Working Memory Practice Depend on Task Demands. *Neuroimage*. Accessed at: http://www.sciencedirect.com/science.

Kahneman, Daniel. 2011. *Thinking, Fast and Slow*. New York: Farrar, Straus and Giroux.

Kahn, Peter. 2011. *Technological Nature: Adaptation and the Future of Human Life*. Cambridge: MIT Press.

Kaiser Family Foundation. 2010. Generation M^2: Media in the Lives of 8- to 18-Year-Olds. Accessed at: https://www.kff.org/other/event/generation-m2-media-in-the-lives-of/.

Kakutani, Michiko. 2018. *The Death of Truth: Notes on Falsehood in the Age of Trump*. New York: Duggan Books.

Kamphuis, Greg. 2017. *A 40 Day Dopamine Fast*. NP: NP.

Kannan, Viji Diane, and Peter J. Veazie. 2023. US Trends in Social Isolation, Social Engagement, and Companionship—Nationally and by Age, Sex, Race/ Ethnicity, Family Income, and Work Hours, 2003–2020. *SSM—Population Health* 21: 101331. Accessed at: https://www.sciencedirect.com/science/article/pii/S235282732200310X.

Kant, Immanuel. 1991. *Political Writings*, 2nd ed. Cambridge: Cambridge University Press.

Kardaras, Nicholas. 2016. *Glow Kids: How Screen Addiction is Hijacking Our Kids—And How to Break the Trance*. New York: St. Martin's Press.

Kardas, Michael, Juliana Schroeder, and O'Brien, ed. 2021. Keep Talking: (Mis)understanding the Hedonic Trajectory of Conversation. *Journal of Personality and Social Psychology*.

Karpf, David. 2016. *Analytic Activism: Digital Listening and the New Political Strategy*. New York: Oxford University Press.

Kayes, D. Christopher, and Jeewhan Yoon. 2022. Cognitive Offloading Strategies and Decrements in Learning: Lessons from Aviation and Aerospace Crises. *Journal of Human Performance in Extreme Environments* 17 (1): 2.

Kellerman, Tom. 2020. COVID-19 and the Need for 'Digital Distancing.' Bank Info Security. Accessed at: https://www.bankinfosecurity.com/kellermann-a-13995; https://www.ptsecurity.com/ww-en/analytics/knowledge-base/work-from-home-digital-distancing-to-keep-your-network-safe/.

Kelly, Kevin. 2010. What Technology Wants. On *Cool Tools*, October 18. Accessed at http://kk.org/cooltools/what-technology/.

Kelly, Yvonne, Afshin Zilanawaia, Cara Booker, and Amanda Sacker. 2018. Social Media Use and Adolescent Mental Health: Findings from the UK Millennium Cohort Study. *eClinicalMedicine* 6: 59–68.

Kendall, Tim. 2020. Testimony at House Committee on Energy and Commerce Hearing on "Mainstreaming Extremism: Social Media's Role in Radicalizing America." Thursday, September 24, 2020. Accessed at: https://energycommerce.house.gov/committee-activity/hearings/hearing-on-mainstreaming-extremism-social-media-s-role-in-radicalizing.

Kerai, Alex. 2023. Cell Phone Usage Statistics. *Reviews.Org*. Accessed at: https://www.reviews.org/mobile/cell-phone-addiction/.

Kessler, David A. 2010. *The End of Overeating: Taking Control of the Insatiable American Appetite*. New York: Rodale Books.

Kierkegaard, Soren. 2000. *The Essential Kierkegaard*. Princeton: Princeton University Press.

Killingsworth, Matthew, and Daniel Gilbert. 2010. A Wandering Mind is an Unhappy Mind. *Science* 330 (6006): 932.

Kim, Victoria. 2024. Australia Has Barred Everyone Under 16 From Social Media. Will It Work? *New York Times,* November 28. Accessed at: https://www.nytimes.com/2024/11/28/world/asia/australia-social-media-ban-law.html.

Kimmerer, Robin. 2023. In David Marchese "You Don't Have to Be Complicit in Our Culture of Destruction." *New York Times Magazine.* Accessed at: https://www.nytimes.com/interactive/2023/01/30/magazine/robin-wall-kimmerer-interview.html.

King, Brett. 2016. *Augmented.* Singapore: Marshall Cavendish International.

King, Cecilia, and Natasha Singer. 2023. Meta Accused by States of Using Features to Lure Children to Instagram and Facebook. *New York Times*, October 24. Accessed at: https://www.nytimes.com/2023/10/24/technology/states-lawsuit-children-instagram-facebook.html.

Kiratli, Osman Sabri. 2023. Social Media Effects on Public Trust in the European Union. *Public Opinion Quarterly* 87 (3): 749–763.

Kirk, Tristan. 2023. Teen Who Plotted to Murder Queen at Windsor Castle with Crossbow 'Encouraged by AI.' *The Standard.* Accessed at: https://www.standard.co.uk/news/crime/queen-crossbow-murder-attempt-windsor-castle-trial-police-b1092313.html.

Kissinger, Henry A., Eric Schmidt, and Daniel Huttenlocher. 2021. *The Age of AI and our Human Future.* New York: Little, Brown and Company.

Kleim, J.A., and T.A. Jones. 2008. Principles of Experience-Dependent Neural Plasticity: Implications for Rehabilitation After Brain Damage. *Journal of Speech, Language, and Hearing Research* 51: S225–S239.

Klein, Ezra. 2022. TikTok May Be More Dangerous Than It Looks. *New York Times*, May 8. Accessed at: https://www.nytimes.com/2022/05/08/opinion/tiktok-twitter-china-bytedance.html.

Klein, Ezra. 2023. America is Dangerously Lonely: Ezra Klein on Her. Accessed at: https://www.nytimes.com/interactive/2023/06/20/opinion/ezra-klein-her-movie.html.

Knausenberger, Judith, Anna Giesen-Leuchter, and Gerald Echterhoof. 2022. Feeling Ostracized by Others' Smartphone Use: The Effect of Phubbing on Fundamental Needs, Mood, and Trust. *Frontiers in Psychology* 13. Accessed at: https://www.frontiersin.org/articles/10.3389/fpsyg.2022.883901/full.

Knibbs, Kate. 2024. AI Slop Is Flooding Medium. *Wired*, October 28. Accessed at: https://www.wired.com/story/ai-generated-medium-posts-content-moderation.

Konrath, Sarah H., Edward H. O'Brien, and Courtney Hsing. 2011. Changes in Dispositional Empathy in American College Students over Time: A Meta-analysis. *Personality and Social Psychology Review* 15 (2): 180–198.

Koivisto, Mika, and Simone Grassini. 2023. Best Humans Still Outperform Artificial Intelligence in a Creative Divergent Thinking Task. *Scientific Reports*

13, 13601. Accessed at: https://www.nature.com/articles/s41598-023-40858-3.

Kosinski, Michal. 2023. Theory of Mind May Have Spontaneously Emerged in Large Language Models. Accessed at: http://export.arxiv.org/abs/2302.02083v1.

Krakauer, David. 2015. *Ingenius: David Krakauer. Nautilus*. Accessed at: https://admin-test.nautil.us/ingenious-david-krakauer-3182/.

Kramer, Adam D., Jamie E. Guillory, and Jeffrey T. Hancock. 2014. Experimental Evidence of Massive-Scale Emotional Contagion Through Social Networks. *Proceedings of the National Academy of Sciences.* 111 (24): 8788–8790.

Kross, E., et al. 2013. Facebook Use Predicts Declines in Subjective Well-Being in Young Adults. *PLoS ONE* 8: e69841.

Kuhn, Reinhard. 1976. *The Demon at Noontide: Ennui in Western Literature.* Princeton: Princeton University Press.

Kuperczko, Diana, Peter Kenyeres, Gergely Darnai, Norbert Kovacs, and Jozsef Janszky. 2022. Sudden Gamer Death: Non-violent Death Cases Linked to Playing Video Games. *BMC Psychiatry* 22: 824. Acccessed at: https://bmcpsychiatry.biomedcentral.com/articles/10.1186/s12888-022-04373-5

Kuznekoff, Jeffrey, and Scott Titsworth. 2013. The Impact of Mobile Phone Usage on Student Learning. *Communication Education* 62 (3): 233–252.

Kurzweil, Ray. 2005. *The Singularity is Near: When Humans Transcend Biology.* New York: Viking.

Kushlev, Kostadin, Jason Proulx, and Elizabeth W. Dunn. 2016. 'Silence Your Phones': Smartphone Notifications Increase Inattention and Hyperactivity Symptoms. *Proceedings of the 2016 CHI Conference on Human Factors in Computing Systems*, 1011–1020. Accessed at: http://dl.acm.org/citation.cfm?d=2858036.2858359.

Kuss, Dara J., and Mark D. Griffiths. 2011. Online Social Networking and Addiction: A Review of the Psychological Literature. *International Journal of Environmental Research and Public Health.* 8 (9): 3528–3552.

Lab, Kaspersky. 2015. *The Rise and Impact of Digital Amnesia.* Moscow: Kapersky Lab.

Lanese, Nicoletta. 2019. Is There Actually Science Behind 'Dopamine Fasting'? *LIVESCIENCE* (November 19). Accessed at: https://www.livescience.com/is-there-science-behind-dopamine-fasting-trend.html.

Lanier, Jaron. 2018. *Ten Arguments for Deleting Your Social Media Accounts Right Now.* New York: Picador.

Larmore, Charles E. 1987. *Patterns of Moral Complexity.* Cambridge: Cambridge University Press.

Larmore, Charles E. 2001. Moral Judgment. In *Judgment, Imagination, and Politics: Themes from Kant and Arendt*, ed. Ronald Beiner and Jennifer Nedelsky, 47–64. New York: Rowman and Littlefield.

LeDoux, Joseph. 2002. *The Synaptic Self: How Our Brains Become Who We Are*. New York: Penguin Books.

Lee, Kai-Fu. 2018. *AI Superpowers: China, Silicon Valley, and the New World Order.* Boston: Houghton Mifflin Harcourt.

Lee, Kai Fu, and Quifan, Chen. 2021. *AI 2041: Ten Visions for our Future*. New York: Currency.

Lembke, Anna. 2021. *Dopamine Nation: Finding Balance in the Age of Indulgence*. New York: Dutton.

Lenkowsky, Leslie. 2023. Americans Are Volunteering Less. *Chronicle of Philanthropy.* https://www.philanthropy.com/article/americans-are-volunteering-less-what-can-nonprofits-do-to-bring-them-back

Lepp, A., J. Li, J.E. Barkley, and S. Salehi-Esfahani. 2015. Exploring the Relationships between College Students' Cell Phone Use, Personality and Leisure. *Computers in Human Behavior* 43: 210–219.

León-Domínguez, Umberto. 2024. Potential Cognitive Risks of Generative Transformer-Based AI Chatbots on Higher Order Executive Functions. *Neuropsychology*. Accessed at: https://doi.org/10.1037/neu0000948.

Leopold, Aldo. 1966. *A Sand County Almanac, with Essays on Conservation from Round River*. New York: Ballantine Books.

Lepp, Andrew, Jacob Barkley, and Aryn Karpinski. 2015. The Relationship Between Cell Phone Use and Academic Performance in a Sample of U.S. College Students. *Sage Open* 5 (1).

Lerman, Rachel. 2019. Q&A: Ex-Googler Harris on How Tech 'Downgrades' Humans. *AP News*. Accessed at: https://apnews.com/article/technology-business-data-privacy-apple-inc-dea7f32d16364c6093f19b938370d600.

Lewis, Cecil Day. 1951. *The Poet's Task.* Oxford: Clarendon Press.

Lewis, Paul. 2017. 'Our Minds Can Be Hijacked': The Tech Insiders Who Fear a Smartphone Dystopia. *The Guardian,* October 6. Accessed at: https://www.theguardian.com/technology/2017/oct/05/smartphone-addiction-silicon-valley-dystopia.

Libet, Benjamin. 1985. Unconscious Cerebral Initiative and the Role of Conscioius Will in Voluntary Action. *Behavioral and Brain Sciences* 8 (4): 529–539.

Lieberman, Daniel, and Long Michael. 2018. *The Molecule of More*. Dallas: BenBella Books.

Lillard, Angeline, and Lori Skibbe. 2005. Theory of Mind: Conscious Attribution and Spontaneious Trait Inference. In *Ran Hassin, James Uleman and John Bargh*, ed. The New Unconscious, 277–305. Oxford: Oxford University Press.

Lin, Fuchun, et al. 2012. Abnormal White Matter Integrity in Adolescents with Internet Addiction Disorder. *PLoS ONE* 7 (1): e30253.

Lin, Luona, Kim Parker, and Juliana Horowitz. 2024. Challenges in the Classroom. Pew Research Center. Accessed at: https://www.pewresearch.org/social-trends/2024/04/04/challenges-in-the-classroom/#policies-around-cellphone-use.

Lindsay, Roddy. 2021. I Designed Algorithms at Facebook. Here's How to Regulate Them. *New York Times,* October 6. Accessed at: https://www.nytimes.com/2021/10/06/opinion/facebook-whistleblower-section-230.html.

Linnet, Jakob, E. Peterson, D.J. Doudet, A. Gjedde, and A. Møller. 2010. Dopamine Release in Ventral Striatum of Pathological Gamblers Losing Money. *Acta Psychiatrica Scandinavica* 122 (4): 326–333.

Liptak, Adam. 2022. Supreme Court Blocks Texas Law Regulating Social Media Platforms. *New York Times,* May 31. Accessed at: https://www.nytimes.com/2022/05/31/us/politics/supreme-court-social-media-texas.html?campaign_id=60&emc=edit_na_20220531&instance_id=0&nl=breaking-news&ref=cta&regi_id=71023177&segment_id=93831&user_id=1b8a7a6c418ba0b24e53c8b7ae5ab191.

Liu, Mingli, et al. 2022. Time Spent on Social Media and Risk of Depression in Adolescents: A Dose-Response Meta-Analysis. *International Journal of Environmental Research and Public Health* 19 (9): 5164.

Loeffler, John. 2018. Why Chinese Artificial Intelligence Will Run The World: How the Chinese tech giants Baidu, Alibaba, and Tencent will Develop the Systems that Will Run the World. *Interesting Engineering*, November 6. Accessed at: https://amp.interestingengineering.com/why-chinese-artificial-intelligence-will-run-the-world.

Loh, Kep Kee, and Ryota Kanai. 2014. Higher Media Multi-Tasking Activity is Associated with Smaller Gray-Matter Density in the Anterior Cingulate Cortex. PLoS One 9 (9). Accessed at: https://doi.org/10.1371/journal.pone.0106698.

Lorenz, Taylor. 2018. Generation Z Is Already Bored by the Internet. *Daily Beast* (April 3). Accessed at: https://www.thedailybeast.com/generation-z-is-already-bored-by-the-internet?ref=scroll.

Lorenz, Taylor. 2022. Doomscrolling Got You Down? Take a Break at a Digital Rest Stop. *The Washington* Post, March 22. Accessed at: https://www.washingtonpost.com/technology/2022/03/22/digital-rest-stop-doomscrolling/?mc_cid=727f278cec&mc_eid=8710c3458f.

Louv, Richard. 2012. *The Nature Principle: Reconnecting with Life in a Virtual Age*. Chapell Hill: Algonquin Books.

Lustig, Robert H. 2017. *The Hacking of the American Mind: The Science Behind the Corporate Takeover of Our Bodies and Brains*. New York: Avery.

Loucaides, Darren. 2022. Message Received. *Wired* (March): 40–57.

Loucaides, Darren. 2023. Intercepted. *Wired* (March): 48–59.

Mac, Ryan, Charlie Warzel, and Alex Kantrowitz. 2018. Growth at Any Cost. *BuzzFeed News,* March 29. Accessed at: https://www.buzzfeednews.com/article/ryanmac/growth-at-any-cost-top-facebook-executive-defended-data#.cnkEEaN0v.

MacIntyre, Alasdair. 1981. *After Virtue: A Study in Moral Theory*. Notre Dame: University of Notre Dame Press.

MacIntyre, Alasdair. 1999. *Dependent Rational Animals*. Chicago: Open Court.

Madrigal, Alexis C. 2018. What Facebook Did to American Democracy. *The Atlantic*, October 12, 2017. Accessed at: https://www.theatlantic.com/technology/archive/2017/10/what-facebook-did/542502/.

Maes, Pattie. 1998. Intelligence Augmentation. *Edge*, January 20. *Accessed at* https://www.edge.org/conversation/pattie_maes-intelligence-augmentation.

Malone, Thomas. 2018. *Superminds: The Surprising Power of People and Computers Thinking Together*. New York: Little, Brown and Company.

Manyika, James and Jacques Bughin. 2018. The Promise and Challenge of the Age of Artificial Intelligence, Briefing Note prepared for the Tallinn Digital Summit, McKinsey Global Institute, New York. Accessed at: https://www.mckinsey.com/featured-insights/artificial-intelligence/the-promise-and-challenge-of-the-age-of-artificial-intelligence.

Marcus, Gary. *Taming Silicon Valley: How We Can Ensure that AI Works for Us*. Cambridge: MIT Press.

Margetts, Helen, Peter John, Scott Hale, and Taha Yasseri. 2016. *Political Turbulence: How Social Media Shape Collective Action*. Princeton: Princeton University Press.

Mark, Gloria. 2015. *Multitasking Int the Digital Age*. NP: Morgan and Claypool.

Mark, Gloria. 2023. *Attention Span: A Groundbreaking Way to Restore Balance, Happiness and Productivity*. New York: Hanover Square Press.

Markoff, John. 2015. *Machines of Loving Grace: The Quest for Common Ground Between Humans and Robots*. New York: HarperCollins.

Marriott, Hannah R., and Valentina Pitardi. 2023. One is the Loneliest Number... Two Can be as Bad as One. The Influence of AI Friendship Apps on Users' Well-being and Addiction. *Psychology and Marketing*, 1–16. https://doi.org/10.1002/mar.21899

Marx, Karl. 1978. *The Marx-Engels Reader.* Robert C. Tucker, 2nd ed. New York: W. W. Norton.

Maslej, Nestor et al. 2024. *The AI Index 2024 Annual Report*. AI Index Steering Committee, Institute for Human-Centered AI, Stanford University. Stanford, CA. Accessed at: https://aiindex.stanford.edu/wp-content/uploads/2024/05/HAI_AI-Index-Report-2024.pdf.

Matsa, Katarina Eva. 2023. More Americans are Getting News on TikTok, Bucking the Trend Seen on Most Other Social Media Sites. Pew Research

Center. Accessed at: https://www.pewresearch.org/short-reads/2023/11/15/more-americans-are-getting-news-on-tiktok-bucking-the-trend-seen-on-most-other-social-media-sites/.

Maushart, Susan. 2010. *The Winter of our Disconnect: How Three Totally Wired Teenagers (and a Mother Who Slept with Her iPhone) Pulled the Plug on Their Technology and Lived to Tell the Tale*. New York: Penguin.

McAfee, Andrew, and Erik Brynjolfsson. 2016. Where Computers Defeat Humans, and Where They Can't. New York Times, March 16, 2016. Accessed at: http://www.nytimes.com/2016/03/16/opinion/where-computers-defeat-humans-and-where-they-cant.html?emc=edit_th_20160316&nl=todaysheadlines&nlid=71023177&_r=0.

McAfee, Andrew, and Erik Brynjolfsson. 2017. *Machine, Platform, Crowd: Harnessing Our Digital Future*. New York: W. W. Norton.

McClure, Paul. 2023. AI Offers Improved Civility for Polarizing Online Conversations. *New Atlas* (October 5). Accessed at: https://newatlas.com/health-wellbeing/ai-offers-improved-civility-for-polarizing-online-conversations/.

McGuinness, Seamus, Konstantinos Pouliakas, and Paul Redmond. 2019. Skills-displacing Technological Change and Its Impact on Jobs: Challenging Technological Alarmism? *Institute of Labor Economics,* No 12541. Accessed at: https://docs.iza.org/dp12541.pdf.

McGuirk, Rod. 2024. Australia Plans a Social Media Ban for Children Under 16. *Associated Press World News,* November 7. Accessed at: https://apnews.com/article/australia-social-media-age-limit-e8259408c0b1456f41967decd474782a.

McKenzie-Mohr, Doug. 2011. *Fostering Sustainable Behavior*. Gabriola, B.C.: New Society Publishers.

McLuhan, Marshall. 1964. *Understanding Media: The Extensions of Man*. New York: McGraw Hill.

McLuhan, Marshall. 1966. Cybernation and Culture. In *The Social Impact of Cybernetic*, ed. Charles R. Dechert, 95–108. Notre Dame: University of Notre Dame Press.

McNamee, Roger. 2019. *Zucked: Waking Up to the Facebook Catastrophe*. New York: Penguin.

Meacham, Jon. 2018. *The Soul of America: The Battle for Our Better Angels*. New York: Random House.

Medium. 2024. Accessed at: https://medium.com/about.

Meet the Press. 2017. January 22. Accessed at: https://www.nbcnews.com/meet-the-press/meet-press-01-22-17-n710491.

Mehl, Matthias R., Simine Vazire, Shannon E. Holleran, and C. Shelby Clark. 2010. Eavesdropping on Happiness: Well-Being is Related to Having Less Small Talk and More Substantive Conversations. *Psychological Science* 21 (4): 539–541.

Melina, Uncapher, Monica Thieu, and Anthony Wagner. 2016. Media Multitasking and Memory: Differences in Working Memory and Long-Term Memory. *Psychonomic Bulletin and Review* 23 (2): 483–490.

Mencken, H.L. 1921. *Prejudices, Second Series*. London: Gape.

Merzenich, M. M., and R.C. deCharms. 1996. Neural Representations, Experience, and Change. In *The Mind-Brain Continuum: Sensory Processes*, eds. R. Llinas and P. Churchland, 61–81. Cambridge: MIT Press.

Messena, Mattia, and Marina Everri. 2023. Unpacking the Relation Between Children's Use of Digital Technologies and Children's Well-being: A Scoping Review. *Clinical Child Psychology and Psychiatry* 28 (1): 161–198.

Michelucci, Pietro, and Janis L. Dickinson. 2016. The Power of Crowds. *Science* 351: 32–33.

Microsoft (Consumer Insights, Canada). 2015. Attention Spans. Accessed at: https://www.scribd.com/document/265348695/Microsoft-Attention-Spans-Research-Report.

Mills, Stuart. 2021. The Future of Nudging Will Be Personal. *Behavioral Scientist* (March 15). Accessed at: https://behavioralscientist.org/the-future-of-nudging-will-be-personal/?utm_source=ActiveCampaign&utm_medium=email&utm_content=How+to+Get+Better+at+Admitting+Your+Wrong++%5BBest+Reads%5D&utm_campaign=Weekly+Digest+%28March+20%29.

Mischel, Walter, Y. Shoda, and P. K. Peake. (1988). The Nature of Adolescent Competencies Predicted by Preschool Delay of Gratification. *Journal of Personality and Social Psychology* 54: 687–696.

Mishan, Ezra J. 1967. *The Costs of Economic Growth*. New York: Praeger.

Misra, Shalili, Lulu Cheng, Jamie Genevie, and Miao Yuan. 2016. The Iphone Effect: The Quality of In-person Social Interactions in the Presence of Mobile Devices. *Environment and Behavior* 48 (2): 275–298.

Mitchell, Amy, and Mason Walker. 2021. More Americans Now Say Government Should Take Steps to Restrict False Information Online than in 2018. Pew Research Center. Accessed at: https://www.pewresearch.org/fact-tank/2021/08/18/more-americans-now-say-government-should-take-steps-to-restrict-false-information-online-than-in-2018/.

Moffitt, T.E., L. Arseneault, D. Belsky, N. Dickson, R.J. Hancox, H.L. Harrington, and A. Caspi. 2011. A Gradient of Childhood Self-Control Predicts Health, Wealth, and Public Safety. *Proceedings of the National Academy of Sciences* 108: 2693–2698.

Mogliner, Cassie. 2012. You'll Feel Less Rushed If You Give Time Away. *Harvard Business Review* 90 (9): 28–29.

Mogliner, Cassie, Zoë Chance, and Michael Norton. 2012. Giving Time Gives You Time. *Psychological Science* 23 (10): 1233–1238.

Moldbug, Mencius. 2008. An Open Letter to Open-Minded Progressives, Chapter 4: Dr. Johnson's Hypothesis. Accessed at: https://www.unqualified-reservations.org/2008/05/ol4-dr-johnsons-hypothesis/.

Monitoring the Future. 2022. Accessed at: https://docs.google.com/spreadsheets/d/1kvIPnbYrYrymQGueld48gACSYXFYfzCfMkdCpkmHKAw/edit?gid=0#gid=0.

Moore, Justine, Bryon Kim, Li Yoko, and Martin Casado. 2024. It's Not a Computer, It's a Companion!. Andreessen Horowitz. Accessed at: https://a16z.com/its-not-a-computer-its-a-companion/.

Morozov, Evgeny. 2011. *The Net Delusion: The Dark Side of Internet Freedom.* New York: PublicAffairs.

Morozov, Evgeny. 2013. *To Save Everything, Click Here: The Folly of Technological Solutionism.* New York: Public Affairs.

Morris, Jonathan S. 2007. Slanted Objectivity? Perceived Media Bias, Cable News Exposure, and Political Attitudes. *Social Science Quarterly* 88: 707–728.

Mohsen Mosleh, Qi Yang, Tauhid Zaman, Gordon Pennycook, and David G. Rand. 2024. Differences in Misinformation Sharing Can Lead to Politically Asymmetric Sanctions. *Nature* 634: 609–624.

Moynihan, Andrew B., Eric R. Igou, and Wijnand A. P. van Tilburg. 2017. Boredom Increases Impulsiveness: A Meaning-Regulation Perspective. *Social Psychology* 48 (5): 293–309.

Moynihan, Andrew B., Eric R. Igou, and Wijnand A. P. van Tilburg. 2018. Lost in the Crowd: Conformity as Escape Following Disbelief in Free Will. *European Journal of Social Psychology* 49 (3): 1–18.

Moynihan, Daniel Patrick. 1996. *Miles to Go.* Cambridge, MA: Harvard University Press.

Mozur, Paul, Aaron Krolik, and Steven Lee Myers. 2024. How Telegram Became a Playground for Criminals, Extremists and Terrorists. *New York Times,* September 7. Accessed at: https://www.nytimes.com/2024/09/07/technology/telegram-crime-terrorism.html.

Mozur, Paul, Muyi Xiao, and John Liu. 2022. 'An Invisible Cage': How China Is Policing the Future. *New York Times,* June 25. Accessed at: https://www.nytimes.com/2022/06/25/technology/china-surveillance-police.html?action=click&module=Well&pgtype=Homepage&section=Business.

Murdoch, Iris. 1970. *The Sovereignty of Good.* UK: Routledge and Kegan Paul.

Murray, Charles. 2020. *Human Diversity: The Biology of Gender, Race, and Class.* New York: Hachette.

Murthy, Vivek. 2020. *Together: The Healing Power of Human Connection in a Sometimes Lonely World.* New York: HarperCollins.

Murthy, Vivek. 2023a. *Social Media and Youth Mental Health: The U.S. Surgeon General's Advisory.* Accessed at: https://www.hhs.gov/sites/default/files/sg-youth-mental-health-social-media-advisory.pdf.

Murthy, Vivek. 2023b. *Our Epidemic of Loneliness and Isolation*. U.S. Department of Health and Human Services. Accessed at: https://www.hhs.gov/sites/default/files/surgeon-general-social-connection-advisory.pdf.

Murthy, Vivek. 2024. Surgeon General: Why I'm Calling for a Warning Label on Social Media Platforms. *New York Times,* June 17. Accessed at: https://www.nytimes.com/2024/06/17/opinion/social-media-health-warning.html.

Nagy Abdelhamid, N.E., A.M. Abdelsamea Mohamed, N.M. Shaker Ibrahim, A. Adel Abdelgawad, M.H. Abd El Moneam, and S.A.I. Marzouk. 2021. Internet Addiction and Its Relation to Emotional Intelligence and Resilience Among a Sample of Egyptian University Students. *Addictive Disorders and Their Treatment* 20 (4): 342–350.

Nass, Clifford. 2009. Interview for *Frontline: Digital Nation*. Accessed at: https://www.pbs.org/wgbh/pages/frontline/digitalnation/interviews/nass.html.

National Institute of Health. 2019. Overdose Death Rates. Accessed at: https://www.drugabuse.gov/related-topics/trends-statistics/overdose-death-rates.

NCVO. 2023. Key Findings from Time Well Spent 2023. London, UK: National Council of Volunteer Organizations. Accessed at: https://www.ncvo.org.uk/news-and-insights/news-index/key-findings-from-time-well-spent-2023/.

Neely Social Media Index. 2023. Accessed at: https://neely.usc.edu/usc-marshalls-neely-center-social-media-index/.

Negroponte, Nicholas. 1995. *Being Digital*. New York: Alfred A. Knopf.

Neophytou, Eliana, Laurie A. Manwell, and Roelof Eikelboom. 2021. Effects of Excessive Screen Time on Neurodevelopment, Learning, Memory, Mental Health, and Neurodegeneration. *International Journal of Mental Health and Addiction* 19: 24–744.

Newport, Cal. 2019. *Digital Minimalism*. New York: Penguin Portfolio.

Newport, Frank. 2015. Americans' Perceived Time Crunch No Worse Than in Past. *Gallup*. Accessed at: https://news.gallup.com/poll/187982/americans-perceived-time-crunch-no-worse-past.aspx.

Nichols, Tom. 2017. *The Death of Expertise: The Campaign against Established Knowledge and Why it Matters*. New York: Oxford University Press.

Nichols, Tom. 2024. Democracy is Not Over. *The Atlantic,* November 6. Accessed at: https://www.theatlantic.com/newsletters/archive/2024/11/trump-victory-democracy/680549/.

Nie, Norman H., Darwin W. Miller, Saar Golde, Daniel M. Butler, and Kenneth Winneg. 2010. The World Wide Web and the U.S. Political News Market. *American Journal of Political Science* 54: 428–439.

Nielsen, Rasmus Kleis, and Sarah Anne Ganter. 2022. *The Power of Platforms*. New York: Oxford University Press.

Nielson. 2018. The Nielsen Total Audience Report: Q3 2108. Accessed at: https://www.nielsen.com/us/en/insights/report/2019/q3-2018-total-audience-report/.

Nietzsche, Friedrich. 1882. *Gesammelte Werke*, vol. 16. Munich: Musarion Verlag.

Nietzsche, Friedrich. 1967. *The Will to Power*. New York: Vintage.

Nietzsche, Friedrich. 1967. *The Genealogy of Morals*. New York: Vintage.

Nietzsche, Friedrich. 1974. *The Gay Science*. New York: Vintage.

Noble, David F. 1998. *The Religion of Technology*. New York: Alfred A. Knopf.

Noake, Rick. 2015. This City Embedded Traffic Lights in the Sidewalks so that Smartphone Users Don't Have to Look Up. *The Washington Post*, April 25. Accessed at: https://www.washingtonpost.com/news/worldviews/wp/2016/04/25/this-city-embedded-traffic-lights-in-the-sidewalks-so-that-smartphone-users-dont-have-to-look-up/.

Oakeshott, Michael. 2001. *The Voice of Liberal Learning*. Indianapolis: Liberty Fund.

Oatley, Keith. 2008. The Mind's Flight Simulator. *In Psychologist* 21: 1030–1032.

Pariser, Eli. 2011. *The Filter Bubble: What the Internet Is Hiding from You*. New York: Penguin Press.

O'Gorman, Marcel. 2018. The Case for Locking Up Your Smartphone. *The Atlantic* (February 2). Accessed at: https://www.theatlantic.com/technology/archive/2018/02/the-case-for-locking-up-your-smartphone/552215/.

Odell, Jenny. 2019. *How to Do Nothing: Resisting the Attention Economy.* Brooklyn: Melville House.

Oeskes, Naomi, and Eik Conway. 2010. *Merchants of Doubt*. New Yok: Bloosmbury.

O'Gieblyn, Meghan. 2022. These Notifications are Killing Me. *Wired* (February), 26–27.

Oggins, Jean, and Jeffrey Sammis. 2012. Notions of Video Game Addiction and Their Relation to Self-Reported Addiction Among Players of World of Warcraft. *International Journal of Mental Health Addiction* 10: 210–230.

O'Neill, Helen. 2016. Deskilling in the Age of Digital Amnesia. *Reader's Digest*. Accessed at: https://www.rdasia.com/true-stories-lifestyle/science-technology/Deskilling-in-the-Age-of-Digital-Amnesia.

Orben, Amy, Andrew K. Przbylski, Blakemore Sarah-Jayne, and Rogier A. Kievit. 2022. Windows of Developmental Sensitivity to Social Media. *Nature Communications* 13 (1649). Accessed at: https://www.nature.com/articles/s41467-022-29296-3#citeas.

Orben, Amy, Tobias Dienlin, and Andrew K. Przybylski. 2019. Social Media's Enduring Effect on Adolescent Life Satisfaction. *PNAS* 116 (21): 10226–10228.

Orben, Amy, and Andrew K. Przybylski. 2019a. The Association Between Adolescent Well-Being and Digital Technology Use. *Nature Human Behaviour* 3: 173–182. Accessed at: https://www.nature.com/articles/s41562-018-0506-1#citeas.

Orben, Amy, and Andrew K. Przybylski. 2019. Screens, Teens, and Psychological Well-Being: Evidence from Three Time-Use-Diary Studies. *Psychological Science* 30 (5): 682–696.

Oremus, Will. 2022. Want to Regulate Social Media? The First Amendment May Stand in the Way. *The Washington Post,* May 30. Accessed at: https://www.washingtonpost.com/technology/2022/05/30/first-amendment-social-media-regulation/?mc_cid=c6290965dd&mc_eid=8710c3458f.

Orr, David. *Down to the Wire: Confronting Climate Collapse*. Oxford: Oxford University Press.

Ortiz-Ospina, Estaban. 2019. Are Facebook and Other Social Media Platforms Bad for Our Well-Being? *Our World in Data*. Accessed at: https://ourworldindata.org/social-media-wellbeing.

Oswald, Tassia, Alice Rumbold, Sophie Kedzior, and Moore Viviene. 2020. Psychological Impacts of "Screen Time" and "Green Time" for Children and Adolescents: A Systematic Scoping Review. PLOS ONE. Accessed at: https://journals.plos.org/plosone/article?id=10.1371/journal.pone.0237725.

Panda Security. 2020. *1/3 of Americans Suffer from Digital Amnesia*. Bilbao: Panda Security.

Paquette, Vincent, et al. 2003. 'Change the Mind and You Change the Brain': Effects of Cognitive Behavioral Therapy on the Neural Correlates of Spider Phobia. *NeuroImage* 18: 401–409.

Pariser, Eli. 2012. *The Filter Bubble*. New York: Penguin.

Paasonen, Susanna. 2021. *Dependent, Distracted, Bored: Affective Formations in Networked Media*. Cambridge: MIT Press.

Penney, Joel. 2017. *The Citizen Marketer: Promoting Political Opinion in the Social Media Age*. New York: Oxford University Press.

Pentland, Alex. 2014. *Social Physics: How Good Ideas Spread—The Lessons from a New Science*. New York: Penguin.

Perrin, Andrew, and Sara Atske. 2021. About Three-In-Ten U.S. Adults Say they are 'Almost Constantly' Online. Pew Research Center. Accessed at: https://www.pewresearch.org/fact-tank/2021/03/26/about-three-in-ten-u-s-adults-say-they-are-almost-constantly-online/.

Pew Research Center. 2018. Teens, Social Media & Technology 2018. Accessed at: https://www.pewinternet.org/2018/05/31/teens-social-media-technology-2018/.

Pew Research Center. 2021. *Digital News Fact Sheet*. Washington, DC: Pew Research Center.

Pew Research Center, Internet and Technology. 2017. Accessed at: http://www.pewinternet.org/2017/03/29/the-future-of-free-speech-trolls-anonymity-and-fake-news-online/; http://www.pewinternet.org/2017/10/19/the-future-of-truth-and-misinformation-online/.

Plous, Scott. 1993. *The Psychology of Judgment and Decision Making*. Philadelphia: Temple University Press.

Poldrack, Russell A. 2021. *Hard to Break: Why our Brains Make Habits Stick*. Princeton: Princeton University Press.

Portwood-Stacer, L. 2012. Media Refusal and Conspicuous Non-consumption: The Performative and Political Dimensions of Facebook Abstention. *New Media and Society* 15 (7): 1041–1057.

Posner, Eric. 2024. The AI Octopus. *Project Syndicate,* January 8. Accessed at: https://www.project-syndicate.org/commentary/ai-will-strengthen-big-tech-oligopoly-market-concentration-and-corporate-political-power-by-eric-posner-2024-01.

Potenza, Marc N., Kyle A. Faust, and Faust David. 2020. *The Oxford Handbook of Digital Technologies and Mental Health*. New York: Oxford University Press.

Prathap, Veena, and S. Swapnil Singh. 2021. Impact of Digital Addiction and Cognitive Offloading on Prospective Memory of Young Adults. *Indian Journal of Health & Wellbeing* 12 (4): 439–445.

Premack, David, and Ann Premack. 2003. *Original Intelligence*. New York: McGraw Hill.

Primack, Brian, et al. 2017. Social Media Use and Perceived Social Isolation Among Young Adults in the U.S. *American Journal of Preventative Medicine* 53 (1): 1–8.

Price, Catherine. 2018. *How to Break Up with Your Phone: The 30-Day Plan to Take Back Your Life*. California: Ten Speed Press.

Przybylski, Andrew, and Neta Weinstein. 2013. Can You Connect With Me Now? How the Presence of Mobile Communication Technology Influences Face-to-Face Conversation Quality. *Journal of Social and Personal Relationships* 30: 3.

Purvis, Jeanette. 2017. Finding Love in a Hopeless Place: Why Tinder Is So 'Evilly Satisfying.' *Salon*. Accessed at: https://www.salon.com/2017/02/12/finding-love-in-a-hopeless-place-why-tinder-is-so-evilly-satisfying/.

Ra, Chaelin, et al. 2018. Association of Digital Media Use With Subsequent Symptoms of Attention-Deficit/Hyperactivity Disorder Among Adolescents. *JAMA* 320 (3): 255–263.

Rabin, Roni Caryn. 2021. Overdoes Deaths Reach Record High as the Pandemic Spread. *New York Times*, November 17. Accessed at: https://www.nytimes.com/2021/11/17/health/drug-overdoses-fentanyl-deaths.html?campaign_id=9&emc=edit_nn_20220213&instance_id=53054&nl=the-morning&regi_id=71023177&segment_id=82588&te=1&user_id=1b8a7a6c418ba0b24e53c8b7ae5ab191.

Ramachandran, V.S., and S. Blakeslee. 1998. *Phantoms in the Brain*. New York: William Morrow and Company.

Rainie, Lee, and Kathryn Zickuhr. 2015. Americans' Views on Mobile Etiquette. Pew Research Center, August 26.

Reid, Rob. 2017. *After On*. New York: Del Rey.

Rheingold, Howard. 2003. *Smart Mobs: The Next Social Revolution*. New York: Perseus.

Riehm, Kira et al. 2019. Associations Between Time Spent Using Social Media and Internalizing and Externalizing Problems Among US Youth. *JAMA Psychiatry* 76 (12): 1266–1273.

Rinta-Kahila, Tapani, T. Penttinen, A. Salovaara, W. Soliman, and J. Ruissalo. 2023. The Vicious Circles of Skill Erosion: A Case Study of Cognitive Automation. *Journal of the Association for Information Systems* 24 (5): 1378–1412.

Rinta-Kahila, Tapani, Esko Penttinen, Antti Salovaara, and Wael Soliman. 2018. Consequences of Discontinuing Knowledge Work Automation—Surfacing of Deskilling Effects and Methods of Recovery. *Proceedings of the 51st Hawaii International Conference on System Sciences*. Accessed at: https://scholarspace.manoa.hawaii.edu/server/api/core/bitstreams/bf98a123-5385-486b-8ecb-76e1e3dbcda0/content.

Roberts, Margaret. 2018. *Censored: Distraction and Diversion inside China's Great Firewall*. Princeton: Princeton University Press.

Rød, E. G, and Weidmann, N. B. 2015. "Empowering activists or autocrats? The Internet in authoritarian regimes." *Journal of Peace Research* 52 (3): 338–351.

Rolland, Romain. 1920. Un Livre de Raymond Lefebvre: *Le Sacrifice d'Abraham*. *L'Humanité*, March 19.

Roose, Kevin. 2024. Can A.I. Be Blamed for a Teen's Suicide? *New York Times*, October 23. Accessed at: https://www.nytimes.com/2024/10/23/technology/characterai-lawsuit-teen-suicide.html.

Rorty, Amelie Oksenberg. 1997. The Social and Political Sources of Akrasia. *Ethics* 107 (July): 644–657.

Rosen, Larry. 2012. *IDisorder: Understanding our Obsession with Technology and Overcoming its Hold on Us*. New York: St. Martin's Griffin.

Rothwell, Jonathan. 2023. Teens Spend Average of 4.8 Hours on Social Media Per Day. *Gallup*. Accessed at: https://news.gallup.com/poll/512576/teens-spend-average-hours-social-media-per-day.aspx.

Russell, Stuart. 2019. *Human Compatible: Artificial Intelligence and the Problem of Control*. New York: Viking.

Ryan, Fergus. 2018. An Orwellian Future Is Taking Shape in China. *The Sydney Morning Herald*, January 8. Accessed at: https://www.smh.com.au/opinion/an-orwellianfuture-is-taking-shape-in-china-20171220-h07vbw.html

Safronova, Valeriya, Neil MacFarquhar, and Adam Satariano. 2022. Where Russians Turn for Uncensored News on Ukraine. *New York Times,* April 16. Accessed at: https://www.nytimes.com/2022/04/16/world/europe/russian-propaganda-telegram-ukraine.html.

Saleem, Nalia, Paul Young, and Saman Yousuf. 2024. Exploring the Relationship Between Social Media Use and Symptoms of Depression and Anxiety Among Children and Adolescents: A Systematic Narrative Review. *Cyberpsychology, Behavior, and Social Networking*. Accessed at: https://www.liebertpub.com/doi/10.1089/cyber.2023.0456.

Sana, Faria, Tina Weston, and Nicholas Cepeda. 2013. Laptop Multitasking Hinders Classroom Learning for Both Users and Nearby Peers. *Computers and Education* 62: 24–31.

Sangani, N. Jangi, and M. R., Razavi. 2024. Investigating the Relationship Between the Use of Social Media, Educational Decline, and Student Mental Health. *Journal of Practice in Clinical Psychology* 12 (2): 123–136.

Santoro, Erik, and Benoît Monin. 2023. The AI Effect: People Rate Distinctively Human Attributes as More Essential to Being Human After Learning About Artificial Intelligence Advances. *Journal of Experimental Social Psychology* 107: 104464.

Sapolsky, Robert. 2017. *Behave: The Biology of Humans at Our Best and Worst*. New York: Penguin.

Sapolsky, Robert. 2023. *Determined: A Science of Life without Free Will*. New York: Penguin.

Satariano, Adam, and Cecilia Kang. 2023. How Nations Are Losing a Global Race to Tackle A.I.'s Harms. *New York Times*, December 6. Accessed at: https://www.nytimes.com/2023/12/06/technology/ai-regulation-policies.html.

Schaeffer, Katherine. 2019. Most U.S. Teens Who Use Cellphones Do It to Pass Time, Connect with Others, Learn New Things. Pew Research Center. Accessed at: https://www.pewresearch.org/fact-tank/2019/08/23/most-u-s-teens-who-use-cellphones-do-it-to-pass-time-connect-with-others-learn-new-things/.

Scharkow, Michael, Frank Mangold, Sebastian Stier, and Johannes Breuer. 2020. How Social Network Sites and Other Online Intermediaires Increase Exposure to News. *PNAS* 117 (6): 2761–2763.

Schmeichel, B.J., and R.F. Baumeister. 2004. Affect Regulation. In *Handbook of Self-Regulation: Research, Theory, and Applications*, ed. R.F. Baumeister and K.D. Vohs, 84–98. New York: Gilford Press.

Schmidt, Eric, and Jared Cohen. 2014. *The New Digital Age: Reshaping the Future of People, Nations and Business*. New York: Vintage.

Schneider, Sabrina, and Michael Leyey. 2019. Me or Information Technology? Adoption of Artificial Intelligence in the Delegation of Personal Strategic Decisions. *Managerial and Decision Economics* 40 (3): 223–231.

Schneier, Bruce, and Nathan Sanders. 2023. The A.I. Wars Have Three Factions, and They All Crave Power. *New York Times,* September 28. Accessed at: https://www.nytimes.com/2023/09/28/opinion/ai-safety-ethics-effective.html.

Schriber, Roberta A., and Richard W. Robins. 2012. Self-Knowledge: An Individual-Differences Perspective. In *Handbook of Self-Knowledge*, ed. Simine Vazire and Timothy D. Wilson, 105–127. New York: The Guilford Press.

Schüll, Natasha Dow. 2012. *Addiction by Design: Machine Gambling in Las Vegas*. Princeton: Princeton University Press.

Schwartz, Barry. 2016. *The Paradox of Choice: Why More is Less*. New York: EccoPress.

Schwartz, J.M., and Sharon Begley. 2002. *The Mind and the Brain: Neuroplasticity and the power of Mental Force*. New York: Harper Collins.

Scott-Dawkins, Kate. 2022. This Year Next Year: Global 2022 End-of-the-Year Forecast. Group M. Accessed at: https://www.groupm.com/longform/this-year-next-year-global-2022-end-of-year-forecast/.

Sen, Amartya. 1999. *Development as Freedom*. New York: Random House.

Sepah, Cameron. 2019. The Definitive Guide to Dopamine Fasting 2.0: The Hot Silicon Valley Trend. *Medium*, October 26. Accessed at: https://medium.com/swlh/dopamine-fasting-2-0-the-hot-silicon-valley-trend-7c4dc3ba2213.

Sherman, Lauren E., Minas Michikyan, and Patricia M. Greenfield. 2013. The Effects of Text, Audio, Video, and In-person Communication on Bonding Between Friends. *Cyberpsychology: Journal of Psychosocial Research on Cyberspace* 7 (2): 3. Accessed at: https://www.cdmc.ucla.edu/wp-content/uploads/sites/170/2018/04/Sherman-et-al.pdf.

Silver, David et al. 2016. Mastering the Game of Go With Deep Neural Networks and Tree Search. *Nature* 529 (28 January): 484–489. https://doi.org/10.1038/nature16961

Silverman, Craig. 2018. *This Analysis Shows How Viral Fake Election News Stories Outperformed Real News On Facebook.* BuzzFeedNews. Accessed at: https://www.buzzfeednews.com/article/craigsilverman/viral-fake-election-newsoutperformed-real-news-on-facebook

Simon, Herbert. 1971. Designing Organizations for an Information-Rich World. In *Computers, Communications, and the Public Interest*, ed. M. Greenberger, 38–52. Baltimore: Johns Hopkins Press.

Singer, Natasha. 2024a. Silicon Valley Battles States Over New Online Safety Laws for Children. *New York Times,* January 31. Accessed at: https://www.nytimes.com/2024/01/31/technology/social-media-free-speech-netchoice.html.

Singer, Natasha. 2024b. Will Chatbots Teach your Children. *New York Times,* January 11. Accessed at: https://www.nytimes.com/2024/01/11/technology/ai-chatbots-khan-education-tutoring.html.

Siu, William. 2022. I Make Video Games. I Won't Let My Daughters Play Them. *New York Times,* October 2. Accessed at: https://www.nytimes.com/2022/10/02/opinion/video-game-addiction.html?campaign_id=9&emc=edit_nn_20221003&instance_id=73583&nl=the-morning&regi_id=71023177&segment_id=108807&te=1&user_id=1b8a7a6c418ba0b24e53c8b7ae5ab191.

Small, Gary W., et al. 2020. Brain Health Consequences of Digital Technology Use. *Dialogues in Clinical Neuroscience* 22 (2): 179–187.

Smith, Adam. 2000. *The Theory of Moral Sentiments*. Amherst: Prometheus Books.

Smith, Adam. 2003. *The Wealth of Nations: An Enquiry into the Nature and Causes of the Wealth of Nations*. New York: Bantam.

Snyder, Timothy. 2021. The American Abyss. *New York* Times, January 9, 2021. Accessed at: https://www.nytimes.com/2021/01/09/magazine/trump-coup.html?mc_cid=d608443470&mc_eid=8710c3458f.

Solon, Olivia. 2017. Ex-Facebook President Sean Parker: Site Made to Exploit Human 'Vulnerability.' *The Guardian*, November 9. Accessed at: https://www.theguardian.com/technology/2017/nov/09/facebook-sean-parker-vulnerability-brain-psychology.

Soon, Chun Siong, Anna Hanxi He, Stefan Bode, and John-Dylan. Haynes. 2013. Predicting Free Choices for Abstract Intentions. *Proceedings for the National Academy of Sciences*. 110 (15): 6217–6222.

Soon, Chun Siong, Marcel Brass, Hans-Jochen. Heinze, and John-Dylan. Haynes. 2008. Unconscious Determinants of Free Decisions in the Human Brain. *Nature Neuroscience* 11 (5): 543–545.

Sophocles. 1993. *The Norton Book of Classical Literature*, ed. Bernard Knox. New York: W. W. Norton.

Soros, George. 2018. Remarks Delivered at the World Economic Forum, Davos Switzerland, January 25, 2108. Accessed at: https://www.georgesoros.com/2018/01/25/remarks-delivered-at-the-world-economic-forum/.

Sparrow, Betsy, Jenny Liu, and Daniel Wegner. 2011. Google Effects on Memory: Cognitive Consequences of Having Information at Our Fingertips. *Science* 333 (6043): 776–778.

Spice, Byron. 2017. Carnegie Mellon Artificial Intelligence Beats Top Poker Pros. Accessed at: https://www.cmu.edu/news/stories/archives/2017/january/AI-beats-poker-pros.html.

Stadler, Felix. 2002. Privacy is Not the Antidote to Surveillance. *Surveillance and Society* 1 (1): 120–124.

Stanford Behavior Design Lab. The Ethical Use of Persuasive Technology. Accessed at: https://behaviordesign.stanford.edu/ethical-use-persuasive-technology.

Statista. 2022. Percentage of Population Volunteering in the U.S. from 2008 to 2017. Accessed at: https://www.statista.com/statistics/189295/percentage-of-population-volunteering-in-the-united-states-since-2003/.

Statista. 2024. Online Advertising Revenue in the United States from 2000 to 2023. Accessed at: https://www.statista.com/statistics/183816/us-online-advertising-revenue-since-2000/.

Stein, David. 2024. The Fundamental Flaws of The Only Meta-Analysis of Social Media Reduction Experiments (And Why It Matters), Part 3. Accessed at: https://www.afterbabel.com/p/fundamental-flaws-part-3.

Stocking, Galen et al. 2022. The Role of Alternative Social Media in the News and Information Environment. Pew Research Center, October 6. Accessed at: https://www.pewresearch.org/journalism/2022/10/06/the-role-of-alternative-social-media-in-the-news-and-information-environment/.

Stokel-Walker, Chris. 2022. TikTok Wants Longer Videos—Whether You Like It or Not. *Wired* (February 21). Accessed at: https://www.wired.com/story/tiktok-wants-longer-videos-like-not/.

Strickland, Eliza. 2024. 15 Graphs That Explain the State of AI in 2024. *IEEE Spectrum*, April 16. Accessed at: https://spectrum.ieee.org/ai-index-2024.

Stroud, Natalie J. 2008. Media Use and Political Predispositions: Revisiting the Concept of Selective Exposure. *Political Behavior* 30: 341–366.

Suleyman, Mustafa. 2023. *The Coming Wave*. New York: Crown.

Sullivan, Andrew. 2016. I Used to be a Human Being. *New York Magazine* (September 19). Accessed at: https://nymag.com/intelligencer/2016/09/andrew-sullivan-my-distraction-sickness-and-yours.html.

Sullivan, Danny. 2015. How Google Now, Siri & Cortana Predict What You Want. *Search Engine Land* (September 18). Accessed at: https://searchengineland.com/how-google-now-siri-cortana-predict-what-you-want-229799.

Sunstein, Cass R. 2013. *Impersonal Default Rules vs. Active Choices vs Personalized Default Rules: A Triptypch. SSRN*. Accessed at: https://ssrn.com/abstract=2171343 or https://doi.org/10.2139/ssrn.2171343.

Sunstein, Cass R. 2017. *#Republic: Divided Democracy in the Age of Social Media*. Princeton: Princeton University Press.

Syvertsen, Trine. 2020. *Digital Detox: The Politics of Disconnecting*. UK: Emerald.

Syvertsen, Trine, and Gunn Enli. 2020. Digital Detox: Media Resistance and the Promise of Authenticity. *Convergence: The International Journal of Research into New Media Technologies* 26: 1269–1283.

Takahashi, Ippei, et al. 2023. Screen Time at Age 1 Year and Communication and Problem-Solving Developmental Delay at 2 and 4 Years. *JAMA Pediatrics*. Accessed at: https://jamanetwork.com/journals/jamapediatrics/fullarticle/2808593.

Taleb, Nassim Nicholas. 2010. *The Black Swan: The Impact of the Highly Improbable*. New York: Random House.

Tam, K. Y. Y., and M. Inzlicht. 2024. Fast-forward to Boredom: How Switching Behavior on Digital Media Makes People more Bored. *Journal of Experimental Psychology: General*. Accessed at: https://psycnet.apa.org/fulltext/2025-11974-001.pdf?auth_token=dbe63f30544cc6199b3c2defe2a8e923e30a8e26.

Tambini, Damian, and Martin Moore, eds. 2022. *Regulating Big Tech*. New York: Oxford University Press.

Taylor, Charles. 1985. *Human Agency and Language: Philosophical Papers 1*. Cambridge: Cambridge University Press.

Tetlock, Philip E. 2005. *Expert Political Judgment*. Princeton: Princeton University Press.

Tetlock, Philip, and Dan Gardner. 2015. *Superforecasting: The Art and Science of Prediction*. New York: Crown.

The Crowd and the Cloud. 2018. Accessed at: http://crowdandcloud.org.

Stanford Digital Economy Lab. *The Digitalist Papers*. 2024. Accessed at: https://www.digitalistpapers.com/essays.

Tetlock, Philip E. 2005. *Expert Political Judgment*. Princeton: Princeton University Press.

The Nation. 1932. Mahatma Gandhi Meets Romain Rolland. (February 10). Accessed at: https://www.thenation.com/article/archive/mahatma-gandhi-meets-romain-rolland/.

Thiele, Leslie Paul. 1990. *Friedrich Nietzsche and the Politics of the Soul*. Princeton: Princeton University Press.

Thiele, Leslie Paul. 1995. *Timely Meditations: Martin Heidegger and Postmodern Politics*. Princeton: Princeton University Press.

Thiele, Leslie Paul. 1997. Postmodernity and the Routinization of Novelty: Heidegger on Boredom and Technology. *Polity* 29: 489–517.

Thiele, Leslie Paul. 1998. *Timely Meditations: Martin Heidegger and Postmodern Politics*. Princeton: Princeton University Press.

Thiele, Leslie Paul. 2006. *The Heart of Judgment: Practical Wisdom, Neuroscience, and Narrative*. Cambridge: Cambridge University Press.

Thiele, Leslie Paul. 2009. The Ontology of Action: Arendt and the Role of Narrative. *Theory and Event*, 12 (December): 4. Johns Hopkins University Press: http://muse.jhu.edu/journals/theory_and_event/.

Thiele, Leslie Paul. 2012. Brain Sculpting as Moral Practice: A Neuro-Aristotelian Approach. In *Frank Vander Valk*, ed. Essays on Neuroscience and Political Theory: Thinking the Body Politic, 119–139. New York: Routledge.

Thiele, Leslie Paul, and Marshall Young. 2016. Practical Judgment, Narrative Experience, and Wicked Problems. *Theoria* 63 (148): 35–52.

Thiele, Leslie Paul. 2018. Digital Politics is the Game: See What Happens When Scholars Play It Well! *Perspectives* 16 (4): 1123–1128.

Thiele, Leslie Paul. 2019. *The Art and Craft of Political Theory*. New York: Routledge.

Thiele, Leslie Paul. 2020. Integrating Political and Technological Uncertainty into Robust Climate Policy. *Climatic Change* 163 (November 2020): 521–538.

Thiele, Leslie Paul. 2021. Rise of the Centaurs: The Internet of Things Intelligence Augmentation. In *Towards an International Political Economy of Artificial Intelligence*, ed. Ryan Kiggins and Tugrul Keskin, 39–61. Palgrave Macmillan.

Thiele, Leslie Paul. 2024. *Sustainability*, 3rd ed. Cambridge: Polity Press.

Thiele, Leslie Paul. 2025. Human Prospects in a World of Smart(er) Machines. In *The Political Economy of Robots: Prospects for Prosperity and Security in the Automated 21st Century*, ed. Ryan Kiggins, 2nd ed. Palgrave Macmillan.

Thorn. 2021. Responding to Online Threats: Minors' Perspectives on Disclosing, Reporting, and Blocking: Findings from 2020 Quantitative Research Among 9–17 year olds. Accessed at: https://info.thorn.org/hubfs/Research/Responding%20to%20Online%20Threats_2021-Full-Report.pdf.

Thornton, Bill, Alyson Faires, Maija Robbins, and E. Eric Rollins. 2014. The Mere Presence of a Cell Phone May be Distracting: Implications for Attention and Task Performance. *Social Psychology* 45 (6): 479–488.

Tice, Dianne, Roy Baumeister, and Liqing Zhang. 2004. The Role of Emotion in Self-Regulation: Differing Roles of Positive and Negative Emotion. In *The Regulation of Emotion*, ed. Pierre Philippot and Robert Feldman, 213–226. Mahwah, NJ: Lawrence Erlbaum.

Tiffany, Kaitlyn. 2022. No, I Will Not BeReal. *The Atlantic* (August 1). Accessed at: https://www.theatlantic.com/technology/archive/2022/08/bereal-app-authenticity-social-media-instagram/671012/?mc_cid=4fc263320c&mc_eid=8710c3458f.

Toffler, Alvin. 1970. *Future Shock*. New York: Bantam Books.

Toolify. 2024. Accessed at: https://www.toolify.ai/ai-news/cancel-plans-with-friends-guiltfree-using-ai-excuse-generator-619560.

Tomasello, Michael, Josep Call, and Brian Hare. 2003. Chimpanzees Understand Psychological States—The Question is Which Ones and to What Extent. *Trends in Cognitive Sciences* 7: 153–156.

Tracii, Ryan, et al. 2014. The Uses and Abuses of Facebook: A Review of Facebook Addiction. *Journal of Behavioral Addiction* 3 (3): 133–148.

Truman, Harry S. 1962. *Public Papers of the Presidents of the United States: Harry S. Truman*, January 1 to December 31, 1946. Washington: United States Government Printing Office.

Trust for America's Health. 2023. Pain in the Nation: The Epidemics of Alcohol, Drug, and Suicide Deaths. Accessed at: https://www.tfah.org/report-details/pain-in-the-nation-2023/.

Tufekci, Zeynep. 2017. *Twitter and Tear Gas: The Power and Fragility of Networked Protest Fragility of Networked Protest.* New Haven: Yale University Press.

Tufekci, Zeynep. 2023. Avert Your Eyes, Avoid Responsibility and Just Blame TikTok. *New York Times,* December 27. Accessed at: https://www.nytimes.com/2023/12/27/opinion/young-voters-tiktok.html.

Tullett-Prado, Deon, R. Doley Jo, Daniel Zarate, Rapson Gomez, and Vasileios Stavropoulos. 2023. Conceptualising Social Media Addiction: A Longitudinal Network Analysis of Social Media Addiction Symptoms and Their Relationships with Psychological Distress In a Community Sample of Adults. *BMC Psychiatry* 23: 509.

Tuquero, Loreben. 2023. PolitiFact FL: Trump Says AI Was Used to Make Him 'Look Bad.' The clips are real. Accessed at: https://www.wmfe.org/politics/2023-12-29/trump-lincoln-project-ai.

Turkle, Sherry. 2015. *Reclaiming Conversation: The Power of Talk in a Digital Age*. New York: Penguin.

Turkle, Sherry. 2017. *Alone Together: Why We Expect More from Technology and Less from Each Other*, 3rd ed. New York: Basic Books.

Tversky, Amos, and Daniel Kahneman. 1982. Judgments of and by Representativeness. In *Judgment under Uncertainty: Heuristics and biases*, eds. Daniel Kahneman, Paul Slovic and A. Tversky. Cambridge: Cambridge University Press.

Twenge, Jean M. 2017. *IGen: Why Today's Super-Connected Kids are Growing Up Less Rebellious, More Tolerant, Less Happy—And Completely Unprepared for Adulthood*. New York: Atria Books.

Twenge, Jean. 2020. Why Increases in Adolescent Depression May be Linked to the Technological Environment. *Current Opinion in Psychology* 32: 89–94.

Twenge, Jean M., and Roy F. Baumeister. 2002. Self-Control: A Limited Yet Renewable Resource. In *Self and Identity: Personal, Social, and Symbolic*, ed.

Yoshihisa Kashima, Margaret Foddy, and Michael Platow, 57–70. Mahwah, NJ: Lawrence Erlbaum Associates.

Twenge, Jean, and W. Keith Campbell. 2009. *The Narcissism Epidemic*. New York: The Free Press.

Twenge, Jean, Brian Spitzberg, and W. Keith Campbell. 2019. Less In-person Social Interaction with Peers Among U.S. Adolescents in the 21st Century and Links to Loneliness. *Journal of Social and Personal Relationships* 36 (6): 1892–1913.

Twenge, Jean M., Jonathan Haidt, Thomas E. Joiner, and W. Keith Campbell. 2020. Underestimating Digital Media Harm. *Nature Human Behavior*. 4: 346–348.

Uhls, Yalda T., et al. 2014. Five Days at Outdoor Education Camp Without Screens Improves Preteen Skills with Nonverbal Emotion Cues. *Computers in Human Behavior* 39: 387–392.

United Kingdom Department for Digital, Culture, Media and Sport. 2018. *A Connected Society: A Strategy for Tackling Loneliness*. Accessed at: https://www.gov.uk/government/publications/a-connected-society-a-strategy-for-tackling-loneliness.

United Kingdom Department of Science, Innovation and Technology. 2023. *AI Regulation: A Pro-innovation Approach*. Accessed at: https://www.gov.uk/government/publications/ai-regulation-a-pro-innovation-approach.

United Nations. 2024a. Resolution A/78/L.49. Accessed at: https://daccess-ods.un.org/access.nsf/Get?OpenAgent&DS=A/78/L.49&Lang=E.

United Nations 2024b. *World Population Prospects 2024*. Accessed at: https://population.un.org/wpp/.

United States Court of Appeals for the Third Circuit. 2024. Tawainna Anderson, Individually and as Administratix of the Estate of n.a., a Deceased Minor, Appellant v. Tiktok, Inc.; Bytedance, Inc. Accessed at: https://www2.ca3.uscourts.gov/opinarch/223061p.pdf.

United States Hospital Fund. 2024. *The Ripple Effects of the Adolescent Behavioral Health Crisis*. New York. Accessed at: https://media.uhfnyc.org/filer_public/61/92/6192cd92-0fc3-4dc7-8a36-eda9531354a3/ripple_effects_adolescent_behavioral_health_chartbook_report.pdf.

Urbinati, Nadia. 2014. *Democracy Disfigured: Opinion, Truth, and the People*. Cambridge: Harvard University Press.

U.S.-China Economic and Security Review Commission. 2024. 2024 Report to Congress. Accessed at: https://www.uscc.gov/sites/default/files/2024-11/2024_Annual_Report_to_Congress.pdf.

U.S. Congress. 2019. S.2314—SMART Act. Accessed at: https://www.congress.gov/bill/116th-congress/senate-bill/2314.

Vaccari, Christina, and Augusto Valeriani. 2018. Digital Political Talk and Participation: Comparing Established and Third Wave Democracies. *Sage Open*

(June): 1–14. Accessed at: https://journals.sagepub.com/doi/10.1177/2158244018784986.

Vallor, Shannon. 2015. Moral Deskilling and Upskilling in New Machine Age. *Philosophy and Technology* 28: 107–124.

Vallor, Shannon. 2016. *Technology and the Virtues: A Philosophical Guide to a Future Worth Wanting*. New York: Oxford University Press.

Vanderklippe, Nathan. 2018. Chinese School Installs Cameras to Monitor Students. *Globe and Mail* (June 2): A3.

Vattimo, Gianni. 1985. *The End of Modernity*. Baltimore: Johns Hopkins University Press.

Vazire, Simine, and Timothy D. Wilson, eds. 2012. *Handbook of Self-Knowledge*. New York: The Guilford Press.

Verdyun, P., et al. 2015. Passive Facebook Usage Undermines Affective Well-Being: Experimental and Longitudinal Evidence. *Journal of Experimental Psychology Gen* 144: 480–488.

Verganti, Roberto, Luca Vendraminelli, and Marco Iansiti. 2020. Innovation and Design in the Age of Artificial Intelligence. *Journal of Product Innovation Management* 37: 212–227.

Verizon. 2019. *Smart Family*. Accessed at: https://www.verizonwireless.com/solutions-and-services/verizon-smart-family/.

Vivanti, Giacomo, and Sally J. Rogers. 2014. Autism and the Mirror Neuron System: Insights from Learning and Teaching. *Philosophical Transactions of the Royal Society London 'B' Biological Science* 369 (1644): 20130184. https://doi.org/10.1098/rstb.2013.0184.

Volkow, Nora D., Gene-Jack Wang, Joanna S. Fowler. 2011. Addiction: Beyond Dopamine Reward Circuitry. *PNAS* 108 (37): 15037–15042. Accessed at: https://www.pnas.org/doi/abs/10.1073/pnas.1010654108.

Vosoughi, Soroush, Deb Roy, and Sinan Aral. 2018. The Spread of True and False News Online. *Science* 359: 1146–1151.

Vuorre, Matti, Amy Orben, and Andrew K. Przybylski. 2021. There Is No Evidence That Associations Between Adolescents' Digital Technology Engagement and Mental Health Problems Have Increased. *Clinical Psychological Science* 9 (5): 823–835.

Wallach, Wendell. 2015. *A Dangerous Master: How to Keep Technology from Slipping Beyond Our Control*. New York: Basic Books.

Walsh, Jeremy, Joel D. Barnes, Jameason D. Cameron, Gary S. Goldfield, Jean-Philippe. Chaput, Katie E. Gunnell, Andrée-Anne. Ledoux, Roger Zemek, and Mark S. Tremblay. 2018. Associations between 24-Hour Movement Behaviours and Global Cognition in a Cross-Sectional Sample of American Children. *The Lancet* 2 (11): 783–791.

Ward, Adrian F., Kristen Duke, Ayelet Gneezy, and Maarten W. Bos. 2017. Brain Drain: The Mere Presence of One's Own Smartphone Reduces Available Cognitive Capacity. *Journal of the Association for Consumer Research* 2 (2): 140–154.

Ward, Zachary J., Sara N. Bleich, Angie L. Cradock, Jessica L. Barrett, Catherine M. Giles, Chasmine Flax, Michael W. Long, and Steven L. Gortmaker. 2019. Projected U.S. State-Level Prevalence of Adult Obesity and Severe Obesity. *New England Journal of Medicine* 381: 2440–2450.

Weber, Max. 1958. *The Protestant Ethic and the Spirit of Capitalism*. New York: Charles Scirbner's Sons.

Wedge, Marilyn. 2017. 'Virtual Autism' May Explain Explosive Rise in ASD Diagnoses. Accessed at: https://www.madinamerica.com/2017/08/virtual-autism-explain-rising-asd-diagnoses/.

Wegmann, Elisa, Sina Ostendorf, and Matthias Brand. 2018. Is It Beneficial to use Internet-Communication for Escaping from Boredom? Boredom Proneness Interacts with Cue-induced Craving and Avoidance Expectancies in Explaining Symptoms of Internet-communication Disorder. *PLoS ONE* 13 (4): e0195742.

Wegner, Daniel. 2018. *The Illusion of Conscious Will*. Cambridge, MA: MIT Press.

Weiser, Benjamin. 2023. Here's What Happens When Your Lawyer Uses ChatGPT. *New York Times*, May 27. Accessed at: https://www.nytimes.com/2023/05/27/nyregion/avianca-airline-lawsuit-chatgpt.html?searchResultPosition=1.

Wells, Georgia, Jeff Horwitz, and Deepa Seetharaman. 2021. Facebook Knows Instagram Is Toxic for Teen Girls, Company Documents Show. *Wall Street Journal*, September 14. Accessed at: https://www.wsj.com/articles/facebook-knows-instagram-is-toxic-for-teen-girls-company-documents-show.

Whitehead, A. N. 1911. *An Introduction to Mathematics*. Cambridge, USA: Henry Holt and Company.

Whitehouse.gov. 2025. *Fact Sheet: Executive Order to Establish United States Leadership in Digital Financial Technology*. Accessed at: https://www.whitehouse.gov/fact-sheets/2025/01/fact-sheet-president-donald-j-trumptakes-action-to-enhance-americas-ai-leadership/

Whitten, Allison. 2022. Brain Chemical Helps Signal to Neurons When to Start a Movement. *Quanta Magazine* (March 22). Accessed at: https://www.quantamagazine.org/brain-chemical-helps-signal-to-neurons-when-to-start-a-movement-20220322/#comments.

Wiener, Norbert. 1954. *The Human Use of Human Beings: Cybernetics and Society*. New York: Houghton Mifflin.

Williams, Florence. 2017. *The Nature Fix: Why Nature Makes us Happier, Healthier, and More Creative*. New York. W.W: Norton.

Williams, James. 2018. *Stand Out of Our Light: Freedom and Resistance in the Attention Economy*. Cambridge: Cambridge University Press.

Wilmer, Henry H., Lauren E. Sherman, and Jason M. Chein. 2017. Smartphones and Cognition: A Review of Research Exploring the Links between Mobile Technology Habits and Cognitive Functioning. *Frontiers in Psychology* 8 (April): 605. Accessed at: https://www.frontiersin.org/articles/10.3389/fpsyg.2017.00605/full.

Wilson, T., D. Reinhard, E. Westgate, D. Gilbert, N. Ellerbeck, C. Hahn, C. Brown, and A. Shaked. 2014. Just Think: The Challenges of the Disengaged Mind. *Science* 345 (6192): 75–77.

Wilson, Timothy. 2002. *Strangers to Ourselves: Discovering the Adaptive Unconscious*. Cambridge, MA: Harvard University Press.

Winner, Langdon. 1977. *Autonomous Technology*. Cambridge, MA: MIT Press.

Winnick, Michael. 2016. Putting a Finger on Our Phone Obsession: Mobile Touches: A Study on How Humans use Technology. *Dscout*. Accessed at: https://pages.dscout.com/hubfs/downloads/dscout_mobile_touches_study_2016.pdf.

Wired. 2024. *The WIRED AI Elections Project*. Accessed at: https://www.wired.com/story/generative-ai-global-elections/.

Wofford, Benjamin. 2022. The Lobbyist Next Door. *Wired* (September), 47–57.

Woodward, Alex. 2020. 'Fake News': A Guide to Trump's Favourite Phrase—And the Dangers It Obscures. *Independent,* October 2. Accessed at: https://www.independent.co.uk/news/world/americas/us-election/trump-fake-news-counter-history-b732873.html.

Woolley, Samuel C., and Philip N. Howard, eds. 2019. *Computational Propaganda*. New York: Oxford University Press.

World Health Organization. 2019. Obesity and Overweight. Accessed at: https://www.cdc.gov/obesity/data/adult.html.

Wu-Ouyang, Biying. 2023. More Mobile Connectedness, Less Well-Being? *Journal of Media Psychology* 35 (5): 291–302.

Wu, Tim. 2016. *The Attention Merchants: The Epic Scramble to Get Inside our Heads*. New York: Alfred A. Knopf.

Wu, Nicole, and Patrick Wu. 2024. Surveying the Impact of Generative Artificial Intelligence on Political Science Education. *PS: Political Science and Politics*, August 9. Accessed at: https://www.cambridge.org/core/journals/ps-political-science-and-politics/article/surveying-the-impact-of-generative-artificial-intelligence-on-political-science-education/D87FE814975D2C1CFA3F9E9E7DEBDB6C.

Yan, Wanling et al. 2022. Does Global Positioning System-Based Navigation Dependency Make Your Sense of Direction Poor? A Psychological Assessment and Eye-Tracking Study. *Frontiers in Psychology* 13.

Accessed at: https://www.frontiersin.org/journals/psychology/articles/10.3389/fpsyg.2022.983019/full.

Yarrow, Jay. 2010. Sergey Brin: 'We Want Google To Be The Third Half Of Your Brain.' *Business Insider* (September 8). Accessed at: https://www.businessinsider.com/sergey-brin-we-want-google-to-be-the-third-half-of-your-brain-2010-9?op=1.

Yeykelis, Leo, James J. Cummings, and Byron Reeves. 2014. Multitasking on a Single Device: Arousal and the Frequency, Anticipation, and Prediction of Switching Between Media Content on a Computer. *Journal of Communication* 64 (1): 167–192.

Zhao, Yihong, et al. 2022. Brain Structural Covariation Linked to Screen Media Activity and Externalizing Behaviors in Children. *Journal of Behavioral Addictions* 11 (2): 417–426.

Zuboff, Shoshana. 2019. *Surveillance Capitalism: The Fight for a Human Future at the New Frontier of Power*. New York: PublicAffairs.

Zuckerberg, Mark. 2015. Post on Facebook. Accessed at: https://www.facebook.com/zuck/posts/10102329188394581?pnref=story.

Index

L. P. Thiele, *Human Agency, Artificial Intelligence, and the Attention Economy*, https://doi.org/10.1007/978-3-031-82086-1

GPSR Compliance
The European Union's (EU) General Product Safety Regulation (GPSR) is a set of rules that requires consumer products to be safe and our obligations to ensure this.

If you have any concerns about our products, you can contact us on

ProductSafety@springernature.com

In case Publisher is established outside the EU, the EU authorized representative is:

Springer Nature Customer Service Center GmbH
Europaplatz 3
69115 Heidelberg, Germany

www.ingramcontent.com/pod-product-compliance
Lightning Source LLC
Chambersburg PA
CBHW071556290626
47387CB00006B/885
* 9 7 8 3 0 3 1 8 2 0 8 5 4 *